Soil Improvement and Water Conservation Biotechnology

Edited By

Israel Valencia Quiroz

Phytochemistry Laboratory, UBIPRO
Superior Studies Faculty (FES)-Iztacala
National Autonomous University of Mexico (UNAM)
Tlalnepantla de Baz, Mexico State
Mexico

Soil Improvement and Water Conservation Biotechnology

Editor: Israel Valencia Quiroz

ISBN (Online): 978-981-5322-43-9

ISBN (Print): 978-981-5322-44-6

ISBN (Paperback): 978-981-5322-45-3

need for a court order if at any point you breach any terms of this License Agreement. In no event will any delay or failure by Bentham Science Publishers in enforcing your compliance with this License Agreement constitute a waiver of any of its rights.

3. You acknowledge that you have read this License Agreement, and agree to be bound by its terms and conditions. To the extent that any other terms and conditions presented on any website of Bentham Science Publishers conflict with, or are inconsistent with, the terms and conditions set out in this License Agreement, you acknowledge that the terms and conditions set out in this License Agreement shall prevail.

Bentham Science Publishers Pte. Ltd.
80 Robinson Road #02-00
Singapore 068898
Singapore
Email: subscriptions@benthamscience.net

BENTHAM SCIENCE

CONTENTS

FOREWORD i

PREFACE ii

LIST OF CONTRIBUTORS iv

CHAPTER 1 SOIL: COMPOSITION, FUNCTION, AND CURRENT CHALLENGES 1
Brenda Karina Guevara Olivar
  INTRODUCTION 1
      Physical Components of Soil 2
      Chemical Components of Soil 3
      Biological Component of Soil 4
      Soil Functions and Environmental Services 7
      Current Challenges in Soil Study and Management 14
  CONCLUDING REMARKS 18
  QUESTIONS RELATED TO THE TEXT 19
  LIST OF ABBREVIATIONS 19
  REFERENCES 20

CHAPTER 2 SOILS OF MEXICAN DESERTS: CHARACTERISTICS AND WATER MANAGEMENT CHALLENGES 34
Blanca González-Méndez and *Elizabeth Chávez-García*
  INTRODUCTION 35
      Factors in Soil Formation 36
      Soils 40
      Rock Weathering by salts 41
      Ferrugination 41
      Oxidation/Reduction Processes 41
      Clay Illuviation 42
      Shrink-Swell Characteristics 42
      Precipitation of secondary minerals 42
      Soil Organic Matter (SOM) and Dust Additions 42
      Organic Matter Decomposition 42
      Diagnostic Horizons 43
      Cambic horizon (Bw, B, or Bq) 43
      Argic horizon (Bt) 43
      Natric horizon (Btn) 43
      Calcic horizon (Bk or Bkk) 43
      Petrocalcic horizon (Bkm or Bkkm) 43
      Gypsic horizon (By or Byy) 43
      Petrogypsic horizon (Bym or Byym) 44
      Salic horizon (Az, Bz, or Cz) 44
      Petroduric horizon (Bqm) 44
      Soils of the Sonoran and Chihuahuan Deserts 44
      Leptsols 45
      Regosols 45
      Calcisols 46
      Vertisols 46
      Arenosols 46
      Solonchaks 47
      Cambisols 47

Phaeozems 47
Luvisols 47
Solonetz 48
Fluvisols 48
Planosols 48
Gypsisols 49
Durisols 49
CASE STUDY: SOIL ASSESSMENT AND RECONDITIONING FOR GREEN INFRASTRUCTURE IN HERMOSILLO, MEXICO 49
CONCLUDING REMARKS 50
AUTHORS' CONTRIBUTION 51
QUESTIONS RELATED TO THE TEXT 52
LIST OF ABBREVIATIONS 52
ACKNOWLEDGEMENTS 53
REFERENCES 53

CHAPTER 3 WATER IN THE ARID AND SEMI-ARID ZONES OF MEXICO 57
Elvia Manuela Gallegos-Neyra, Francisco José Torner-Morales, América Patricia Garcia-Garcia, Cesar Alejandro Zamora-Barrios, Iván Andrés Arredondo-Fragoso and Israel Valencia Quiroz
INTRODUCTION 58
Current Water Situation in Mexico 59
Water Sources in Arid and Semi-arid Zones 59
Water Use And Management in Arid and Semi-Arid Zones of Mexico 60
Problems and Challenges in Water Management 62
Strategies, Public Policies, and Legal Frameworks for Sustainable Water Management in Mexico 62
Case Studies and Successful Experiences in Water Management in Mexico 63
Challenges and Future Perspectives of Water in the Arid and Semi-Arid Zones of Mexico   64
CONCLUDING REMARKS 66
AUTHORS' CONTRIBUTION 66
QUESTIONS RELATED TO THE TEXT 67
CONSENT FOR PUBLICATION 67
CONFLICT OF INTEREST 67
LIST OF ABBREVIATIONS 67
REFERENCES 68

CHAPTER 4 'OMICS' STUDIES ON RHIZOSPHERE-MICROORGANISM INTERACTIONS IN SOILS 77
Edgar Antonio Estrella-Parra, José G. Avila-Acevedo, Adriana Montserrat Espinosa González, Ana M. García-Bores, Jessica Hernández-Pineda, Nallely Alvarez-Santos, José Cruz Rivera-Cabrera and Erick Nolasco Ontiveros
INTRODUCTION 78
Trash Landfills, Microplastics, and Antibiotics: Soil Pollution 79
Agricultural-type Soils: Food Challenges and Overexploitation 82
Study of Soils and/or Microbiota in Contaminated Soils Through OMICS Tools 84
DNA Extraction 84
Metabolomics Analyses 84
Bacterial Metagenomics 85
Exposure of Soil to Nano-plastic Contaminants by Earthworm Analysis 85
CONCLUDING REMARKS 88
AUTHORS' CONTRIBUTION 88

QUESTIONS RELATED TO THE TEXT 88
LIST OF ABBREVIATIONS 89
REFERENCES 89

CHAPTER 5 BIOENGINEERING TECHNIQUES FOR SOIL EROSION PREVENTION 93
Iván E. Barrera and *Paris O. Gonzalez*
INTRODUCTION 94
    Soil Erosion 94
    Bioengineering as a Soil Erosion Prevention Strategy 95
    Soil Erosion Prevention Techniques on Slopes and Embankments 95
        Bush Mattress 95
        Wattle Fences 96
        Log Brush Barrier 98
        Fascines (Bush Wattles) 99
        Fences 99
        Live Cribwall 101
        Jute Netting 102
        Jute-mat Log 104
        Brush Layering 105
        Live Reinforced Earth Walls 106
    Considerations 107
CONCLUDING REMARKS 109
AUTHORS' CONTRIBUTION 109
QUESTIONS RELATED TO THE TEXT 109
ACKNOWLEDGEMENT 110
REFERENCES 110

**CHAPTER 6 BIOFERTILIZERS AND BIOPESTICIDES: SUSTAINABLE ALTERNATIVES
FOR AGRICULTURE** 113
*Rafael Torres-Martínez, Yesica R. Cruz-Martinez, Ana K. Villagómez-Guzmán, Olivia
Pérez-Valera, Héctor M. Arreaga-González* and *Tzasna Hernández-Delgado*
INTRODUCTION 114
BIOFERTILIZERS 115
    Classification and Mechanism of Action of Biofertilizers 115
    Advantages and Drawbacks of Biofertilizers Compared to Chemical Fertilizers 116
    Biofertilizer Market, Development Costs, and Input Economic Impact 118
    Biotechnology in the Development of Biofertilizers 119
    Formulation and Biofertilizer Application 120
    Biopesticides 122
    Classification and Mechanisms of Biopesticide Action 123
    Microbial Pesticides (Bacterium, Fungus, Virus or Protozoan) 124
    Biochemicals Pesticides 125
    Plant-Based Extracts and Essential Oils 125
    Insect Pheromones 126
    Plant-Incorporated-Protectants (PIPs) 127
    Nanobiopesticides 127
    The Market, Impact, and Advantages of Biopesticide over Synthetic Pesticides 127
    Formulation and Biopesticide Application 128
CONCLUDING REMARKS 129
AUTHORS' CONTRIBUTION 130
QUESTIONS RELATED TO THE TEXT 130
ACKNOWLEDGMENT 130

LIST OF ABBREVIATIONS 130
REFERENCES 131

**CHAPTER 7 MICROALGAE AND BIOTECHNOLOGY: WATER PURIFICATION AND
BIOMASS PRODUCTION** 137
Yuri Córdoba Campo and *Israel Valencia Quiroz*
INTRODUCTION 138
    Importance of Water Purification 139
MICROALGAE REMEDIATION MECHANISMS 141
    Biosorption 141
    Bioaccumulation 143
    Biodegradation 143
ENVIRONMENTAL APPLICATIONS OF MICROALGAE IN WATER PURIFICATION 145
    Domestic Effluents 145
    Industrial Effluents 146
    Tannery Effluent 147
    Effluents from the Pharmaceutical Industry 147
BIOMASS PRODUCTION BY MICROALGAE 148
    Organic Materials 149
    Nutritional Supplement 150
    Biodiesel Production 150
    Bioethanol and Biomethane Production 151
CONCLUDING REMARKS 151
AUTHORS' CONTRIBUTION 151
QUESTIONS RELATED TO THE TEXT 151
LIST OF ABBREVIATIONS 152
REFERENCES 152

**CHAPTER 8 BIOLOGICAL DESALINATION: BIOTECHNOLOGICAL ALTERNATIVES
FOR FRESHWATER EXTRACTION** 157
Ana K. Villagómez-Guzmán, Héctor M. Arreaga-González and *Tzasna Hernández-
Delgado*
INTRODUCTION 158
    Current Distribution State of Water Worldwide 160
    Salinity and Conventional Desalination Process 161
        Reverse Osmosis (RO) 163
        Multi-stage Flash Distillation (MSF) 163
        Multiple-effect Distillation (MED) 163
        Vapor-compression Distillation 163
    Biodesalination Concept 164
    Biodesalination through Salt-tolerant Organisms 164
        Halophytes Plants 164
    Algae 167
    Microalgae 168
    Bacterial 170
    Cyanobacterial 171
    Microbial Desalination Cells 172
    Biological Membranes and Biomaterials for Desalination 174
CONCLUDING REMARKS 174
AUTHORS' CONTRIBUTION 175
QUESTIONS RELATED TO THE TEXT 175
ACKNOWLEDGMENTS 175

LIST OF ABBREVIATIONS 176
REFERENCES 177

CHAPTER 9 BIOLOGICAL TREATMENT OF WASTEWATER: USE OF MICROORGANISMS FOR PURIFICATION 183
Román Adrián González Cruz, Alonso Ezeta Miranda and *Israel Valencia Quiroz*
INTRODUCTION 184
    Types of Wastewater 185
    Biological Treatment Methods 185
        Aerobic Treatment Processes 186
        Anaerobic Treatment Processes 187
    Role of Microorganisms in Wastewater Treatment 187
        Types of Microorganisms Used 189
    Microorganisms used in Water Bioremediation 190
        Bacteria 190
        Fungi 190
        Algae 190
        Archaea 191
    Overview of Microbial Species that Play a Pivotal Role in Bioremediation. 191
        Bacteria 191
        Fungi 192
        Algae 192
        Archaea 192
    Role of Microbial Consortia 193
    Technological and Research Advances 193
    Case Study: the use of Phytoremediation in Radionuclide-contaminated Sites 193
        Application of Phytoremediation 194
    Procedure: Implementing Microbial Bioremediation 195
    Challenges and Future Directions in Biological Treatment 195
    Emerging Technologies 196
CONCLUDING REMARKS 196
AUTHORS' CONTRIBUTION 197
QUESTIONS RELATED TO THE TEXT 197
ACKNOWLEDGEMENTS 198
LIST OF ABBREVIATIONS 198
REFERENCES 198

CHAPTER 10 BIOREMEDIATION OF CONTAMINATED WATERS: STRATEGIES AND SUCCESS CASES 205
Nichdaly Ortiz Chacón, Aliana Zacaria Vital and *Israel Valencia Quiroz*
INTRODUCTION 206
    Sources and Types of Contaminants 207
    Bioremediation: an Overview 207
        Definition and Principles 208
    Microbial Bioremediation Techniques 209
        Biostimulation 209
    Plant-based Bioremediation Techniques 210
        Phytoremediation 210
    Successful Cases of Bioremediation 211
        Case Study: Bioremediation in Various Contexts 212
    Procedure Steps: Implementing Microbial Bioremediation 220
    Challenges and Future Directions 221

Key Challenges in Bioremediation 221
CONCLUDING REMARKS 222
AUTHORS' CONTRIBUTION 222
QUESTIONS RELATED TO THE TEXT 223
LIST OF ABBREVIATIONS 223
REFERENCES 223

**CHAPTER 11 LUMINESCENT (BIO)SENSORS FOR PESTICIDE DETECTION: AN
INNOVATIVE TOOL FOR WATER MONITORING** 230
*María K. Salomón-Flores, Iván J. Bazany-Rodríguez, Helen Paola Toledo-Jaldin,
Juan Pablo León-Gómez and Alejandro Dorazco-González*
    INTRODUCTION 231
        Fluorescent Metal-free Organic Sensors for Pesticides in Real Samples 233
            Macrocycles 233
            Cyclodextrins 233
            Calixarenes 233
            Curcubiturils 233
            Pillararenes 233
        Relevant Examples of Macrocycle-based Sensors for Pesticides in Real Samples 234
        Relevant Examples of Organic Acyclic Sensors for Pesticides in Real Samples 236
        Relevant Examples of Organic Polymers for Pesticides in Real Samples 238
        Luminescent Metal Molecular Complexes for Chemodetection of Organophosphorus
        Pesticides 239
        Luminescent Metal-organic Frameworks (LMOFS) for the Sensing of Pesticides in Real
        Samples 244
        Luminescent Biosensors for Pesticides in Real Samples 249
        Fluorescent Enzymatic-based Biosensors 249
    CONCLUDING REMARKS 252
    QUESTIONS RELATED TO THE TEXT 253
    AUTHORS' CONTRIBUTION 254
    ACKNOWLEDGMENTS 254
    LIST OF ABBREVIATIONS 254
    REFERENCES 256

CHAPTER 12 NANOTECHNOLOGY FOR WATER AND SOIL CONSERVATION 263
María Alejandra Istúriz-Zapata and Alfredo Jiménez-Pérez
    INTRODUCTION 263
        Water Conservation 266
        Soil Conservation 274
        Nanoencapsulation 280
    CONCLUDING REMARKS 284
    QUESTIONS 285
    ACKNOWLEDGEMENTS 285
    AUTHORS' CONTRIBUTION 285
    LIST OF ABBREVIATIONS 285
    REFERENCES 286

**CHAPTER 13 CONSERVATION AND REUSE OF WATER IN AGRICULTURE:
BIOTECHNOLOGICAL TECHNIQUES FOR EFFICIENT USE** 297
*Israel Valencia Quiroz, Diana Violeta Sánchez Oropeza, María Fernanda Trujillo
Lira, Miriam Arlette López Pérez, Casandra Rosales García and Susana Rafael Maya*
    INTRODUCTION 298

Importance of Water in Agriculture 298
Challenges in Water Management in Agriculture 301
    Scarcity of Freshwater Resources 301
Biotechnological Approaches for Water Conservation 302
    Drought-Tolerant Crop Varieties 303
Biotechnological Techniques for Water Reuse 303
    Wastewater Treatment and Recycling 306
Case Studies and Success Stories 306
    The Use of Hydroponics Techniques to Reduce the Amount of Water Used in the
    Growth of Forage Crops in Qatar 306
    Low-pressure Center Pivot: A Case Study on Water Savings in Sugarcane
    Commercial Plantations 307
    Implementation of Biotechnological Solutions in Real-world Settings 308
Future Directions and Innovations 308
    Emerging Technologies in Water Conservation 309
New Technologies for Agriculture: Optimization of Water Resources 309
    Rule-Based Agriculture System (RBAS) 310
    Precision Agriculture and Data-Driven Farming 311
    Data Collection in Smart Farming 312
    Machine Learning in Agriculture 312
Advantages and Challenges of Precision Agriculture 312
    Advantages 312
    Challenges 313
Artificial Intelligence (AI) and Smart Agriculture for Sustainable Development ... 313
CONCLUDING REMARKS 313
AUTHORS' CONTRIBUTION 315
QUESTIONS RELATED TO THE TEXT 315
LIST OF ABBREVIATIONS 316
REFERENCES 316

CHAPTER 14 ETHICAL AND SOCIAL CHALLENGES OF ENVIRONMENTAL
BIOTECHNOLOGY 322
Israel Valencia Quiroz
INTRODUCTION 323
Ethical Considerations in Environmental Biotechnology 323
Principles of Bioethics 325
Social Implications of Environmental Biotechnology 326
    Impact on Traditional Practices 326
Regulatory Frameworks and Policies 327
    International Agreements and Conventions 328
Bioremediation of contaminated sites 328
    Case Study: The Application of Biochar in Soil Amendment 330
    Future Directions and Emerging Issues 336
    Technological Innovations 336
CONCLUDING REMARKS 337
AUTHORS' CONTRIBUTION 338
QUESTIONS RELATED TO THE TEXT 338
ACKNOWLEDGEMENTS 338
LIST OF ABBREVIATIONS 338
REFERENCES 339

SUBJECT INDEX 344

FOREWORD

More than one hundred years have passed since the Hungarian agricultural engineer Károly Ereki introduced the term "Biotechnology" to the world. This combination, perhaps strange at the time, between biology and technology, resulted from the observation and empirical knowledge of processes that provided us with goods, mainly associated with food products such as cheese, wine, bread, and beer. Products that, to this day, remain fundamental and in high demand.

However, according to historical and archaeological information, there is evidence from around 100 B.C. that the Teotihuacan culture already knew various biotechnological processes such as fermentation for the production of pulque (a traditional alcoholic drink from central Mexico), nixtamalization of corn, algae cultivation (spirulina), the use of natural dyes, as well as the medicinal use of plants and mushrooms. This shows that biotechnology, without being described as such, already played an important role in using living organisms and their metabolic processes for food production, as well as in medical, religious, and cultural matters.

Considering that humans have substantially increased their knowledge, a product of empirical understanding since the dawn of humanity, it is only natural that this knowledge has been perfected over time thanks to the progressive accumulation of observations and discoveries. Thanks to the development of the scientific method, which revolutionized the way we conduct scientific research in the modern era, significant growth and refinements have been achieved in the field of biotechnology, which includes a wide range of fields such as industry, agriculture, and the environment.

This compilation "Soil Improvement and Water Conservation Biotechnology" exemplifies, in a compelling manner, through perfectly structured examples, how the biotechnological approach applied to natural resources such as soil and water in Mexico has been successfully implemented. Although biotechnology is a controversial topic of great scientific and social interest, it is essential to apply a multidisciplinary and transdisciplinary approach given the complexities of biological systems for better understanding and resolution of environmental problems.

Finally, this book represents the culmination of an important and significant collective challenge: to face the complexity of addressing problems related to soil and water through a biotechnological approach in Mexico. We hope this work inspires future researchers and professionals to continue exploring and applying biotechnology for the well-being of our society and the environment.

Fernando Ayala Niño
Laboratory of Applied Edaphology and Environmental Services
UBIPRO, FES-Iztacala
National Autonomous University of Mexico (UNAM)
Tlalnepantla de Baz, Mexico State
Mexico

PREFACE

The significant issues of soil degradation and water scarcity represent some of the most crucial challenges currently confronting our planet. With the expansion of global populations and the escalating impact of climate change, the urgency for sustainable resolutions to these issues has never been more pronounced. The publication titled "Soil Improvement and Water Conservation Biotechnology" endeavors to tackle these challenges by investigating state-o--the-art biotechnological processes that have the potential to improve soil quality and optimize water utilization, particularly in ecosystems that are susceptible to adverse impacts.

The structure of this manuscript is designed to offer a comprehensive comprehension of the various facets of soil and water conservation. The initial sections establish the groundwork by analyzing the composition, functionality, and existing obstacles related to soil, with a specific emphasis on the distinctive attributes of soils located in the arid and semi-arid regions of Mexico. These areas, renowned for their severe conditions and limited water reserves, present distinct challenges and opportunities for the implementation of biotechnological interventions.

Subsequently, the publication explores sophisticated biotechnological strategies for soil and water conservation. Sections concerning 'omics' investigations uncover the intricate relationships between rhizosphere microorganisms and polluted soils, providing insights into how these relationships can be utilized for environmental rehabilitation. Techniques in bioengineering and the utilization of biofertilizers and biopesticides are presented as sustainable substitutes for mitigating soil degradation and enhancing agricultural output.

The discussion on water conservation biotechnology takes a prominent position in the latter segments of the book. Chapters dedicated to microalgae biotechnology, biological desalination, and the employment of microorganisms in wastewater treatment showcase inventive approaches for purifying and preserving water resources. The utilization of luminescent biosensors for pesticide detection and the significance of nanotechnology in water and soil conservation underscore the capacity of advanced technologies to transform environmental monitoring and governance.

In addition to furnishing a comprehensive theoretical insight, this publication is purposed to serve as an academic resource, providing practical methodologies directly linked to the topics covered in each chapter. Every chapter contains detailed explanations of practical methods and case studies that exemplify real-world applications of the technologies discussed. To further reinforce the material, there are review queries at the conclusion of each chapter, prompting readers to contemplate and engage with the content.

Acknowledging that technological progressions must be accompanied by ethical and societal considerations, the manuscript culminates with a section devoted to the ethical and societal dilemmas surrounding environmental biotechnology. This discourse stresses the significance of fair access to biotechnological resolutions and the necessity for community involvement in decision-making processes that impact their surroundings.

"Soil Improvement and Water Conservation Biotechnology" stands as a valuable reference for researchers, professionals, and scholars aiming to grasp and implement biotechnological innovations for sustainable soil and water administration. By amalgamating scientific expertise with pragmatic applications, this publication strives to contribute to the global endeavor to conserve our natural resources for forthcoming generations while equipping individuals with the essential tools for academic and occupational advancement.

Israel Valencia Quiroz
Phytochemistry Laboratory, UBIPRO
Superior Studies Faculty (FES)-Iztacala
National Autonomous University of Mexico (UNAM)
Tlalnepantla de Baz, Mexico State
Mexico

List of Contributors

América Patricia Garcia-Garcia	Research Laboratory on Emerging Pathogens, Interdisciplinary Research Unit in Health and Education Sciences (UIICSE), Faculty of Higher Studies Iztacala (FES)-Iztacala, Division of Research and Graduate Studies, National Autonomous University of Mexico, (UNAM), Tlalnepantla de Baz, Mexico State, Mexico
Adriana Montserrat Espinosa González	Phytochemistry Laboratory, UBIPRO, Superior Studies Faculty (FES)-Iztacala, National Autonomous University of Mexico (UNAM), Tlalnepantla de Baz, Mexico State, Mexico
Ana M. García-Bores	Phytochemistry Laboratory, UBIPRO, Superior Studies Faculty (FES)-Iztacala, National Autonomous University of Mexico (UNAM), Tlalnepantla de Baz, Mexico State, Mexico
Ana K. Villagómez-Guzmán	Natural Products Bioactivity Laboratory, UBIPRO, Superior Studies Faculty (FES)-Iztacala, National Autonomous University of Mexico (UNAM), Tlalnepantla de Baz, Mexico State, Mexico
Alonso Ezeta Miranda	Phytochemistry Laboratory, UBIPRO, Superior Studies Faculty (FES)-Iztacala, National Autonomous University of Mexico (UNAM), Tlalnepantla de Baz, Mexico State, Mexico
Aliana Zacaria Vital	Phytochemistry Laboratory, UBIPRO, Superior Studies Faculty (FES)-Iztacala, National Autonomous University of Mexico (UNAM), Tlalnepantla de Baz, Mexico State, Mexico
Alejandro Dorazco-González	Institute of Chemistry, National Autonomous University of Mexico (UNAM), Mexico City, Mexico
Alfredo Jiménez-Pérez	Centro de Desarrollos de Productos Bióticos, Instituto Politécnico Nacional (IPN), Yautepec, Morelos, Mexico
Brenda Karina Guevara Olivar	Faculty of Higher Studies Aragón, National Autonomous University of Mexico (UNAM), Mexico State, Mexico
Blanca González-Méndez	CONAHCYT, Insurgentes Sur 1582, Crédito Constructor, Benito Juárez, CP 03940 Mexico City, Mexico Northwest Regional Station, Institute of Geology, National Autonomous University of Mexico, Colosio and Madrid s/n, ZIP Code 83000, Hermosillo, Sonora, Mexico
Cesar Alejandro Zamora-Barrios	Environmental Conservation and Improvement Project, Interdisciplinary Research Unit in Health and Education Sciences (UIICSE), Faculty of Higher Studies Iztacala (FES)-Iztacala, Division of Research and Graduate Studies, National Autonomous University of Mexico, (UNAM), Tlalnepantla de Baz, Mexico State, Mexico
Casandra Rosales García	Phytochemistry Laboratory, UBIPRO, Superior Studies Faculty (FES)-Iztacala, National Autonomous University of Mexico (UNAM), Tlalnepantla de Baz, Mexico State, Mexico
Diana Violeta Sánchez Oropeza	Phytochemistry Laboratory, UBIPRO, Superior Studies Faculty (FES)-Iztacala, National Autonomous University of Mexico (UNAM), Tlalnepantla de Baz, Mexico State, Mexico

Elizabeth Chávez-García

Faculty of Philosophy and Literature, National Autonomous University of Mexico, University City, ZIP Code 04510, Mexico City, Mexico

Elvia Manuela Gallegos-Neyra

Research Laboratory on Emerging Pathogens, Interdisciplinary Research Unit in Health and Education Sciences (UIICSE), Faculty of Higher Studies Iztacala (FES)-Iztacala, Division of Research and Graduate Studies, National Autonomous University of Mexico, (UNAM), Tlalnepantla de Baz, Mexico State, Mexico

Edgar Antonio Estrella-Parra

Phytochemistry Laboratory, UBIPRO, Superior Studies Faculty (FES)-Iztacala, National Autonomous University of Mexico (UNAM), Tlalnepantla de Baz, Mexico State, Mexico

Erick Nolasco Ontiveros

Phytochemistry Laboratory, UBIPRO, Superior Studies Faculty (FES)-Iztacala, National Autonomous University of Mexico (UNAM), Tlalnepantla de Baz, Mexico State, Mexico

Francisco José Torner-Morales

Environmental Conservation and Improvement Project, Interdisciplinary Research Unit in Health and Education Sciences (UIICSE), Faculty of Higher Studies Iztacala (FES)-Iztacala, Division of Research and Graduate Studies, National Autonomous University of Mexico, (UNAM), Tlalnepantla de Baz, Mexico State, Mexico

Héctor M. Arreaga-González

Biotechnology Laboratory, Institute of Agroindustries, Technological University of Mixteca, Heroic City of Huajuapan de Leon, Oaxaca, Mexico

Helen Paola Toledo-Jaldin

Institute of Metallurgy, Autonomous University of San Luis Potosi, San Luis Potosi, Mexico
National Technological of Mexico, Technological of Superior Studies of Tianguistenco, Mechanical Engineering Division, Tenango-La Marquesa Km22, Santiago Tilapa, Santiago Tianguistenco, Mexico

Iván E. Barrera

Postgraduate in Biological Sciences, Postgraduate Studies Unit, National Autonomous University of Mexico (UNAM), Coyoacan, Mexico City, Mexico

Israel Valencia Quiroz

Phytochemistry Laboratory, UBIPRO, Superior Studies Faculty (FES)-Iztacala, National Autonomous University of Mexico (UNAM), Tlalnepantla de Baz, Mexico State, Mexico

Iván J. Bazany-Rodríguez

Faculty of Chemistry, National Autonomous University of Mexico (UNAM), Mexico City, Mexico

José G. Avila-Acevedo

Phytochemistry Laboratory, UBIPRO, Superior Studies Faculty (FES)-Iztacala, National Autonomous University of Mexico (UNAM), Tlalnepantla de Baz, Mexico State, Mexico

Jessica Hernández-Pineda

Department of Infectious Diseases and Immunology, INPer, SSA, Mexico City, Mexico

José Cruz Rivera-Cabrera

Liquid Chromatography Laboratory, Department of Pharmacology, Military School of Medicine, CDA, Palomas S/N, Lomas de San Isidro, Mexico City, Mexico

Juan Pablo León-Gómez

Institute of Chemistry, National Autonomous University of Mexico (UNAM), Mexico City, Mexico

María K. Salomón-Flores	Institute of Chemistry, National Autonomous University of Mexico (UNAM), Mexico City, Mexico
María Alejandra Istúriz-Zapata	Centro de Desarrollos de Productos Bióticos, Instituto Politécnico Nacional (IPN), Yautepec, Morelos, Mexico
María Fernanda Trujillo Lira	Phytochemistry Laboratory, UBIPRO, Superior Studies Faculty (FES)-Iztacala, National Autonomous University of Mexico (UNAM), Tlalnepantla de Baz, Mexico State, Mexico
Miriam Arlette López Pérez	Phytochemistry Laboratory, UBIPRO, Superior Studies Faculty (FES)-Iztacala, National Autonomous University of Mexico (UNAM), Tlalnepantla de Baz, Mexico State, Mexico
Nallely Alvarez-Santos	Phytochemistry Laboratory, UBIPRO, Superior Studies Faculty (FES)-Iztacala, National Autonomous University of Mexico (UNAM), Tlalnepantla de Baz, Mexico State, Mexico
Nichdaly Ortiz Chacón	Phytochemistry Laboratory, UBIPRO, Superior Studies Faculty (FES)-Iztacala, National Autonomous University of Mexico (UNAM), Tlalnepantla de Baz, Mexico State, Mexico
Olivia Pérez-Valera	Institute of Chemistry, National Autonomous University of Mexico (UNAM), Mexico City, Mexico
Paris O. Gonzalez	Phytochemistry Laboratory, UBIPRO, Superior Studies Faculty (FES)-Iztacala, National Autonomous University of Mexico (UNAM), Tlalnepantla de Baz, Mexico State, Mexico
Rafael Torres-Martínez	Chemical Ecology and Agroecology Laboratory, Research Institute for Ecosystems and Sustainability, National Autonomous University of Mexico (UNAM), Morelia, Michoacan, Mexico
Román Adrián González Cruz	Natural Products Bioactivity Laboratory, UBIPRO, Superior Studies Faculty (FES)-Iztacala, National Autonomous University of México (UNAM), Tlalnepantla de Baz, México State, México
Susana Rafael Maya	Phytochemistry Laboratory, UBIPRO, Superior Studies Faculty (FES)-Iztacala, National Autonomous University of Mexico (UNAM), Tlalnepantla de Baz, Mexico State, Mexico
Tzasna Hernández-Delgado	Natural Products Bioactivity Laboratory, UBIPRO, Superior Studies Faculty (FES)-Iztacala, National Autonomous University of Mexico (UNAM), Tlalnepantla de Baz, Mexico State, Mexico
Ván Andrés Arredondo-Fragoso	Environmental Conservation and Improvement Project, Interdisciplinary Research Unit in Health and Education Sciences (UIICSE), Faculty of Higher Studies Iztacala (FES)-Iztacala, Division of Research and Graduate Studies, National Autonomous University of Mexico, (UNAM), Tlalnepantla de Baz, Mexico State, Mexico
Yesica R. Cruz-Martinez	Institute of Chemistry, National Autonomous University of Mexico (UNAM), Mexico City, Mexico
Yuri Córdoba Campo	Fundación Universitaria Navarra, Uninavarra, Neiva, Huila, Colombia

CHAPTER 1

Soil: Composition, Function, and Current Challenges

Brenda Karina Guevara Olivar[1,*]

[1] *Faculty of Higher Studies Aragón, National Autonomous University of Mexico (UNAM), Mexico State, Mexico*

Abstract: Soil embodies the fundamental basis for the sustenance of human existence on the planet. We are linked to its origin, evolution, and the set of processes that occur within its matrix. It forms close links with the water, energy, and food of all living beings that inhabit the planet. While it is customary to periodically conduct assessments of its physical, chemical, and biological characteristics, it is imperative to evaluate and identify the most effective research methodology to enhance the utility of the existing data. The current environmental challenges are setting the tone for the paradigm that must be followed for developing the tools and forming the working groups that will allow us to face the challenges that arise from an objective, scientific, and technological basis. In the current research endeavor, an examination is conducted on the fundamental components that confer significance to the soil, alongside an investigation of various concepts such as resilience and sustainability, which are integral to the foundational guidelines that ought to be considered as the preliminary framework in assessing the properties of the soil. Furthermore, the study emphasizes the development of bioindicators that facilitate the standardization of evaluative processes regarding the capabilities of ecosystem services and their associated attributes. The planet faces a dramatic environmental scenario with climate change as the most significant challenge. Along with water scarcity, soil salinity, and the acceleration of aridification, the scientific community must reorient efforts and work together; we can no longer be spectators.

Keywords: Bioindicators, Climate change, Ecosystems, Environmental services, Land aridification, Soil, Sustainability, Soil properties.

INTRODUCTION

One of the paramount characteristics of soil is its considerable variability, which consequently renders it exceedingly intricate due to the extensive interrelationship of systems that coalesce within its matrix. The genesis of soil can be attributed to

---

[*] **Corresponding author Brenda Karina Guevara Olivar:** Faculty of Higher Studies Aragón, National Autonomous University of Mexico (UNAM), Mexico State, Mexico; Tel: +525516814412;
E-mail: brendaguevaraguo@aragon.unam.mx

Israel Valencia Quiroz (Ed.)

the intricate interplay of various environmental determinants, microbial organisms, geological constituents termed "parent material", and topographical features, which are interconnected over extensive temporal spans; consequently, sedimentary layers of diverse depths are generated, consistently shaped by a predominant soil-forming agent. This agent imparts certain characteristics to the newly synthesized sediments, thereby endowing them with novel physical, chemical, and biological attributes.

As soil-forming factors are never homogeneous and are governed by the laws of thermodynamics, the variability of their properties is also influenced by the environmental conditions in which they develop. In this way, when the parent material that gives rise to the same specific type of sediment develops in two different environments, it forms two different types of soils; consequently, anisotropic processes and conditions vary according to the direction that the initial materials take as they continue to differentiate into several horizons parallel to the surface, which is called "Horizonization". Although knowledge about pedogenesis and formation processes is extensive, the concept of soil continues to evolve as we attribute new meanings to it in our lives, regardless of the context or discipline in question. A study [1] initially proposed that there are two forms of components from which soils can be considered: i) based on the nature of their properties, and ii) based on their specific functions or land use. This elucidates our current assertion that fundamentally, soil constitutes a natural entity consisting of stratified layers (horizons) that are formed from weathered mineral constituents, organic substances, gaseous elements, and hydric components, having evolved as a vital element of both the "Earth" and the "ecosystems" [2].

Physical Components of Soil

Soil physical attributes, such as texture, structure, and moisture regime, are closely linked to climate and significantly impact plant growth and crop production [3, 4]. Drought, affecting about 57% of global soils, is influenced by soil texture, depth, and water-holding capacity [5]. Soil coarsening can alleviate precipitation constraints on vegetation growth in drylands [6]. Soil cracking, which is related to mineralogy, is another physical restriction affected by soil-water-atmosphere interactions [7]. Climate change is expected to increase aridity, leading to soil degradation and altered soil functions [8]. Arid soils, in particular, face challenges, such as low organic carbon, poor structure, and reduced biodiversity [9, 10].

Among physical soil properties, structure is the parameter that is considered most relevant to evaluate the productivity, sustainability, and degradation of ecosystems or agroecosystems [11]. Soil structures consist of a set of data, including aggregate stability and porosity [12]. The stability of aggregates is ascertained through an evaluation of a range of chemical, physical, and biological characteristics, along with agricultural management or conservation methodologies [13]. Soil porosity and structure significantly influence water retention and movement, which in turn affect soil quality and agricultural productivity [14, 15]. Land use and management practices impact soil structure, pore size distribution, and water retention properties [16, 17]. For instance, no-tillage practices generally improve soil aggregation and microporosity while reducing macroporosity [17]. Wetting and drying cycles can alter soil pore architecture and water retention characteristics [18, 19] and liming affects pore geometry and water retention in acid soils [20]. Soil structure has a significant impact on water infiltration, with correlations established between infiltration rates and structural properties [21]. Understanding these relationships is crucial for sustainable soil management and optimizing agricultural practices to maintain soil health and water availability for crops [22]. Studies carried out [23] on the micromorphology of aggregates and their relationship with environmental factors showed that different species of arbuscular mycorrhizal fungi have the capacity to generate different types of aggregates, so these tend to reconfigure indefinitely as the conditions vary. Arbuscular mycorrhizal fungi (AMF) play a crucial role in improving soil fertility and reducing erosion in agricultural ecosystems. It enhances nutrient uptake, particularly phosphorus, and improves soil structure [24]. AMF inoculation has been shown to significantly reduce erosion-induced soil nutrient losses, especially nitrogen and phosphorus, by enhancing plant nutrient uptake and decreasing runoff and sediment loss [25]. The synergistic interaction between AMF and associated microbiota contribute to soil fertility and plant health [26, 27]. Utilizing AMF and microbial consortia in agriculture can promote sustainable practices and improve crop yields [28]. In summary, the aforementioned attributes endow the soil with essential qualities necessary for preserving the optimal standards of its environmental functions.

Chemical Components of Soil

Chemical weathering of primary minerals is a fundamental process in soil formation, transforming rock components into stable substances in the surface environment [29]. This process is influenced by factors such as climate, topography, and parent material [29, 30]. Soil composition varies with weathering intensity, with clay-rich soils typically forming in warm climates where chemical weathering dominates [31]. In this manner, the byproducts of chemical weathering exhibit fundamental stability, provided that they persist within an environment analogous to that in which their formation transpired [32]. Currently, soil science considers three predominant origins of weathered materials, which then become soil: geological, organic, and anthropic. In a study [33], it is proposed to classify mineral and organogenic materials based on their characteristics, as well as their chemical and mineralogical composition. This system identifies five classes related to the ferromagnesian content and the sixth one related to the presence of carbonates. The seventh and eighth classifications are conditioned by salt content, and the last considers the proportion of organic matter. It should be noted that even this classification generates different conceptual conflicts with established geological definitions. In this way, igneous, sedimentary, and metamorphic rocks are broadly acknowledged as the fundamental progenitor materials in geological discourse.

Peat is an organic material derived from plant remains that accumulates in water-saturated environments, where decomposition is inhibited [34]. The main parent materials of origin currently considered by soil science are made up of bryophyte peats in boreal, arctic, and subarctic regions, peats originating in temperate rainy regions, and peat from swampy areas in humid tropical regions. In Mexico, such parent material is distributed mainly in old lacustrine or swamp areas, and the genus Sphagnum constitutes the most abundant element in the deposits [35].

Based on the WRB [36], two types of anthropic materials are recognized. The first is made up of soil that has been subjected, for decades, to prolonged anthropogenic processes, including intensive fertilization, both organic and inorganic, liming, permanent annual irrigation and the addition of significant amounts of sediments and deep subsoil. All these processes, when repeatedly applied to the soil, create diagnostic horizons, new parental materials, and a distinct epipedon. The second type of anthropogenic parent material is made up of various unconsolidated mineral or organic materials that result from the technogenic activities of humanity, such as urban and sanitary landfills, mine slag (tailings), and dredging materials, among others. One of the elements that greatly transforms the chemical properties of soils is water. It plays a crucial role in chemical weathering processes, acting as the primary solvent and reaction medium [37, 38]. Its effectiveness stems from its ability to wet mineral surfaces, penetrate small openings, and form sheaths around dissolved ions [39]. The main chemical weathering processes include dissolution, hydrolysis, oxidation, and carbonatization [40, 41], all of which are influenced by factors such as pH, Eh, and the presence of complexes [42].

Biological Component of Soil

One of the most important biogeochemical processes that occur within the soil is the nitrogen cycle, which is related to microbial activity and edaphic fauna such as earthworms, nematodes, protozoans, fungi, bacteria, and arthropods. Soil biology plays a fundamental role in soil composition and characteristics.

However, being a newly discovered science, much remains to be investigated about how it affects the nature of soils. Soil organisms decompose organic matter from plant and animal remains, in turn releasing nutrients to be assimilated by plants. Nutrients stored within soil organisms prevent their loss through leaching. Soil microorganisms are instrumental in preserving the structural integrity of the soil, whereas earthworms facilitate the removal of soil materials. Bacteria are integral to the nitrogen cycle, participating in various biochemical processes that are essential for nutrient cycling [2].

Nitrogen fixation in the soil occurs when the mineralization of nitrogen occurs from impregnation with ammonia or the ammonia component (NH_3). A process where pure forms of nitrogen are transformed into ammonium (NH_4^+) with the help of decomposers or bacteria. Nitrogen mineralization, or ammonification, is a crucial process in the nitrogen cycle where organic nitrogen is converted to inorganic forms, primarily ammonium (NH_4^+), by soil microorganisms [43, 44]. This process is essential for plant nutrition and is influenced by various factors, including soil temperature, moisture, organic matter composition, and microbial activity [45, 46]. Subsequently, the nitrification process occurs, which is divided into three stages. In the first, bacteria transform nitrogen into ammonium (NH_4^+) so it can be absorbed by the roots of the plants. In the second stage, the ammonium is oxidized, and nitrite NO_2^- is formed. Finally, in the third stage, nitrate $NO_3^-$is formed through oxidation. Nitrogen fixation is facilitated by bacteria or algae in the soil capable of fixing atmospheric nitrogen, incorporating it into their organisms, and depositing it in the soil once upon their death. *Azobacter* and *Clostridium* bacteria are named as non-symbiotic nitrogen fixers. Bacteria that carry out symbiotic attachment include *Rhizobia*. Its habitat is found around legume roots, which form nodules in the cortical cells inhabited by bacteria. On the other hand, the denitrification process returns nitrogen to the atmosphere. Anaerobic bacteria *Achromobacter* and *Pseudomonas* convert nitrates and nitrites to nitrogen oxide N_2O or molecular N (N_2). In excess, the process tends to lead to total losses of available nitrogen in the soil and, consequently, its fertility.

Soil microorganisms play a crucial role in carbon cycling and organic matter dynamics [47]. They actively participate in decomposition processes, influencing soil carbon storage and turnover [48]. Microbial compounds, rather than plant residues, predominate in long-term soil organic matter [49, 50]. Soil fauna significantly impacts carbon turnover, sometimes altering carbon sinks into sources [51]. Increased agricultural productivity is associated with greater soil carbon stocks and faster turnover rates [52]. A large part of the organic matter originating from the annual decomposition of plant residues accumulates on the soil surface or in the root zone, where it is almost completely consumed by soil organisms, creating a carbon reserve with a rapid rate of renewal, in numerous instances, between 1 and 3 years. The byproducts of microbial consumption result in emissions of carbon dioxide (CO_2), water (H_2O), and various organic compounds known as humus. Humus is composed of substances that are difficult to degrade and, therefore, have a slow decomposition rate. Formed in superficial soil horizons, generally, a fraction precipitates towards lower profiles as clay-humic complexes. In deeper soil profiles, the oxygen content is usually lower, which makes it difficult for organisms to decompose humus. However, over time, due to several natural processes that remove the soil, the humus is returned to higher horizons, where it can decompose and release additional CO_2. This is why humus constitutes a more stable reserve for soil carbon with a long duration, persisting for hundreds to thousands of years [53]. In most soils, rapid and slow humus decomposition leads to a residence time of about 20 to 30 years [54]. Soil microorganisms are highly sensitive to organic carbon and water content, as well as temperature, so they increase respiration in high carbon content, elevated temperatures, and humid conditions [55]. Recent studies highlight the critical role of soil organic matter (SOM) and soil organic carbon (SOC) in maintaining soil health, productivity, and ecosystem services. SOM enhances soil physical, chemical, and biological properties, supporting microbial activity and nutrient cycling [56, 57]. Conservation agriculture practices can increase SOM content, improving soil functions and crop productivity [58]. SOC dynamics strongly impact soil health, food security, and climate change mitigation [59]. Forest ecosystems can store significant amounts of carbon, with seasonal variations affecting SOC and microbial biomass carbon levels [60]. Overall, maintaining and enhancing SOM is crucial for sustainable agriculture and ecosystem health [61].

Soil microbiota plays a crucial role in maintaining soil fertility, structure, and health through various mechanisms. They contribute to nutrient cycling, organic matter decomposition, and soil aggregate formation [62 - 64]. Microorganisms like bacteria, fungi, and algae form associations with plant roots, promoting growth and protecting against pathogens [64]. The core microbiota is particularly important in maintaining complex connections between bacterial taxa and is associated with multinutrient cycling [65]. Additionally, soil microbiomes also enhance plant resilience to abiotic stresses and suppress diseases [66]. Understanding and harnessing soil microbiomes offer promising opportunities for sustainable agriculture and ecosystem management [67 - 69]. Soil aggregate stability plays a crucial role in various physical and biogeochemical processes, significantly influencing soil quality and ecosystem functioning [70, 71]. Aggregate stability is affected by pore system characteristics, with smaller aggregates exhibiting higher porosity [72]. Microaggregates within macroaggregates contribute to overall stability and pore space architecture [73]. Soil organic matter and its humic fraction modify pore size distribution and hydrophobicity, impacting aggregate stability [74]. Microbial processing of organic matter drives aggregate formation and stability [75]. The pore network within aggregates influences water and soluble carbon transport [76]. Aggregate-associated microbial communities, described as "microbial villages," play a significant role in biogeochemical cycling [77, 78]. Soil aggregation is a complex process involving biotic and abiotic interactions that contribute to soil structure and stability [79]. Plant roots, mycorrhizal fungi, and soil biota play crucial roles in aggregate formation and stabilization through various mechanisms [80, 81]. Spatial aggregation patterns are influenced by root architecture, hydrologic fluctuations, and carbon inputs [82]. The dynamics of soil aggregation occur at multiple scales, from micro to macro, and are affected by factors such as soil moisture, organic matter, and microbial activity [79, 83]. Understanding these processes is essential for soil carbon sequestration and ecosystem functioning [83]. While progress has been made in documenting the contributions of soil biota to aggregation, quantitative models are still in the early stages of development [79]. Further research is needed to elucidate the mechanisms of temporal and spatial aggregation across different scales [84 - 86]. For example, Arbuscular Mycorrhizal Fungi (AMF) play a crucial role in soil aggregation and structure. AMF increases the formation of large macroaggregates and slows down their disintegration while enhancing microaggregate turnover [87]. They improve soil structure by increasing organic matter content and glomalin-related soil protein, with hyphal density having the most significant direct effect [88]. AMF binds mineral particles and organic materials to form stable microaggregates [89]. Studies using split-root devices have shown that mycorrhizal hyphae have a greater effect on soil aggregation than roots alone but less than mycorrhizal roots [90, 91]. Different AMF species vary in their ability to improve soil structure [23, 92].

Soil Functions and Environmental Services

Historically, soil science has been connected to environmental challenges as they arise, with agriculture being the activity most dependent on variations in soil properties. This dependence has led agriculture to face different challenges over the decades. In the mid-1980s, scholarly investigations pertaining to pedology predominantly concentrated on themes including climatic variations, governance of natural resources, and the provisioning of ecosystem services [93, 94], substantiating the importance of soil in the face of environmental problems [95, 96]. In a study [96], it was articulated that one method to elucidate the significance of soil science is through the acknowledgment of its multifaceted roles and the array of services that this critical resource provides within both ecosystems and agroecosystems. This is particularly pertinent to issues concerning food security, water scarcity, climate change, biodiversity loss, and various health threats, all of which, when managed effectively, can substantially contribute to the achievement of certain Sustainable Development Goals.

Soil functions are essential capacities that soil provides for agriculture, the environment, and human society. These functions include biomass production, nutrient cycling, water regulation, environmental interactions, biological habitat, genetic reserve, raw material provision, and support for anthropic structures [97 - 99]. Soil functions are linked to processes that represent the system's capacity to provide services valued by individuals and society.

Based on the available information [100], the authors of a study grouped the set of key soil properties that are most importance to evaluate the functions that soil has in ecosystems (Fig. **1**). They emphasize the importance of soil information, advocating for the inclusion of its properties in all ecosystem service (ES) modeling studies. They also propose forming multidisciplinary work groups to guide the study of soil functions in alignment with the sustainability objectives of the United Nations.

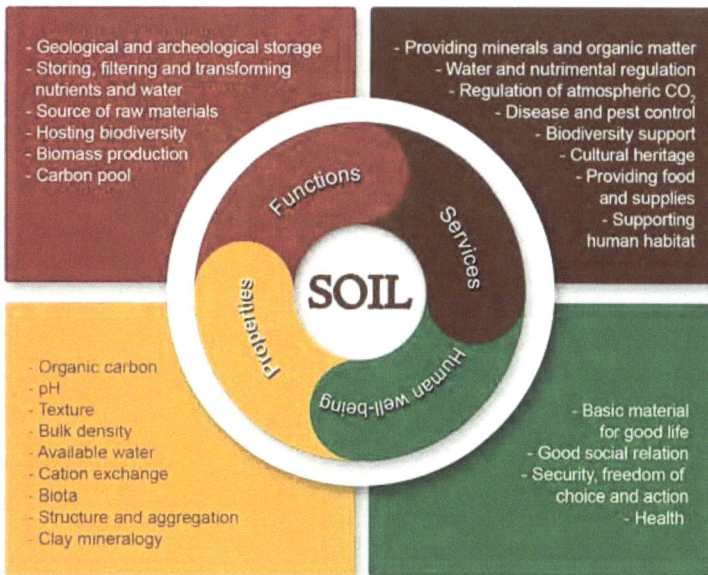

Fig. (1). A conceptual framework delineating the interrelationships among essential soil characteristics, ecosystem services, and soil functionalities contributing to human well-being (adapted from [100]).

The provision of ecosystem services by soil is contingent upon the inherent characteristics of soil and the interactions that are predominantly shaped by its utilization and stewardship [101].

Soil degradation is a critical global challenge threatening food security and ecosystem sustainability [102, 103]. Factors such as landslides, erosion, declines in soil carbon, and biodiversity are leading to soil degradation. The contribution of soils to human well-being beyond food production requires appreciation, which can be addressed by incorporating soils into the ecosystem services framework and linking them to the multitude of functions they provide [104, 105]. Currently, there is significant competition for land resources, with primary uses focused on the production of energy (agrofuels), human food (greater demand), and animal food. This competition jeopardizes fundamental aspects at risk, such as protection of biodiversity, ecosystem services, energy security, water security, and climate change mitigation [106, 107]. In light of this extensive overview, there has been a notable escalation in initiatives aimed at safeguarding the resource recently, particularly emphasizing the establishment of legal regulatory frameworks [108].

While many studies link soil properties to ecosystem functionalities, direct connections between soil properties and services are limited [101]. Most of these studies focus on economic evaluations and valuations, but these approaches cannot capture the complexity of nature and the services it provides [109]. Nonetheless, these studies are needed to provide an understanding of the functionalities of the ecosystem and the services it offers, as these are evolving. Recently, there have been a series of articles in which the potential role of soils has been analyzed in a holistic analysis of the ecosystem.

Mapping spatial aspects of soil ecosystem services is crucial for sustainable environmental management, but challenges remain in data quality, model uncertainty, and user needs assessment [110, 111]. The quality of the data is a bottleneck regarding the reliability of the map. This issue could be addressed using model uncertainty and accuracy assessments. Additionally, understanding soil microbial diversity and its role in soil functional variability should receive more attention, as it is essential for translating biodiversity into soil services [112, 113].

Soil science plays a crucial role in achieving the UN Sustainable Development Goals (SDGs). To address these challenges effectively, soil scientists must adopt transdisciplinary approaches, collaborating with experts from various fields [114, 115]. Future studies should focus on soil carbon and water-related services [116]. An example of an interdisciplinary approach is where soil scientists collaborate proactively with crop breeders, agronomists, and irrigation engineers, demonstrating the value of soil information on crop yield in the context of climate change. Recent research emphasizes the need to rethink the fundamentals of soil science and effectively communicate their value to policymakers [96]. It is important to understand the limitations of the term "soil contribution" for ecosystem services because the current knowledge base may be limited, and an incomplete understanding can lead to poor decisions. We should influence the ecological and socioeconomic aspects, services, and their provision to ensure a sustainable future of the soil and the global ecosystem. Advanced techniques like digital soil mapping and geomorphometry can enhance soil information collection and processing [117, 118].

Assessing soil health requires holistic approaches using bioindicators that reflect soil functionality and resilience [119, 120]. Bioindicators are measurable surrogates for environmental endpoints that are themselves too complex to assess or too difficult to interpret in terms of ecological significance [121]. The challenge posed by a growing population is the increased and intensified use of ecosystem services provided by soils. To ensure sustainable agriculture, soil management must simultaneously include different aspects of the multiple soil functions and ecosystem services linked to this nexus [122]. The United Nations Sustainable Development Goals (SDGs) present a significant challenge and opportunity for soil science to contribute to sustainable development [123, 124]. The search for relevant bioindicators for arable soils has given rise to an enormous production of studies at a national and international level over the last decades, currently driven by the specification of the SDGs towards the integration of more soil-related indicators [116].

Soil quality assessment is crucial for sustainable land management, but establishing standardized indicators remains challenging [125]. The two main approaches used to establish sustainability indicators are the development of a single composite index or the development of a set of indicators. A single indicator is often deemed insufficient to provide a complete picture of a complex system [126]. However, too many indicators make it difficult to collect and process data at a reasonable cost and time. Conversely, relying on very few indicators can miss crucially important events. To evaluate an indicator, reference values are first required to be compared and contextualized; otherwise, a real interpretation is not possible [127]. In general, most current approaches reflect natural dimensions, including soil physical, chemical, and biological parameters [128]. However, these approaches often lack social dimensions that are relevant, especially for production ecosystems. Some promising methods have emerged to address these gaps. Several studies [129, 130] calculated the maximum ecological potential of arable soils by comparing them with a representative reference soil sample (equal to 100%). Additionally, a research study [131] reported the estimation of their so-called biological index of soil quality (BSQ), based on soil microarthropods in relation to land use and land cover (LULC) data, to perform a state-level sustainability assessment. Furthermore, references [132, 133] highlights the need for holistic and interdisciplinary collaboration in sustainability research and to encourage ecologists to engage in increasingly important new interfaces between disciplines. Soil health assessment is crucial for sustainable agriculture and ecosystem services. Traditional approaches focus on physical, chemical, and biological parameters [134, 135]. However, these approaches often lack social dimensions that are relevant, especially for production ecosystems. Recent studies emphasize the importance of soil microarthropods as biological indicators of soil quality [136, 137]. The Biological Soil Quality (BSQ) index, based on microarthropods, has been developed to evaluate soil quality in relation to land use [138]. In sustainability research, there is a need for holistic and interdisciplinary collaboration to address these issues [120, 135]. Promising approaches can be found that attempt to solve at least some of these problems, including the calculation of the maximum ecological potential of arable soils by comparing them with a representative reference soil sample (equal to 100%) and the estimation of biological indices of soil quality based on soil microarthropods in relation to land use and land cover (LULC) data [131].

Resilience in ecosystems is defined as the capacity to absorb disturbances while maintaining essential functions and structures [139, 140]. It is not a single parameter but rather a holistic metafunction of a community, a soil, or an entire ecosystem, derived from all its individual properties, in interaction with ongoing processes driven by interactions of the biota. It is a measure of the (pre)adaptive potential to cope with future perturbations and represents the multifunctionality of the system [140, 141]. Soil resilience is increasingly recognized as a crucial indicator of soil sustainability and quality [119, 142]. It reflects the soil's ability to maintain functional integrity and recover from disturbances, integrating aspects of past management, current status, and future adaptive potential [96, 143]. Resilience encompasses all levels of indication important for assessing soil status: it is derived from past soil management as part of the soil memory, represents the current status of the soil under given pressures and disturbances, and provides possible recommendations for the future improvement of holistic soil management through the evaluation of intrinsic adaptive potential. Consequently, resilience is considered a crucial part of sustainability or is even equated with the term sustainability [119].

The concept of panarchy in agroecosystems emphasizes the interconnectedness of field-scale processes with soil structural units and biodiversity across multiple spatial and temporal scales [144, 145]. This multiscale approach involves understanding how legacies affect slower cycles (farm, landscape) at larger scales and faster cycles (pedon, biogenic structure) at smaller scales [146]. Agricultural management practices aimed at homogenizing soil can have significant impacts on soil structure. The homogenization of pedons affects soil's physical and chemical properties, leading to changes in biogenic structures, such as the alteration of aggregates and biogenic pores. These changes can vary from crop cycle to crop cycle. In turn, biogenic structures contribute to the formation of the soil profile, resulting in structurally diverse pedons and increasing heterogeneity within an agricultural field [147, 148].

Agricultural ecosystems provide essential services and rely on natural ecosystem services. A farm benefits from the ecosystem services provided by soil biota, such as regulating water balance, aeration, and promoting soil fertility [149]. Conversely, a farm can cause damage to the ecosystem, such as root infestation by pests and soil-borne pathogens. These impacts are controlled by agricultural management practices [150, 151]. Agricultural landscapes are shaped by multiple farms, resulting in what is known as a cultivated landscape. Farm emissions, such as agrochemicals through drainage and climate-relevant trace gases through livestock farming, can cause environmental stress on the landscape [152]. However, a high degree of diversification improves the aesthetic value of a cultivated landscape and provides multiple habitat structures for beneficial organisms on farms [153].

Soil biodiversity plays a crucial role in ecosystem functioning and stability across scales. At smaller scales, the genetic diversity of soil biota drives soil processes, while at larger scales, species diversity ensures soil functioning and the provision of ecosystem services [154, 155]. In both cases, redundancy allows for functional stabilization and makes the system more robust against adverse impacts. This is similar to the relevance of large-scale ecosystem diversity. To assess the health of the soil ecosystem, soil resilience can be considered an additional dimension of soil functioning [119].

The concept of resilience has evolved across multiple disciplines, originating in engineering and psychology before being adopted in ecology [156]. In psychology, resilience is viewed as a dynamic process of positive adaptation to adversity [157]. For ecosystems, there are two distinct definitions of resilience, both of which address aspects of system stability: engineering resilience, focusing on efficiency, control, consistency, and predictability [158]. The concept of ecological resilience emphasizes the dynamic characteristics of ecosystems, particularly the idea that there is not a single equilibrium state [139]. Resilience is characterized by the time required for a system to return to an equilibrium state after a disturbance, which defines stability by the attributes of persistence, adaptability, variability, and unpredictability [159].

In ecological systems, resilience encompasses two distinct aspects: engineering resilience, which focuses on maintaining function efficiency, and ecological resilience, which emphasizes the preservation of function existence [160, 161].

These two contrasting aspects of system stability have fundamentally different consequences for understanding, evaluating, and managing complexity and change, especially regarding a multifunctional system like soil [162]. The resilience approach challenges the dominant eco-efficiency paradigm in sustainability, arguing that optimizing for short-term efficiency can reduce long-term ecological resilience [163]. This paradox arises because, while short-term efficiency may be beneficial, it often hinders long-term sustainability by diminishing ecological resilience [164].

Soils are complex, dynamic systems that face increasing anthropogenic pressures [165, 166]. In this sense, we can say that anthropogenic management creates additional pressure and pulse disturbances for the soil ecosystem. Therefore, soils need adaptive management that leaves opportunities for positive legacy effects (*e.g.*, novelty, redundancy) and provide time for self-organized restoration periods that contribute to soil resilience [167, 168].

Soil sustainability and multifunctionality are complex concepts requiring holistic approaches to measurement and management [119, 142]. Resilience has emerged as a promising indicator for soil sustainability, integrating ecological and social components across scales [119, 169]. As a metafunction of soil, resilience incorporates the system's multifunctionality, making it a promising parameter for assessing soil management sustainability. It reflects the highly interrelated ecological and social components, while its importance for ecosystem functioning is derived from all levels of the functional hierarchy, on all scales [170]. Soil resilience, crucial for ecosystem stability and sustainability, is derived from functional redundancy and reinforcement across different levels [119]. The measurable part of soil resilience is redundancy within scales represented by soil response diversity. This allows the identification of soil ecosystem resistance and resilience thresholds and potential tipping points that could lead to a regime shift away from irreversible or permanent unfavorable soil states (*e.g.*, rigidity or poverty traps) [171].

Arable soils (*e.g.*, in contrast to forest soils) are posited to exhibit, on the one hand, elevated ecological resilience, as they have been conditioned to endure perturbations (recurring structural failures, mechanical and chemical stresses resultant from tillage, harvesting, and agrochemical applications). Consequently, arable soils manifest a broader spectrum of responses compared to forest soils [165]. The water-energy-food nexus further highlights the interconnectedness of these systems and the need for holistic approaches to address sustainability challenges [172]. Agroecosystems prioritize production efficiency, which only supports what is directly and immediately beneficial, leading to drastic losses in soil resilience [119]. Therefore, to find management solutions for challenges linked to the water-energy-food nexus, we need holistic approaches that explicitly incorporate meta-functions such as resilience towards any sustainability goal.

In this regard, effective measures are essential for monitoring the resilience of arable soil effectively and adjusting our adaptive management accordingly to sustain the functional integrity of the soil. Resilience, defined as the ability to recover from disturbances, can serve as a bioindicator for soil sustainability [173]. Using the multi-omics approach in this context allows higher-level traits, *i.e.*, resilience, to be resolved alongside very specific and discriminating traits, such as key functions and organisms that have a disproportionate effect on soil functioning, making them crucial to overall soil health [174, 175].

Current Challenges in Soil Study and Management

Soil health is crucial for achieving the UN Sustainable Development Goals (SDGs), but it has been underrepresented in policy and research efforts [176]. Soil scientists have not adequately participated in the debate on SDG targets and indicators [177]. While there are four SDG targets that specifically mention soil and others that indirectly relate to it, only one explicit soil indicator has been established [123, 124]. Soil scientists have been minimally involved in SDG discussions and indicator development, possibly due to their strong focus on pure soil science rather than engaging in interdisciplinary discussions on soil-related issues with other stakeholders [123]. To help achieve the UN Sustainable Development Goals (SDGs), it is essential to encourage interdisciplinary soil research among soil scientists and researchers in fields related to social sciences, climate sciences, ecology, and environmental sciences [178, 179]. When national and local governments formulate policies in line with the UN SDGs, soil scientists should be encouraged to play a more active role, and decision-makers should seek their advice [180]. For example, nominating soil scientists to key steering committees can ensure their involvement in policy development related to soil management and sustainability. Soil systems are recognized as complex and heterogeneous, presenting challenges for sustainable land management due to their inherent spatial heterogeneity, which occurs across scales from micro to global levels [181, 182]. This variability complicates the prediction of non-linear relationships between various soil processes and system behaviors [182]. Representing small-scale processes in broader models remains challenging [183].

Soil organic carbon (SOC) stocks exhibit significant spatial heterogeneity and scale-dependent variability, posing challenges for accurate estimation and modeling. Regional estimates of SOC stocks can differ by up to 60% at different scales due to this heterogeneity [184]. Furthermore, little is known about the vertical distribution of organic carbon in the subsurface, and since large amounts of carbon are stored in deep soils, it is essential to understand the state and mechanisms of soil carbon cycling throughout the lithosphere [185].

Long-term studies are crucial for understanding soil dynamics and developing sustainable agricultural practices, as many prominent issues, such as the climate crisis, require insightful, long-term, evidence-based solutions. However, many existing studies, especially those on emerging issues, are based on short-term findings. For example, a recent study on pasture systems suggested that several species could be planted to mitigate nitrogen leaching associated with cow urine [186]. Yet, this promising finding was based on data collected over less than a year. Longer-term studies are needed to verify the effectiveness of such strategies. Additionally, greater efforts should be made in the research and development of accelerated aging techniques [187].

Soil health assessment is crucial for sustainable land management, requiring holistic indicators that reflect various processes in a concise, quantifiable, reliable, and meaningful way [135, 188]. These assessments should be site-specific, taking into account local geological, climatic, and social conditions [188]. This approach would be particularly valuable in assisting farmers in decision-making and translating soil science into practical applications for sustainable soil use and management practices. Recent research highlights the importance of sustainable soil management and the need for improved soil mapping at various scales. High-resolution mapping and clustering of soil properties on regional, national, or global scales would allow specific recommendations for sustainable soil management [189]. While many existing soil sustainability studies have focused on the impacts of socioeconomic activities (*i.e.*, soil management) on soil systems (*i.e.*, soil health), studies exploring the impacts of soil systems on socioeconomic systems are less common [190].

Sustainable land management requires effective information sharing and knowledge building among scientists, land managers, and stakeholders [191, 192]. Emerging technologies like 5G, big data, and machine learning offer opportunities to improve information management and knowledge sharing [192]. However, challenges remain for knowledge creators in making their information visible in a vast ocean of data and for users in distinguishing whether the information is valuable or not [193, 194]. Interdisciplinary studies initiated by, or in collaboration with, communications and computer engineers have great potential to improve our capacity for the sustainable use and management of land resources.

Recent studies on landscape connectivity in Latin America underscore its importance for guiding ecosystem conservation and restoration efforts. While these studies represent an initial approach to conservation, they also present numerous challenges due to the complex integration of cultural, economic, social, environmental, and political factors. Despite the limited number of studies registered on the topics of connectivity, ecosystem services, land use, and restoration, these studies highlight significant scientific collaboration, providing diverse examples and the functioning of complex systems in various regions of Latin America over the last ten years [195, 196].

Research on landscape connectivity in Latin America has increased, but significant biases and gaps remain. There are still biases that need to be resolved, particularly in the six countries that did not register efforts to develop studies on landscape connectivity. Few studies were identified on the topics of connectivity in ecosystem services and connectivity and climate change, particularly the latter [195, 196]. The need for more research on these topics is evident [197, 198]. Addressing these biases is crucial for effective conservation planning and climate change resilience [199, 200].

For instance, the Mesoamerican Biological Corridor, comprising seven nations in Central America and Mexico, was designed as a holistic framework aimed at curbing the proliferation of the agricultural frontier to safeguard remnants of tropical ecosystems and forests. This initiative aimed to address challenges posed by climate change by diminishing fragmentation and enhancing the connectivity of ecosystems within the landscape by fostering sustainable productive methodologies that elevate the living standards of local communities to preserve biological diversity [201, 202]. Another illustration executed through the Convention on Biological Diversity in 2010 pertains to Aichi Target 11, which aimed to implement effective and immediate actions to halt the decline of biological diversity, guaranteeing that by the year 2020, ecosystems would be resilient and persist in delivering essential services. This would safeguard the diversity of life on the planet and enhance human well-being while facilitating the alleviation of poverty [203, 204].

Particularly in Mexico since the 1980s, following its accession to the General Agreement on Tariffs, Customs, and Trade (GATT), the government has progressively retreated from its involvement in agricultural production, storage and price regulation, as well as the framework of credit and subsidy provisions. As a result, food production ceased to be a key component of the national development strategy [205]. The industrial agriculture system that emerged has led to numerous environmental and social issues, including soil degradation, water contamination, loss of agro-biodiversity, greenhouse gas emissions, poverty, migration, monopolization of the agro-food system by transnational corporations, and public health concerns such as chronic diseases, obesity, and malnutrition [206, 207]. This system, dominated by large corporations, has marginalized small farmers and sustainable practices, resulting in the monopolization of the agro-food system [208, 209]. Additionally, the shift towards monocultures has contributed to the rise of chronic-degenerative diseases, obesity, and malnutrition [210, 211].

Recent agricultural policies in Mexico have shifted focus towards small producers and agro-ecological practices. Programs like Sembrando Vida and Production for Well-being promote agro-diversity and some agro-ecological practices, offering ongoing training for small producers [212]. Evaluating the impact of these programs on soil quality and environmental conservation is crucial and should be monitored through transdisciplinary studies and research to improve the design of these programs [213, 214]. On the other hand, programs such as Livestock Credit to the Word and Fertilizers for Well-being may also impact social and environmental areas, specifically soil quality, which must be monitored.

The "4 per 1000" initiative, launched at COP21 in Paris, aims to increase soil organic carbon (SOC) stocks to contribute to the increase in organic matter in the soil and mitigate high concentrations of carbon dioxide (CO_2) in the atmosphere [215, 216]. While this initiative provides a collaborative platform for stakeholders to address climate change, the specific actions taken by Mexico under this commitment remain unclear.

Soil organic matter (SOM) is essential for maintaining soil sustainability and protecting soil structure. Adding organic matter to soil increases porosity, improves structure, enhances water retention, and stabilizes soil aggregates. This, in turn, helps prevent erosion, provides energy to soil biota, and supplies nutrients to plants [217, 218]. Sustainable soil management through organic matter conservation supports agricultural productivity by maintaining soil quality and reducing erosion [219]. There are no official projects in Mexico related to this initiative.

The Initiative 20x20, launched at COP20 in Lima, aims to restore 20 million hectares of degraded and deforested lands in Latin America and the Caribbean [220]. Mexico has committed to restoring approximately 8.5 million hectares of degraded lands. However, the agreement only included areas under the care of public programs and not field projects operated by other entities, which limits the initiative's scope [221].To enhance its effectiveness and better address local needs, it has been suggested that the initiative be implemented at the municipal level.

Mexico's National Forest Commission (CONAFOR) has allocated nearly 600 million dollars to reforest one million hectares of degraded soils in the country, a goal initially set to be achieved by 2018 [221]. Additionally, almost half a million dollars were allocated to a private company for reforesting the Lake Texcoco area [222].

The Mexican Society of Soil Science (SMCS) has significant potential to contribute to improving the design of national public policy programs and monitoring international agreements, ensuring the correct use of resources [223, 224]. The magnitude of the deterioration of soils in Mexico demands that the SMCS diversify its academic activities to engage more closely with Mexican society and the National Government. This could include promoting forums, generating dissemination publications, participating in current policy discussions, and acting as consultants in relevant programs [225]. These efforts can be facilitated through networks and media while proactively engaging directly with farmers, rural schools, associations, cooperatives, and political-administrative authorities responsible for public policies and programs [178, 179]. These actions would align with international efforts, such as the World Soil Agenda promoted by the International Union of Soil Sciences, which focuses on science, policy, and implementation.

CONCLUDING REMARKS

This chapter emphasizes the critical importance of soil as a dynamic system that supports life and sustains ecosystems. Soil's complexity arises from its diverse physical, chemical, and biological properties, which are shaped by both environmental conditions and human activities. Understanding these properties is essential to managing soil sustainably and ensuring it continues to provide vital ecosystem services, such as water regulation, carbon sequestration, and nutrient cycling.

Current challenges, including climate change, water scarcity, and soil degradation, demand innovative and integrated approaches to soil management. Developing bioindicators and standardized methods for assessing soil health is crucial for monitoring its ability to sustain these services. Additionally, enhancing soil resilience is key to maintaining its functionality amid growing environmental pressures.

To address these challenges, soil science must become more deeply integrated into global sustainability frameworks, such as the United Nations Sustainable Development Goals. This integration will require active engagement from soil scientists in policy-making and interdisciplinary discussions to ensure that soil health and management are prioritized.

In summary, sustainable soil management is essential for food security, ecosystem health, and climate resilience. By fostering interdisciplinary collaboration and advancing research and innovative practices, we can protect and enhance the vital functions of soils for future generations.

QUESTIONS RELATED TO THE TEXT

1.- What are the main physical components of soil, and how do they influence soil productivity and sustainability?

2.- How do chemical properties of soil, such as mineralogy and organic matter content, affect its fertility and ability to support plant growth?

3.- Explain the role of biological components in soil health, specifically focusing on the nitrogen cycle and the contribution of soil microorganisms.

4.- Discuss the ethical and social considerations of implementing biotechnological interventions like biochar in agriculture, particularly for smallholder farmers.

5.- What are the current environmental challenges facing soil management, and how can biochar and other biotechnological solutions address these challenges?

6.- How do soil functions contribute to ecosystem services, and why is it important to integrate soil science into broader environmental and agricultural policies?

LIST OF ABBREVIATIONS

AMF	Arbuscular Mycorrhizal Fungi
BSQ	Biological Index of Soil Quality
CO_2	Carbon Dioxide
CONAFOR	Mexico´s National Forest Commission
Eh	Redox Potential
GATT	General Agreement on Tariffs Customs and Trade
H_2O	Water
LULC	Land Use and Land Cover
N	nitrogen
N_2	Molecular Nitrogen
N_2O	Nitrogen Oxide
NH_3	Ammonia
NH_4^+	Ammonium
NO_2^-	Nitrite
NO_3^-	Nitrate
pH	Hydrogen Potential

SDGs	Sustainable Development Goals
SMCS	Mexican Society of Soil Science
SOC	Soil Organic Carbon
SOM	Soil Organic Matter
UN	United Nations
WRB	World Reference Base

REFERENCES

[1] Strzemski M. Ideas underlying soil systematics. Poland: National Center for Scientific, Technical and Economic Information, Foreign Scientific Publications Dept 1975.

[2] Food and Agriculture Organization of the United Nations. All definitions | FAO Soils Portal [Internet]. [cited 2024 Aug 25]. Available from: https://www.fao.org/soils-portal/about/all-definitions/en/

[3] Várallyay G. Soil-water stress. Cereal Research Communications 2009; 37(2): 315-9.
[http://dx.doi.org/10.1556/CRC.37.2009.Suppl.7]

[4] Gardner C. Soil Physical Constraints to Plant Growth and Crop production. Land and Water Development Division, Food and Agriculture Organization. 1999; p. 96.

[5] Jones HG. Drought and other abiotic stresses. In: Plants and Microclimate: A Quantitative Approach to Environmental Plant Physiology. Cambridge University Press 2013; 255-89.

[6] Zhu X, Liu H, Xu C, Wu L, Shi L, Liu F. Soil coarsening alleviates precipitation constraint on vegetation growth in global drylands. Environ Res Lett 2022; 17(11): 114008.
[http://dx.doi.org/10.1088/1748-9326/ac953f]

[7] Ribeiro Filho JC, Andrade EM, Guerreiro MS, Palácio HAQ, Brasil JB. Soil–Water–Atmosphere Effects on Soil Crack Characteristics under Field Conditions in a Semiarid Climate. Hydrology 2023; 10(4): 83.
[http://dx.doi.org/10.3390/hydrology10040083]

[8] Várallyay G. The impact of climate change on soils and their water management. Agronomy Research 2010; 8(Special II): 385-96.

[9] Naorem A, Jayaraman S, *et al.* Soil Constraints in an Arid Environment—Challenges, Prospects, and Implications. Agronomy 2023; 13(1): 220.

[10] United States Department of Agriculture. Soil taxonomy: a basic system of soil classification for making and interpreting soil surveys 1975.

[11] Torres-Guerrero CA, Gutierrez-Castorena MC, Solorio CAO, Gutierrez-Castorena EV, Herrera JM. Quantification of soil components in thin sections: High resolution mosaics *versus* individual images. Bol Soc Geol Mex 2020; 72(1): 1-17.

[12] Watteau F, Jangorzo NS, Schwartz C. A micromorphological analysis for quantifying structure descriptors in a young constructed technosol. Bol Soc Geol Mex 2019; 71(1): 11-20.
[http://dx.doi.org/10.18268/BSGM2019v71n1a2]

[13] Totsche KU, Amelung W, Gerzabek MH, *et al.* Microaggregates in soils. J Plant Nutr Soil Sci 2018; 181(1): 104-36.
[http://dx.doi.org/10.1002/jpln.201600451]

[14] Liu C, Tong F, Yan L, Zhou H, Hao S. Effect of Porosity on Soil-Water Retention Curves: Theoretical and Experimental Aspects. Geofluids. 2020.

[15] Sekucia F, Dlapa P, Kollár J, Cerdà A, Hrabovský A, Svobodová L. Land-use impact on porosity and water retention of soils rich in rock fragments. CATENA 2020; 195: 104807.

[http://dx.doi.org/10.1016/j.catena.2020.104807]

[16]   Dlapa P, Hriník D, Hrabovský A, *et al.* The Impact of Land-Use on the Hierarchical Pore Size Distribution and Water Retention Properties in Loamy Soils. Water. 2020; 12(2): 339.
[http://dx.doi.org/10.3390/w12020339]

[17]   Mondal S, Chakraborty D. Global meta-analysis suggests that no-tillage favourably changes soil structure and porosity. Geoderma 2022; 405: 115443.
[http://dx.doi.org/10.1016/j.geoderma.2021.115443]

[18]   Pires LF, Auler AC, Roque WL, Mooney SJ. X-ray microtomography analysis of soil pore structure dynamics under wetting and drying cycles. Geoderma 2020; 362: 114103.
[http://dx.doi.org/10.1016/j.geoderma.2019.114103] [PMID: 32184497]

[19]   Ng C, Peprah-Manu D. Pore structure effects on the water retention behaviour of a compacted silty sand soil subjected to drying-wetting cycles. Eng Geol 2022.

[20]   Ferreira TR, Pires LF, Auler AEC, Brinatti AEM, Ogunwole JO. Water retention curve to analyze soil structure changes due to liming. An Acad Bras Ci Encias. 2019.

[21]   Basset C, Abou Najm M, Ghezzehei T, Hao X, Daccache A. How does soil structure affect water infiltration? A meta-data systematic review. Soil Tillage Res 2023; 226: 105577.
[http://dx.doi.org/10.1016/j.still.2022.105577]

[22]   Pagliai M, Vignozzi N. Image Analysis and Microscopic Techniques to Characterize Soil Pore System. In: Physical Methods in Agriculture [Internet]. 2002. p. 13-38. Available from: 10.1007/978-1-4615-0085-8_2
[http://dx.doi.org/10.1007/978-1-4615-0085-8_2]

[23]   Jiménez-Martínez A, Gutiérrez-Castorena MC, Montaño NM, Gutiérrez-Castorena EV, Alarcón A, Gavito ME. Micromorphology and thematic micro-mapping reveal differences in the soil structuring traits of three arbuscular mycorrhizal fungi. Pedobiologia (Jena) 2024; 104: 150953.
[http://dx.doi.org/10.1016/j.pedobi.2024.150953]

[24]   Khade SW, Rodrigues BF. Applications of arbuscular mycorrhizal fungi in agroecosystems. Tropical and Subtropical Agroecosystems 2009; 10: 337-54.

[25]   Gou X, Ni H, Sadowsky MJ, Chang X, Liu W, Wei X. Arbuscular mycorrhizal fungi alleviate erosion-induced soil nutrient losses in experimental agro-ecosystems. Catena 2023; 220: 106687.
[http://dx.doi.org/10.1016/j.catena.2022.106687]

[26]   Giovannini L, Palla M, Agnolucci M, *et al.* Arbuscular Mycorrhizal Fungi and Associated Microbiota as Plant Biostimulants: Research Strategies for the Selection of the Best Performing Inocula. Agronomy 2020; 10(1): 106.

[27]   Fall A, Nakabonge G, Ssekandi J, *et al.* Roles of Arbuscular Mycorrhizal Fungi on Soil Fertility: Contribution in the Improvement of Physical, Chemical, and Biological Properties of the Soil. Front Fungal Biol 2022; 3: 723892.

[28]   Aguilar-Paredes A, Valdés G, Nuti M. Ecosystem Functions of Microbial Consortia in Sustainable Agriculture. Agronomy (Basel) 2020; 10(12): 1902.
[http://dx.doi.org/10.3390/agronomy10121902]

[29]   Wilson MJ. Weathering of the primary rock-forming minerals: processes, products and rates. Clay Miner 2004; 39(3): 233-66.
[http://dx.doi.org/10.1180/0009855043930133]

[30]   Velbel MA. Weathering and Soil-Forming Processes.Ecological Studies. Springer New York 1988; pp. 93-102.

[31]   Earle S. Chapter 5: Weathering and the Formation of Soil. In: Physical Geology. 2015; 104-30.

[32]   Tarbuck EJ. Earth: an introduction to physical geology, global edition. 12th ed., Pearson Education Limited 2016.

[33]   Fitzpatrick RW. Iron Compounds as Indicators of Pedogenic Processes: Examples from the Southern Hemisphere. In: Iron in Soils and Clay Minerals [Internet]. 1988. p. 351-96.
[http://dx.doi.org/10.1007/978-94-009-4007-9_13]

[34]   Moore P. The ecology of peat-forming processes: a review. Int J Coal Eco 1989; 12(1-4): 89-103.
[http://dx.doi.org/10.1016/0166-5162(89)90048-7]

[35]   Sánchez MR, Rangel OC, Aguirre C. Ecological studies in the Cordillera Oriental IV: Sinecological aspects of the bryophyte flora of paramo peat deposits around Bogotá. Caldasia 1989; 16(76): 41-57.

[36]   World reference base for soil resources: a framework for international classification, correlation and communication. 2nd ed. Rome: Food and Agriculture Organization of the United Nations; (World soil resources reports). 2006; 128.

[37]   Cawsey D, Mellon P. A review of experimental weathering of basic igneous rocks. Geol Soc Spec Publ 1983.
[http://dx.doi.org/10.1144/GSL.SP.1983.011.01.03]

[38]   Depetris P, Pasquini AI, Lecomte K. Chemical Weathering Processes on the Earth's Surface 2014.
[http://dx.doi.org/10.1007/978-94-007-7717-0_4]

[39]   Speight J. Water Chemistry. Nat Water Remediat 2020.

[40]   Kužvart M. The Weathering Processes in Geography. Geografie (Utrecht) 1969; 74(2): 101-8.
[http://dx.doi.org/10.37040/geografie1969074020101]

[41]   Formoso M. Some topics on geochemistry of weathering: a review. An Acad Bras Ci Encias.M 2006.

[42]   Stumm W, Wollast R. Coordination chemistry of weathering: Kinetics of the surface-controlled dissolution of oxide minerals. Reviews of Geophysics 1990; 28(1): 53-69.
[http://dx.doi.org/10.1029/RG028i001p00053]

[43]   Romillac N. Ammonification.Encyclopedia of Ecology. Elsevier 2019; pp. 256-63.
[http://dx.doi.org/10.1016/B978-0-12-409548-9.10889-9]

[44]   Deenik J. Nitrogen Mineralization Potential in Important Agricultural Soils of Hawai'i 2006.

[45]   Cabrera ML, Kissel DE, Vigil MF. Nitrogen mineralization from organic residues: research opportunities. J Environ Qual 2005; 34(1): 75-9.
[http://dx.doi.org/10.2134/jeq2005.0075] [PMID: 15647536]

[46]   Persson T, Rudebeck A, Jussy JH, Colin-Belgrand M, Priemé A, Dambrine E, *et al.* Soil Nitrogen Turnover — Mineralisation, Nitrification and Denitrification in European Forest Soils.Ecological Studies. Springer Berlin Heidelberg 2000; pp. 297-311.

[47]   Wu H, Cui H, Fu C, *et al.* Unveiling the crucial role of soil microorganisms in carbon cycling: A review. Sci Total Environ 2024; 909: 168627.
[http://dx.doi.org/10.1016/j.scitotenv.2023.168627] [PMID: 37977383]

[48]   Basile-Doelsch I, Balesdent J, Pellerin S. Reviews and syntheses: The mechanisms underlying carbon storage in soil. Biogeosciences 2020; 17: 5223-42.

[49]   Paul E, Kravchenko A, Grandy AS, Morris S. Soil organic matter dynamics: Controls and management for sustainable ecosystem functioning. In: Hamilton SK, Doll JE, Robertson GP (Eds.). The Ecology of Agricultural Landscapes: Long-Term Research on the Path to Sustainability. Oxford University Press, New York, New York, USA. 2015; 104-34.

[50]   Gleixner G. Soil organic matter dynamics: a biological perspective derived from the use of compound-specific isotopes studies. Ecol Res 2013; 28(5): 683-95.
[http://dx.doi.org/10.1007/s11284-012-1022-9]

[51]   Filser J, Faber JH, Tiunov AV, *et al.* Soil fauna: key to new carbon models. Soil (Gottingen) 2016; 2(4): 565-82.
[http://dx.doi.org/10.5194/soil-2-565-2016]

[52]   Sanderman J, Creamer C, Baisden WT, Farrell M, Fallon S. Greater soil carbon stocks and faster turnover rates with increasing agricultural productivity. Soil (Gottingen) 2017; 3(1): 1-16. [http://dx.doi.org/10.5194/soil-3-1-2017]

[53]   Liski J, Ilvesniemi H, Annikki Mäkelä, Westman CJ. CO_2 Emissions from Soil in Response to Climatic Warming Are Overestimated: The Decomposition of Old Soil Organic Matter Is Tolerant of Temperature. Ambio 1999; 28(2): 171-4. http://www.jstor.org/stable/4314871

[54]   Fierer N, Colman BP, Schimel J, Jackson RB. Predicting the temperature dependence of microbial respiration in soil: A continental-scale analysis. Global Biogeochemical Cycles 2006; 20(3): GB3026. [http://dx.doi.org/10.1029/2005GB002644]

[55]   Metcalfe DB. Microbial change in warming soils. Science 2017; 358(6359): 41-2. [http://dx.doi.org/10.1126/science.aap7325] [PMID: 28983036]

[56]   Gurmu G. Soil Organic Matter and its Role in Soil Health and Crop Productivity Improvement. For Ecol Manage 2019; 7: 475-83.

[57]   Biswas T, Kole SC. Soil Organic Matter and Microbial Role in Plant Productivity and Soil Fertility.Microorganisms for Sustainability. Springer Singapore 2017; pp. 219-38.

[58]   Kassam A, Gonzalez-Sanchez E, Carbonell-Bojollo RM, Friedrich T, Derpsch R. Global Spread of Conservation Agriculture for Enhancing Soil Organic Matter, Soil Health, Productivity, and Ecosystem Services. In: Soil Organic Matter and Feeding the Future. CRC Press; 2021; pp. 91-126.

[59]   Lal R. Soil health and carbon management. Food Energy Secur 2016; 5(4): 212-22. [http://dx.doi.org/10.1002/fes3.96]

[60]   Babur E, Dindaroglu T. Seasonal Changes of Soil Organic Carbon and Microbial Biomass Carbon in Different Forest Ecosystems.Environmental Factors Affecting Human Health. IntechOpen 2020. [http://dx.doi.org/10.5772/intechopen.90656]

[61]   Voltr V, Menšík L, Hlisnikovský L, Hruška M, Pokorný E, Pospíšilová L. The Soil Organic Matter in Connection with Soil Properties and Soil Inputs. Agronomy (Basel) 2021; 11(4): 779. [http://dx.doi.org/10.3390/agronomy11040779]

[62]   Abinandan S, Subashchandrabose SR, Venkateswarlu K, Megharaj M. Soil microalgae and cyanobacteria: the biotechnological potential in the maintenance of soil fertility and health. Crit Rev Biotechnol 2019; 39(8): 981-98. [http://dx.doi.org/10.1080/07388551.2019.1654972] [PMID: 31455102]

[63]   Dhall M, Mishra B, Barman S, Boddana P. Role of soil microbiota in soil fertility. Int J Agric Sci 2021; 17(2): 729-39. [http://dx.doi.org/10.15740/HAS/IJAS/17.2/729-739]

[64]   Jamir E, Kangabam RD, Borah K, Tamuly A, Deka Boruah HP, Silla Y. Role of Soil Microbiome and Enzyme Activities in Plant Growth Nutrition and Ecological Restoration of Soil Health. In: Kumar, A., Sharma, S. (eds) Microbes and Enzymes in Soil Health and Bioremediation. Microorganisms for Sustainability, vol 16. Springer, Singapore. 2019; 99-132. [http://dx.doi.org/10.1007/978-981-13-9117-0_5]

[65]   Jiao S, Xu Y, Zhang J, Hao X, Lu Y. Core Microbiota in Agricultural Soils and Their Potential Associations with Nutrient Cycling. mSystems 2019; 4(2): e00313-18. [http://dx.doi.org/10.1128/mSystems.00313-18] [PMID: 30944882]

[66]   Khan AU, Ahmad H, Khan Z, Noor M, Bibi M, Khan AM. Unraveling the Complexities of Soil Microbiomes: A Review of Their Role in Crop Production and Health. EPH - *Int J Agric*. Environ Res 2024; 3702-10.

[67]   Six J, Bossuyt H, Degryze S, Denef K. A history of research on the link between (micro)aggregates, soil biota, and soil organic matter dynamics. Soil Tillage Res 2004; 79(1): 7-31. [http://dx.doi.org/10.1016/j.still.2004.03.008]

[68] Lehmann J, Bossio DA, Kögel-Knabner I, Rillig MC. The concept and future prospects of soil health. Nat Rev Earth Environ 2020; 1(10): 544-53.
[http://dx.doi.org/10.1038/s43017-020-0080-8] [PMID: 33015639]

[69] Prasad S, Malav L, Choudhary J, *et al.* Soil microbiomes for healthy nutrient recycling. In: Yadav AN, Singh J, Singh C, Yadav N (Eds.). Current Trends in Microbial Biotechnology for Sustainable Agriculture. Springer: Singapore. 2020; 1-21.

[70] Igwe C, Obalum SE. Microaggregate Stability of Tropical Soils and its Roles on Soil Erosion Hazard Prediction. Advances in Agrophysical Research. InTech. 2013.
[http://dx.doi.org/10.5772/52473]

[71] Amezketa E. Soil aggregate stability: a review. Journal of Sustainable Agriculture 1999; 14(2–3): 83-151.
[http://dx.doi.org/10.1300/J064v14n02_08]

[72] Menon M, Mawodza T, Rabbani A, *et al.* Pore system characteristics of soil aggregates and their relevance to aggregate stability. Geoderma 2020; 366: 114259.
[http://dx.doi.org/10.1016/j.geoderma.2020.114259]

[73] Yudina AV, Klyueva VV, Romanenko KA, Fomin DS. Micro- within macro: How micro-aggregation shapes the soil pore space and water-stability. Geoderma 2022; 415: 115771.
[http://dx.doi.org/10.1016/j.geoderma.2022.115771]

[74] Ferro N, Berti A, Francioso O, Ferrari E, Matthews GP, Morari F. Investigating the effects of wettability and pore size distribution on aggregate stability: the role of soil organic matter and the humic fraction. Eur J Soil Sci 2012; 63(2): 152-64.

[75] Rabbi S, Minasny B, McBratney A, Young I. Microbial processing of organic matter drives stability and pore geometry of soil aggregates. Geoderma 2020; 360: 114033.
[http://dx.doi.org/10.1016/j.geoderma.2019.114033]

[76] Smucker AJM, Park EJ, Dorner J, Horn R. Soil Micropore Development and Contributions to Soluble Carbon Transport within Macroaggregates. Vadose Zone J 2007; 6(2): 282-90.
[http://dx.doi.org/10.2136/vzj2007.0031]

[77] Yudina A, Kuzyakov Y. Dual nature of soil structure: The unity of aggregates and pores. Geoderma 2023; 434: 116478. [Internet].
[http://dx.doi.org/10.1016/j.geoderma.2023.116478]

[78] Wilpiszeski RL, Aufrecht JA, Retterer ST, *et al.* Towards Understanding Microbiome Interactions at Biologically Relevant Scales. Appl Environ Microbiol 2019; 85: e00324-19.

[79] Lehmann A, Zheng W, Rillig MC. Soil biota contributions to soil aggregation. Nat Ecol Evol 2017; 1(12): 1828-35.
[http://dx.doi.org/10.1038/s41559-017-0344-y] [PMID: 29038473]

[80] Rillig MC, Aguilar-Trigueros CA, Bergmann J, Verbruggen E, Veresoglou SD, Lehmann A. Plant root and mycorrhizal fungal traits for understanding soil aggregation. New Phytol 2015; 205(4): 1385-8.
[http://dx.doi.org/10.1111/nph.13045] [PMID: 25231111]

[81] Degens B. Macro-aggregation of soils by biological bonding and binding mechanisms and the factors affecting these: a review. Australian Journal of Soil Research 1997; 35(3): 431-60.
[http://dx.doi.org/10.1071/S96016]

[82] Ghezzehei T, Or D. Root architecture and hydrologic fluctuations explain spatiotemporal soil aggregation patterns. 22nd EGU General Assembly, held online 4-8 May 2020; id.21957.
[http://dx.doi.org/10.5194/egusphere-egu2020-21957]

[83] Kumar R, Rawat KS, Singh J, Singh A, Rai A. Soil aggregation dynamics and carbon sequestration. J Appl Nat Sci 2013; 5(1): 250-67.
[http://dx.doi.org/10.31018/jans.v5i1.314]

[84]   Grear J, Schmitz O. Effects of grouping behavior and predators on the spatial distribution of a forest floor arthropod. Ecology 2005; 86(4): 960-71.
[http://dx.doi.org/10.1890/04-1509]

[85]   Murphy J, Giller P, Horan M. Spatial scale and the aggregation of stream macroinvertebrates associated with leaf packs. Freshwater Biol 1998; 39(2): 325-37.
[http://dx.doi.org/10.1046/j.1365-2427.1998.00284.x]

[86]   Fiene JG, Sword GA, VanLaerhoven SL, Tarone AM. The role of spatial aggregation in forensic entomology. J Med Entomol 2014; 51(1): 1-9.
[http://dx.doi.org/10.1603/ME13050] [PMID: 24605447]

[87]   Morris EK, Morris DJP, Vogt S, *et al.* Visualizing the dynamics of soil aggregation as affected by arbuscular mycorrhizal fungi. ISME J 2019; 13(7): 1639-46.
[http://dx.doi.org/10.1038/s41396-019-0369-0] [PMID: 30742058]

[88]   Tao G. Impacts of arbuscular mycorrhizal fungi on soil aggregation dynamics of neutral purple soil 2011.

[89]   Tisdall JM, Smith SE, Rengasamy P. Aggregation of soil by fungal hyphae. Soil Res 1997; 35(1): 55.
[http://dx.doi.org/10.1071/S96065]

[90]   PENG Sili , SHEN Hong , ZHANG Yuting , GUO Tao . Compare different effect of arbuscular mycorrhizal colonization on soil structure. Acta Ecol Sin 2012; 32(3): 863-70.
[http://dx.doi.org/10.5846/stxb201104080458]

[91]   Liang T, Shi X, Guo T, Peng S. Arbuscular Mycorrhizal Fungus Mediate Changes in Mycorrhizosphere Soil Aggregates. Agric Sci 2015; 6(12): 1455-63.

[92]   Barbosa MV, Pedroso DF, Curi N, Carneiro MA. Do different arbuscular mycorrhizal fungi affect the formation and stability of soil aggregates?. Cienc Agrotecnologia 2019; p. 43.

[93]   Tinker PB, Buckwell A, Lynch JM, Blaxter KL, Fowden L. Crop nutrients: control and efficiency of use. Philos Trans R Soc Lond B Biol Sci 1985; 310(1144): 175-91.
[http://dx.doi.org/10.1098/rstb.1985.0106]

[94]   Bridges EM, Catizzone M. Soil science in a holistic framework: discussion of an improved integrated approach. Geoderma 1996; 71(3-4): 275-87.
[http://dx.doi.org/10.1016/0016-7061(96)00015-8]

[95]   Herrick JE. Soil quality: an indicator of sustainable land management? Appl Soil Ecol 2000; 15(1): 75-83.
[http://dx.doi.org/10.1016/S0929-1393(00)00073-1]

[96]   Keesstra SD, Bouma J, Wallinga J, *et al.* The significance of soils and soil science towards realization of the United Nations Sustainable Development Goals. Soil (Gottingen) 2016; 2(2): 111-28.
[http://dx.doi.org/10.5194/soil-2-111-2016]

[97]   Benedetti A, Dell'Abate MT, Napoli R. Soil Functions and Ecological Services. In: Costantini E, Dazzi C (eds) The Soils of Italy. World Soils Book Series. Springer, Dordrecht. 2013; 179-203.
[http://dx.doi.org/10.1007/978-94-007-5642-7_7]

[98]   Blum W. Soil Resources - The Basis of Human Society and the Environment. Die Bodenkultur Journal of Land Management Food and Environment 2006; 57(1): 197-202.

[99]   Blum WEH, Warkentin B, Frossard E. Soil, human society and the environment. In: Function of Soils for Human Societies and the Environment. Geological Society, London, Special Publications 2006; Vol. 266: pp. 1-8.
[http://dx.doi.org/10.1144/GSL.SP.2006.266.01.01]

[100]   Adhikari K, Hartemink AE. Linking soils to ecosystem services — A global review. Geoderma 2016; 262: 101-11.
[http://dx.doi.org/10.1016/j.geoderma.2015.08.009]

[101] Janků J, Jehlička J, Heřmanová K, *et al.* An overview of land evaluation in the context of ecosystem services. Soil & Water Res 2022; 17(1): 1-14.
[http://dx.doi.org/10.17221/136/2021-SWR]

[102] Koch A, McBratney A, Adams M, *et al.* Soil Security: Solving the Global Soil Crisis. Glob Policy 2013; 4(4): 434-41.
[http://dx.doi.org/10.1111/1758-5899.12096]

[103] Zhu Y, Meharg AA. Protecting global soil resources for ecosystem services. Ecosyst Health Sustain 2015; 1(3): 1-4.
[http://dx.doi.org/10.1890/EHS15-0010.1]

[104] Robinson DA, Hockley N, Dominati E, *et al.* Natural Capital, Ecosystem Services, and Soil Change: Why Soil Science Must Embrace an Ecosystems Approach. Vadose Zone J 2012; 11(1): vzj2011.0051.
[http://dx.doi.org/10.2136/vzj2011.0051]

[105] Comerford NB, Franzluebbers AJ, Stromberger ME, Morris L, Markewitz D, Moore R. Assessment and Evaluation of Soil Ecosystem Services. Soil Horiz (Madison) 2013; 54(3): 0.
[http://dx.doi.org/10.2136/sh12-10-0028]

[106] McBratney A, Field DJ, Koch A. The dimensions of soil security. Geoderma 2014; 213: 203-13.
[http://dx.doi.org/10.1016/j.geoderma.2013.08.013]

[107] Hartemink AE, McBratney A. A soil science renaissance. Geoderma 2008; 148(2): 123-9.
[http://dx.doi.org/10.1016/j.geoderma.2008.10.006]

[108] Bone J, Head M, Barraclough D, *et al.* Soil quality assessment under emerging regulatory requirements. Environ Int 2010; 36(6): 609-22.
[http://dx.doi.org/10.1016/j.envint.2010.04.010] [PMID: 20483160]

[109] Bartkowski B, Bartke S, Helming K, Paul C, Techen AK, Hansjürgens B. Potential of the economic valuation of soil-based ecosystem services to inform sustainable soil management and policy. PeerJ 2020; 8: e8749.
[http://dx.doi.org/10.7717/peerj.8749] [PMID: 32231877]

[110] Andrew ME, Wulder MA, Nelson TA, Coops NC. Spatial data, analysis approaches, and information needs for spatial ecosystem service assessments: a review. GIsci Remote Sens 2015; 52(3): 344-73.
[http://dx.doi.org/10.1080/15481603.2015.1033809]

[111] Schulp CJE, Burkhard B, Maes J, Van Vliet J, Verburg PH. Uncertainties in ecosystem service maps: a comparison on the European scale. PLoS One 2014; 9(10): e109643.
[http://dx.doi.org/10.1371/journal.pone.0109643] [PMID: 25337913]

[112] Maron PA, Mougel C, Ranjard L. Soil microbial diversity: Methodological strategy, spatial overview and functional interest. C R Biol 2011; 334(5-6): 403-11.
[http://dx.doi.org/10.1016/j.crvi.2010.12.003] [PMID: 21640949]

[113] Smith P, Cotrufo MF, Rumpel C, *et al.* Biogeochemical cycles and biodiversity as key drivers of ecosystem services provided by soils. Soil (Gottingen) 2015; 1(2): 665-85.
[http://dx.doi.org/10.5194/soil-1-665-2015]

[114] Bouma J. Soil science contributions towards Sustainable Development Goals and their implementation: linking soil functions with ecosystem services. J Plant Nutr Soil Sci 2014; 177(2): 111-20.
[http://dx.doi.org/10.1002/jpln.201300646]

[115] Rodrigo-Comino J, López-Vicente M, Kumar V, *et al.* Soil Science Challenges in a New Era: A Transdisciplinary Overview of Relevant Topics. Air Soil Water Res 2020; 13: 1178622120977491.
[http://dx.doi.org/10.1177/1178622120977491]

[116] Bouma J. Contributing pedological expertise towards achieving the United Nations Sustainable Development Goals. Geoderma 2020; 375: 114508.

[http://dx.doi.org/10.1016/j.geoderma.2020.114508]

[117] Sánchez PA, Ahamed S, Carré F, *et al.* Environmental science. Digital soil map of the world. Science 2009; 325(5941): 680-1.
[http://dx.doi.org/10.1126/science.1175084] [PMID: 19661405]

[118] Hengl T, Reuter H. Geomorphometry: Concepts, software, applications. Elsevier 2009; 33: 1-765.

[119] Ludwig M, Wilmes P, Schrader S. Measuring soil sustainability *via* soil resilience. Sci Total Environ 2018; 626: 1484-93.
[PMID: 29054651]

[120] Griffiths BS, Faber J, Bloem J. Applying Soil Health Indicators to Encourage Sustainable Soil Use: The Transition from Scientific Study to Practical Application. Sustainability (Basel) 2018; 10(9): 3021.
[http://dx.doi.org/10.3390/su10093021]

[121] Cairns J Jr, McCormick PV, Niederlehner BR. A proposed framework for developing indicators of ecosystem health. Hydrobiologia 1993; 263(1): 1-44.
[http://dx.doi.org/10.1007/BF00006084]

[122] Doran JW, Zeiss MR. Soil health and sustainability: managing the biotic component of soil quality. Appl Soil Ecol 2000; 15(1): 3-11.
[http://dx.doi.org/10.1016/S0929-1393(00)00067-6]

[123] Bouma J, Montanarella L, Evanylo G. The challenge for the soil science community to contribute to the implementation of the UN Sustainable Development Goals. Soil Use Manage 2019; 35(4): 538-46.
[http://dx.doi.org/10.1111/sum.12518]

[124] Tóth G, Hermann T, da Silva MR, Montanarella L. Monitoring soil for sustainable development and land degradation neutrality. Environ Monit Assess 2018; 190(2): 57.
[http://dx.doi.org/10.1007/s10661-017-6415-3] [PMID: 29302746]

[125] Zinck JA, Farshad A. Issues of sustainability and sustainable land management. Can J Soil Sci 1995; 75(4): 407-12.
[http://dx.doi.org/10.4141/cjss95-060]

[126] Zornoza R, Acosta JA, Bastida F, Domínguez SG, Toledo DM, Faz A. Identification of sensitive indicators to assess the interrelationship between soil quality, management practices and human health. Soil (Gottingen) 2015; 1(1): 173-85.
[http://dx.doi.org/10.5194/soil-1-173-2015]

[127] Syers JK, Pushparajah E, Hamblin A. Indicators and thresholds for the evaluation of sustainable land management. Can J Soil Sci 1995; 75(4): 423-8.
[http://dx.doi.org/10.4141/cjss95-062]

[128] Jónsson JÖG, Davíðsdóttir B, Jónsdóttir EM, Kristinsdóttir SM, Ragnarsdóttir KV. Soil indicators for sustainable development: A transdisciplinary approach for indicator development using expert stakeholders. Agric Ecosyst Environ 2016; 232: 179-89.
[http://dx.doi.org/10.1016/j.agee.2016.08.009]

[129] Bastida F, Zsolnay A, Hernández T, García C. Past, present and future of soil quality indices: A biological perspective. Geoderma 2008; 147(3-4): 159-71.
[http://dx.doi.org/10.1016/j.geoderma.2008.08.007]

[130] Rutgers M, van Wijnen HJ, Schouten AJ, *et al.* A method to assess ecosystem services developed from soil attributes with stakeholders and data of four arable farms. Sci Total Environ 2012; 415: 39-48.
[http://dx.doi.org/10.1016/j.scitotenv.2011.04.041] [PMID: 21704358]

[131] Rüdisser J, Tasser E, Peham T, Meyer E, Tappeiner U. The dark side of biodiversity: Spatial application of the biological soil quality indicator (BSQ). Ecol Indic 2015; 53: 240-6.
[http://dx.doi.org/10.1016/j.ecolind.2015.02.006]

[132] Hinckley ELS, Bonan GB, Bowen GJ, *et al.* The soil and plant biogeochemistry sampling design for The National Ecological Observatory Network. Ecosphere 2016; 7(3): e01234.
[http://dx.doi.org/10.1002/ecs2.1234]

[133] McCormick R, Kapustka LA. The answer is 42 … What is THE question? J Environ Stud Sci 2016; 6(1): 208-13.
[http://dx.doi.org/10.1007/s13412-016-0376-7]

[134] Rinot O, Levy GJ, Steinberger Y, Svoray T, Eshel G. Soil health assessment: A critical review of current methodologies and a proposed new approach. Sci Total Environ 2019; 648: 1484-91.
[http://dx.doi.org/10.1016/j.scitotenv.2018.08.259] [PMID: 30340293]

[135] Cardoso EJBN, Vasconcellos RLF, Bini D, *et al.* Soil health: looking for suitable indicators. What should be considered to assess the effects of use and management on soil health? Sci Agric 2013; 70(4): 274-89.
[http://dx.doi.org/10.1590/S0103-90162013000400009]

[136] Menta C. Use of microarthropods as biological indicators of soil quality: the BSQ synthetic indicator. 2005.

[137] Gardi C, Menta C. Use of microarthropods as biological indicators of soil quality: the BSQ synthetic indicator. 2002.

[138] Jacomini C, Menta C, Parisi V. Evaluation of land use and crop management impacts on soil quality: application of QBS methods 2003.

[139] Gunderson LH. Ecological Resilience—In Theory and Application. Annu Rev Ecol Syst 2000; 31(1): 425-39.
[http://dx.doi.org/10.1146/annurev.ecolsys.31.1.425]

[140] Walker B, Holling CS, Carpenter S, Kinzig A. Resilience. Adaptability and Transformability in Social–ecological Systems. Ecology and Society 2003; 9(2): 5.

[141] Ingrisch J, Bahn M. Towards a Comparable Quantification of Resilience. Trends Ecol Evol 2018; 33(4): 251-9.
[http://dx.doi.org/10.1016/j.tree.2018.01.013] [PMID: 29477443]

[142] Ludwig M, Schrader S. Conceptual design for measuring soil management sustainability. In: Jahrestagung der DBG 2017: Horizonte des Bodens Bodens, 02.-07.09.2017, Göttingen.

[143] Davis AG, Huggins DR, Reganold JP. Linking soil health and ecological resilience to achieve agricultural sustainability. Front Ecol Environ 2023; 21(3): 131-9.
[http://dx.doi.org/10.1002/fee.2594]

[144] Apeldoorn D. Panarchy rules?: rethinking resilience of agroecosystems Wageningen University 2014.

[145] Apeldoorn D, Kok K, Sonneveld M, Veldkamp T. Panarchy rules: rethinking resilience of agroecosystems, evidence from Dutch dairy-farming. Ecology and Society 2011; 16(1): 39.
[http://dx.doi.org/10.5751/ES-03949-160139]

[146] Holling CS. Understanding the Complexity of Economic, Ecological, and Social Systems. Ecosystems (N Y) 2001; 4(5): 390-405.
[http://dx.doi.org/10.1007/s10021-001-0101-5]

[147] Slager S. Morphological studies of some cultivated soils. Pudoc: Wageningen 1966.
[http://dx.doi.org/10.18174/187953]

[148] Bouma J. Microstructure and stability of two sandy loam soils with different soil management. Wageningen: Centrum voor Landbouwpublikaties en Landbouwdocumentatie (Pudoc) 1969; vi, [110] p ill.

[149] Power AG. Ecosystem services and agriculture: tradeoffs and synergies. Philos Trans R Soc Lond B Biol Sci 2010; 365(1554): 2959-71.

[http://dx.doi.org/10.1098/rstb.2010.0143] [PMID: 20713396]

[150] Bommarco R, Kleijn D, Potts SG. Ecological intensification: harnessing ecosystem services for food security. Trends Ecol Evol 2013; 28(4): 230-8.
       [http://dx.doi.org/10.1016/j.tree.2012.10.012] [PMID: 23153724]

[151] Schrader S, Banse M, van Capelle C. Ecology and economy of ecosystem services provided by the diversity of soil organisms in agricultural systems. Johann Heinrich von Thünen-Institut 2021.

[152] Collins SJ, Bellingham L, Mitchell GW, Fahrig L. Life in the slow drain: Landscape structure affects farm ditch water quality. Sci Total Environ 2019; 656: 1157-67.
       [http://dx.doi.org/10.1016/j.scitotenv.2018.11.400] [PMID: 30625647]

[153] Latimer CE, Smith OM, Taylor JM, *et al.* Landscape context mediates the physiological stress response of birds to farmland diversification 2020.
       [http://dx.doi.org/10.1111/1365-2664.13583]

[154] Elmqvist T, Folke C, Nyström M, *et al.* Response diversity, ecosystem change, and resilience 2003.
       [http://dx.doi.org/10.1890/1540-9295(2003)001[0488:RDECAR]2.0.CO;2]

[155] Yang G, Wagg C, Veresoglou SD, Hempel S, Rillig MC. How Soil Biota Drive Ecosystem Stability. Trends Plant Sci 2018; 23(12): 1057-67.
       [http://dx.doi.org/10.1016/j.tplants.2018.09.007] [PMID: 30287162]

[156] Rose A. Defining Resilience Across Disciplines. In: Defining and Measuring Economic Resilience from a Societal, Environmental and Security Perspective. Integrated Disaster Risk Management. Springer; Singapore. 2017; 19-27.
       [http://dx.doi.org/10.1007/978-981-10-1533-5_3]

[157] Blair JM, Mabee WE. Resilience.International Encyclopedia of Human Geography. Elsevier 2020; pp. 451-6.
       [http://dx.doi.org/10.1016/B978-0-08-102295-5.10754-1]

[158] Angeler DG, Allen CR, Persson ML. Resilience concepts in psychiatry demonstrated with bipolar disorder. Int J Bipolar Disord 2018; 6(1): 2.
       [http://dx.doi.org/10.1186/s40345-017-0112-6] [PMID: 29423550]

[159] Ludwig D, Walker B, Holling CS. Sustainability, Stability, and Resilience. Conserv Ecol 1997; 1(1): art7.
       [http://dx.doi.org/10.5751/ES-00012-010107]

[160] Bundschuh M, Schulz R, Schäfer RB, Allen CR, Angeler DG. Resilience in ecotoxicology: Toward a multiple equilibrium concept. Environ Toxicol Chem 2017; 36(10): 2574-80.
       [http://dx.doi.org/10.1002/etc.3845] [PMID: 28493505]

[161] Dakos V, Kéfi S. Ecological resilience: what to measure and how. Environ Res Lett 2022; 17(4): 043003.
       [http://dx.doi.org/10.1088/1748-9326/ac5767]

[162] Davies J, Batista P, Janes-Bassett V, O'Riordan R, Quinton J, Yumashev D. Soil systems as critical infrastructure: do we know enough about soil system resilience and vulnerability to secure our soils? 2020.
       [http://dx.doi.org/10.5194/egusphere-egu2020-20273]

[163] Korhonen J, Seager TP. Beyond eco-efficiency: a resilience perspective. Bus Strategy Environ 2008; 17(7): 411-9.
       [http://dx.doi.org/10.1002/bse.635]

[164] Gunderson LH. Adaptive dancing: interactions between social resilience and ecological crises.Navigating Social-Ecological Systems. Cambridge University Press 2001; pp. 33-52.
       [http://dx.doi.org/10.1017/CBO9780511541957.005]

[165] Smith P, House JI, Bustamante M, *et al.* Global change pressures on soils from land use and

management. Glob Change Biol 2016; 22(3): 1008-28.
[http://dx.doi.org/10.1111/gcb.13068] [PMID: 26301476]

[166] Lin H. A New Worldview of Soils. Soil Science Society of America Journal 2014; 78(6): 1831-44.
[http://dx.doi.org/10.2136/sssaj2014.04.0162]

[167] Birgé HE, Bevans RA, Allen CR, Angeler DG, Baer SG, Wall DH. Adaptive management for soil ecosystem services. J Environ Manage 2016; 183(Pt 2): 371-8.
[http://dx.doi.org/10.1016/j.jenvman.2016.06.024] [PMID: 27344211]

[168] Curran M, Maynard D, Heninger R, *et al.* An adaptive management process for forest soil conservation. The Forestry Chronicle 2005; 81(5): 717-22.
[http://dx.doi.org/10.5558/tfc81717-5]

[169] Hodbod J, Barreteau O, Allen C, Magda D. Managing adaptively for multifunctionality in agricultural systems. J Environ Manage 2016; 183(Pt 2): 379-88.
[http://dx.doi.org/10.1016/j.jenvman.2016.05.064] [PMID: 27349502]

[170] Delgado-Baquerizo M, Maestre FT, Reich PB, *et al.* Microbial diversity drives multifunctionality in terrestrial ecosystems. Nat Commun 2016; 7(1): 10541.
[http://dx.doi.org/10.1038/ncomms10541] [PMID: 26817514]

[171] Smith C, Jayathunga S, Gregorini P, Pereira FC, McWilliam W. Using Soil Sustainability and Resilience Concepts to Support Future Land Management Practice: A Case Study of Mt Grand Station, Hāwea, New Zealand. Sustainability (Basel) 2022; 14(3): 1808.
[http://dx.doi.org/10.3390/su14031808]

[172] Hogeboom RJ, Borsje BW, Deribe MM, *et al.* Resilience Meets the Water–Energy–Food Nexus: Mapping the Research Landscape. Front Environ Sci 2021; 9: 630395.
[http://dx.doi.org/10.3389/fenvs.2021.630395]

[173] Ludwig M, Wilmes P, Schrader S. Measuring soil sustainability *via* soil resilience. Sci Total Environ 2018; 626: 1484-93.
[http://dx.doi.org/10.1016/j.scitotenv.2017.10.043] [PMID: 29054651]

[174] Djemiel C, Dequiedt S, Karimi B, *et al.* Potential of Meta-Omics to Provide Modern Microbial Indicators for Monitoring Soil Quality and Securing Food Production. Front Microbiol 2022; 13: 889788.
[http://dx.doi.org/10.3389/fmicb.2022.889788] [PMID: 35847063]

[175] Bertola M, Ferrarini A, Visioli G. Improvement of Soil Microbial Diversity through Sustainable Agricultural Practices and Its Evaluation by -Omics Approaches: A Perspective for the Environment, Food Quality and Human Safety. Microorganisms. 2021.

[176] Wall DH, Six J. Give soils their due. Science 2015; 347(6223): 695.
[http://dx.doi.org/10.1126/science.aaa8493] [PMID: 25678633]

[177] Smith P, Keesstra SD, Silver WL, *et al.* Soil-derived Nature's Contributions to People and their contribution to the UN Sustainable Development Goals. Philos Trans R Soc Lond B Biol Sci 2021; 376(1834): 20200185.
[http://dx.doi.org/10.1098/rstb.2020.0185] [PMID: 34365826]

[178] Bouma J, Montanarella L. Facing policy challenges with inter- and transdisciplinary soil research focused on the UN Sustainable Development Goals. Soil (Gottingen) 2016; 2(2): 135-45.
[http://dx.doi.org/10.5194/soil-2-135-2016]

[179] Keesstra S, Bouma J, Wallinga J, Tittonell P, Smith P, Cerdà A, *et al.* The significance of soils and soil science towards realization of the United Nations sustainable development goals 2016.
[http://dx.doi.org/10.5194/soil-2-111-2016]

[180] Bouma J. Soil Security in Sustainable Development. Soil Syst 2019; 3(1): 5.
[http://dx.doi.org/10.3390/soilsystems3010005]

[181] Medina N, Vandermeer J. Developing systems theory in soil agroecology: incorporating heterogeneity and dynamic instability. Front Environ Sci 2023; 11: 1171194.
[http://dx.doi.org/10.3389/fenvs.2023.1171194]

[182] Görres J. Developing systems theory in soil agroecology: incorporating heterogeneity and dynamic instability. Front Environ Sci. 2020; 11

[183] Or D. The Tyranny of Small Scales—On Representing Soil Processes in Global Land Surface Models. Water Resour Res 2020; 56(6): 2019WR024846.
[http://dx.doi.org/10.1029/2019WR024846]

[184] Homann P, Sollins P, Fiorella M, Thorson T, Kern J. Regional Soil organic carbon storage estimates for western Oregon by multiple approaches 1998.
[http://dx.doi.org/10.2136/sssaj1998.03615995006200030036x]

[185] Jobbágy EG, Jackson RB. The vertical distribution of soil organic carbon and its relation to climate and vegetation. Ecol Appl 2000; 10(2): 423-36.
[http://dx.doi.org/10.1890/1051-0761(2000)010[0423:TVDOSO]2.0.CO;2]

[186] Cusser S, Bahlai C, Swinton SM, Robertson GP, Haddad NM. Long-term research avoids spurious and misleading trends in sustainability attributes of no-till. Glob Change Biol 2020; 26(6): 3715-25.
[http://dx.doi.org/10.1111/gcb.15080] [PMID: 32175629]

[187] Richter DD, Markewitz D, Heine PR, et al. Legacies of agriculture and forest regrowth in the nitrogen of old-field soils. For Ecol Manage 2000; 138(1-3): 233-48.
[http://dx.doi.org/10.1016/S0378-1127(00)00399-6]

[188] Zornoza R, Acosta JA, Bastida F, Domínguez SG, Toledo DM, Faz A. Identification of sensitive indicators to assess the interrelationship between soil quality, management practices and human health 2014.
[http://dx.doi.org/10.5194/soild-1-463-2014]

[189] Huynh TN, Lobry de Bruyn LA, Wilson B, Knox O. Insights, implications and challenges of studying local soil knowledge for sustainable land use: a critical review 2020.
[http://dx.doi.org/10.1071/SR19227]

[190] Amundson R. The policy challenges to managing global soil resources. Geoderma 2020; 379: 114639.
[http://dx.doi.org/10.1016/j.geoderma.2020.114639]

[191] Bosch OJH, Ross AH, Beeton RJ. Integrating science and management through collaborative learning and better information management 2003.
[http://dx.doi.org/10.1002/sres.536]

[192] Hou D, Bolan NS, Tsang DCW, Kirkham MB, O'Connor D. Sustainable soil use and management: An interdisciplinary and systematic approach. Sci Total Environ 2020; 729: 138961.
[http://dx.doi.org/10.1016/j.scitotenv.2020.138961] [PMID: 32353725]

[193] Lobry de Bruyn L, Ingram J. Soil information sharing and knowledge building for sustainable soil use and management: insights and implications for the 21 st Century. Soil Use Manage 2019; 35(1): 1-5.
[http://dx.doi.org/10.1111/sum.12493]

[194] Bouma J. How to communicate soil expertise more effectively in the information age when aiming at the UN Sustainable Development Goals. Soil Use Manage 2019; 35(1): 32-8.
[http://dx.doi.org/10.1111/sum.12415]

[195] Correa Ayram CC, Mendoza ME, Etter A, Pérez Salicrup DR. Habitat connectivity in biodiversity conservation 2016.
[http://dx.doi.org/10.1177/0309133315598713]

[196] Leija EG, Mendoza ME. Estudios de conectividad del paisaje en América Latina: retos de investigación. Madera Bosques 2021; 27(1)
[http://dx.doi.org/10.21829/myb.2021.2712032]

[197] Balbuena-Serrano Á, Zarco-González MM, Monroy-Vilchis O. Biases and information gaps in the study of habitat connectivity in the Carnivora in the Americas. Mammal Rev 2023; 53(2): 99-115.
[http://dx.doi.org/10.1111/mam.12312]

[198] LaPoint S, Balkenhol N, Hale J, Sadler J, van der Ree R. Ecological connectivity research in urban areas. Funct Ecol 2015; 29(7): 868-78.
[http://dx.doi.org/10.1111/1365-2435.12489]

[199] Rudnick DA, Ryan SJ, Beier P, Cushman SA, Dieffenbach F, Epps CW, *et al.* The Role of Landscape Connectivity in Planning and Implementing Conservation and Restoration Priorities. Issues in Ecology 2012.

[200] Kool JT, Moilanen A, Treml EA. Population connectivity: recent advances and new perspectives. Landsc Ecol 2013; 28(2): 165-85.
[http://dx.doi.org/10.1007/s10980-012-9819-z]

[201] Bodin Ö. Collaborative environmental governance: Achieving collective action in social-ecological systems. Science 2017; 357(6352): eaan1114.
[http://dx.doi.org/10.1126/science.aan1114] [PMID: 28818915]

[202] World Bank. Mexico - Mesoamerican Biological Corridor Project [Internet]. 2024. Available from: https://documents.worldbank.org/en/publication/documents-reports/documentdetail/674191468774634551/Mexico-Mesoamerican-Biological-Corridor-Project

[203] Woodley S, Bertzky B, Crawhall N, Dudley N, Londoño JM, Mackinnon K, *et al.* Meeting Aichi Target 11: What does success look like for protected area systems? 2012.

[204] Rands MRW, Adams WM, Bennun L, *et al.* Biodiversity conservation: challenges beyond 2010. Science 2010; 329(5997): 1298-303.
[http://dx.doi.org/10.1126/science.1189138] [PMID: 20829476]

[205] Cotler H, Merino L, Martinez-Trinidad S. Forest Soil Management: A Mexican Experience. Open J Soc Sci 2020; 10(9): 374-90.
[http://dx.doi.org/10.4236/ojss.2020.109020]

[206] Horrigan L, Lawrence RS, Walker P. How sustainable agriculture can address the environmental and human health harms of industrial agriculture. Environ Health Perspect 2002; 110(5): 445-56.
[http://dx.doi.org/10.1289/ehp.02110445] [PMID: 12003747]

[207] Gomiero T, Pimentel D, Paoletti MG. Is there a need for a more sustainable agriculture? Crit Rev Plant Sci 2011; 30(1-2): 6-23.
[http://dx.doi.org/10.1080/07352689.2011.553515]

[208] Cabeza M. El sistema agroalimentario globalizado: imperios alimentarios y degradación social y ecológica 2010.

[209] González H. What socioenvironmental impacts did 35 years of export agriculture have in Mexico? (1980–2014): A transnational agri-food field analysis. J Agrar Change 2020; 20(1): 163-87.
[http://dx.doi.org/10.1111/joac.12343]

[210] Moreno Thompson JA. Environmental and social impacts of agricultural biodiversity degradation from the ultra-processed food industry. InterNaciones 2023; 24: 141-64.
[http://dx.doi.org/10.32870/in.vi24.7235]

[211] Burigo AC, Porto MF. Agenda 2030, saúde e sistemas alimentares em tempos de sindemia: da vulnerabilização à transformação necessária. Cien Saude Colet 2021; 26(10): 4411-24.
[http://dx.doi.org/10.1590/1413-812320212610.13482021] [PMID: 34730632]

[212] Sarmiento Escobar A, Renard MC. Priority programs of the 4T government in attention to the coffee sector in Chiapas. Mexico: EDUCATECONCIENCIA 2024.

[213] Marín CM, Martínez Paz JM, Alcón Provencio FJ. Evaluación de políticas públicas sobre la diversificación de cultivos. Proc 8th Workshop Agri-Food Res Young Res WiA.

[214] Rivas García T, Espinosa-Calderón A, Hernández-Vázquez B, Schwentesius-Rinderman R. Participatory agroecological diagnosis in small and medium-sized producers in Michoacán and Morelos, Mexico. Agro Product. 2022.

[215] Lal R. Beyond COP 21: Potential and challenges of the "4 per Thousand" initiative. J Soil Water Conserv 2016; 71(1): 20A-5A.
[http://dx.doi.org/10.2489/jswc.71.1.20A]

[216] Rumpel C, Amiraslani F, Chenu C, Garcia Cardenas M, Kaonga M, Koutika L, *et al.* The 4 per 1000 initiative: Opportunities, limitations and challenges for implementing soil organic carbon sequestration as a sustainable development strategy. Ambio 2019.
[PMID: 30905053]

[217] Fageria NK. Role of Soil Organic Matter in Maintaining Sustainability of Cropping Systems. Commun Soil Sci Plant Anal 2012; 43(16): 2063-113.
[http://dx.doi.org/10.1080/00103624.2012.697234]

[218] Yadav SS, McNeil DL, Andrews M, Chen CC, Brand J, Singh GS, *et al.* Soil nutrient management.The lentil: botany, production and uses. CABI 2009; pp. 194-212.
[http://dx.doi.org/10.1079/9781845934873.0194]

[219] White PJ, Crawford JW, Díaz Álvarez MC, García Moreno R. Soil Management for Sustainable Agriculture 2013. Appl Environ Soil Sci 2014; 2014: 1-2.
[http://dx.doi.org/10.1155/2014/536825]

[220] Romijn E, Coppus R, De Sy V, Herold M, Roman-Cuesta RM, Verchot L. Land Restoration in Latin America and the Caribbean: An Overview of Recent, Ongoing and Planned Restoration Initiatives and Their Potential for Climate Change Mitigation. Forests 2019; 10(6): 510.
[http://dx.doi.org/10.3390/f10060510]

[221] Zamora-Cristales R, Gonzalez M, Rachmaninoff V, Franco Chuaire M, Vergara W, De Camino R, *et al.* Curando as Feridas da Terra: o papel dos incentivos financeiros públicos para ampliar os esforços de restauração em seis países da América Latina. World Resour Inst 2022.

[222] Mallén Rivera C. A decade of planting the seeds of forestry knowledge. Rev Mex Cienc For 2018; 3(13): 3-8.

[223] Bouma J. The new role of soil science in a network society. Soil Sci 2001; 166(12): 874-9.
[http://dx.doi.org/10.1097/00010694-200112000-00002]

[224] Montanarella L. Translating Soil Science Knowledge to Public Policy. In: Field DJ, Morgan CLS., McBratney AB (eds) Global Soil Security. Progress in Soil Science. Springer: Cham. 2017; 451-6.
[http://dx.doi.org/10.1007/978-3-319-43394-3_42]

[225] Null. A world soils agenda: Discussing international actions for the sustainable use of soils. Centre for Development and Environment | Bern, Switzerland 2002; 63.

CHAPTER 2

Soils of Mexican Deserts: Characteristics and Water Management Challenges

Blanca González-Méndez[1,2,*] and **Elizabeth Chávez-García**[3]

[1] *CONAHCYT, Insurgentes Sur 1582, Crédito Constructor, Benito Juárez, CP 03940 Mexico City, Mexico*

[2] *Northwest Regional Station, Institute of Geology, National Autonomous University of Mexico, Colosio and Madrid s/n, ZIP Code 83000, Hermosillo, Sonora, Mexico*

[3] *Faculty of Philosophy and Literature, National Autonomous University of Mexico, University City, ZIP Code 04510, Mexico City, Mexico*

Abstract: Mexico is home to two major deserts, the Chihuahuan and the Sonoran, which date back to the late Miocene. Both cover a vast region of northern Mexico and parts of the south of the USA. These deserts have been primarily shaped by recent tectonic forces and are characterized by tall mountain ranges that rise abruptly from alluvial plains. The landscapes include piedmont slopes, basin floors, alluvial deposits, closed drainage systems, and beaches. The soils are predominantly composed of alluviums, volcanic rocks, and isolated patches of sedimentary marine rocks. Intense wind erosion shapes the topography, creating sand dunes and impacting the biological productivity of ecosystems. The Sonoran Desert's climate is influenced by the surrounding mountains, which block moisture-laden air, resulting in rare winter frosts. In contrast, the Chihuahuan Desert experiences near- or below-freezing temperatures in its higher mountains. The vegetation in both deserts varies significantly due to differences in soil composition, topography, weather conditions, and soil age. Biological crusts play a crucial role in reducing erosion, trapping water and nutrients, and aiding in soil formation. Soils in desert environments typically consist of bare rock, varnished stone pavements, and coarse-weathered mantles. They undergo minimal weathering and leaching, resulting in coarser textures, shallow soil profiles, high concentrations of salts, and the accumulation of aeolian dust. These conditions pose challenges for providing ecosystem services such as flood regulation, water purification, climate regulation, and soil contaminant retention in urban areas located in deserts, thereby increasing vulnerability to climate change. Consequently, reconditioning soils in desert cities is essential before implementing any infrastructure to enhance ecosystem services.

---

* **Corresponding author Blanca González-Méndez:** Northwest Regional Station, Institute of Geology, National Autonomous University of Mexico, Colosio and Madrid s/n, ZIP Code 83000, Hermosillo, Sonora, Mexico; Tel: +52 662 217 5019; E-mail: blancagm@geologia.unam.mx

Israel Valencia Quiroz (Ed.)

Keywords: Aeolian erosion, Biocrusts, Chihuahuan Desert, Dryness, Low profile development, Patchy vegetation, Sonoran Desert, Salts accumulation.

INTRODUCTION

Solar radiation strikes our planet most intensely near the Equator. Due to the Earth's tilt of approximately 23.5° relative to its orbit, the zone of maximum solar intensity shifts seasonally northwards towards the Tropic of Cancer and southwards towards the Tropic of Capricorn. Thus, the tropics (23°N to 23°S latitudes) form a warm belt where heat generates rising and unstable air. This warm air leads to the condensation of evaporated moisture from the oceans, resulting in heavy tropical rains. The dry air continues its movement away from the tropical belt and cools and descends around 25-30° latitude, where it forms dry conditions with a predominance of calm air and rare rainstorms. These corridors of stable atmosphere are known as the "horse" latitudes, where most of the world's major deserts are located [1].

Part of the Mexican territory is located around 32°N latitude, which is affected by the horse's northern latitude (30°). Additionally, Mexico's topography contributes significantly to the formation of drylands. When moisture-laden winds coming from the Pacific Ocean and the Gulf of Mexico encounter the continental mountain ranges of the Sierra Madre Occidental, they cool and condense into fog, feeding the montane cloud forests. After losing their moisture, these winds descend over the mountain crests, where they compress and warm, becoming hot and dry, which inhibits precipitation. In addition, Mexico's widening northern territory creates a barrier that prevents humid winds from penetrating inland, resulting in extreme climate conditions conducive to desert formation [2].

As a result, Mexico has 2 major deserts: the Chihuahuan and the Sonoran, which are relatively young on the continent, dating back to around 8 million years (late Miocene). Over time, the Sonoran Desert has fluctuated in size, expanding and contracting in response to climates (Pleistocene glacial and interglacial periods). Additionally, about 6 million years ago, the Baja California Peninsula (described as part of the Sonoran Desert) moved northwestward along the East Pacific Rise (later the San Andreas Fault). This isolation gave rise to much endemism [3].

The Sonoran Desert extends from the southwestern USA into northwestern Mexico (Sonora and Baja California), covering up to 324 300 km^2. As most of the desert lies on Mexican territory (71% or 230 635 km^2), it gets its name from the Mexican state of Sonora [4]. This Desert is subtropical, with complex geology, diverse soil types, and a wide diversity of species. It receives rainfall twice a year: light, prolonged rains in the winter and heavier downpours in the summer. This seasonal pattern helps alleviate drought stress, contributing to the desert's rich biodiversity. Frosting is virtually absent, allowing succulents and cacti to thrive [5].

The Chihuahuan Desert spans a vast region of northern Mexico (including territories from the states of Chihuahua, Coahuila, Durango, Zacatecas, San Luis Potosí, Nuevo León, and Tamaulipas), as well as parts of the southwestern USA, including west Texas, New Mexico, and Arizona. It accounts for 13% of the Mexican territory. The region is predominantly sedimentary, with extensive alluvial plains. Limestone is the predominant bedrock, with gypsum and igneous rocks in some areas. The desert contains distinctive white gypsum dunes, the largest of which is found in the White Sands National Monument in New Mexico [6].

The Chihuahuan Desert is located between two orographic barriers: the Sierra Madre Occidental to the West and the Sierra Madre Oriental to the East. It is characterized by elevations ranging from 400 meters along the Rio Grande to as high as 2,000 meters. This higher elevation contributes to cooler and more moist conditions compared to the Sonoran Desert. The combination of climate diversity and geological features contributes to a wide range of animal and plant life, particularly in grassland areas, although it is not as biologically diverse as the Sonoran Desert.

Factors in Soil Formation

Soils in the Sonoran and Chihuahuan Deserts are generally made of alluviums from the Pleistocene and Pliocene alluviums, volcanic rocks from the Cenozoic and Pleistocene (mainly granites, andesites, basalts, rhyolites, and tuffs), and, in isolated patches, sedimentary rocks of marine origin from the Cenozoic and the Mesozoic (mainly limestones, shales, marls, and calcareous sandstones) [7]. In areas close to the Sonoran Desert, rhyolitic alluvium, mixed alluvium derived from volcanic lithologies with subordinate metamorphic and limestone clasts, and mixed alluvium derived from rhyolite and monzonite have been identified as parent materials [8].

Parent material has a major influence on soil properties such as texture, composition, water retention, infiltration, aeration, and nutrient availability. For instance, soils derived from rhyolite (a slow-disintegrating extrusive igneous rock) contain more gravel and, therefore, have less plant-available water than soils derived from monzonite. Furthermore, under similar environmental conditions, limestone parent material inhibits the formation of argillic horizons, contrasting with soils derived from igneous rocks, where the argillic horizons tend to be common [9]. Regarding alluvium, when it is derived from limestone, it is usually rich in silt and has both a high carbonate content and higher proportions of mixed-layer clays. In contrast, alluvium from igneous bedrock contains feldspar ($\approx 25\%$) and quartz as the dominant material, as well as more kaolinite and smectite. Both types of alluvium contain similar amounts of illite [10].

As we mentioned before, the Mexican deserts were primarily shaped by recent tectonic forces and are characterized by tall mountain ranges, or sierras, that rise abruptly from alluvial plains. At the foothills of these mountains, sediment-laden streams often form alluvial fans that spread out into the plain or bajada zone. Many of these desert basins have closed drainage systems, creating geomorphic features known as bolsons, which typically contain salty beaches at their lowest points. One prominent example in the Chihuahuan Desert is the Cuatro Ciénegas Basin [11].

The following are the major landscape components in the Mexican deserts [12, 13]:

1. Mountains and hills: Defined by tall mountains like the Sierra Madre Occidental, these mountains contrast with the surrounding flat plains, showcasing pine-covered areas such as the Copper Canyon (Barranca del Cobre) in the Sonora-Chihuahua region [14].
2. Piedmont slopes (bajadas): At the foothills of the mountains, sediment-laden streams form alluvial fans that spread into the plain or bajada zone. The bajadas could exhibit alluvial deposits from different geological ages, which result in distinct levels of soil development or erosion. The alluvial fans generally consist of gravelly to stony alluvium deposited in areas near the mountain front and finer alluvium in more distal locations [13].
3. Basin floors: Eroded sediments from ranges filled adjacent basins, forming alluvium deposits more than 3 km deep in some basins of the Sonoran Desert [15]. Coalescing basins are common in the Chihuahuan Desert [6].
4. Low relief alluvial deposits with poorly developed channels: Due to the scarcity of through-flowing rivers, there are few areas of erosional lowlands. Therefore, in stable areas (*i.e.*, Pleistocene fan remnants), pedogenesis occurs over time. These older soils are characterized by strong profile development with distinct illuvial clay-enriched horizons (argillic) and cemented calcium carbonate layers (caliche, calcic, and petrocalcic horizons) formed from wind-blown dust and precipitation deposits [16].
5. Closed drainage systems and beaches: Interior drainage systems form geomorphic features known as bolsons, which often contain salty beaches at their lowest points. Mapimí, de los Muertos, and Cuatro Ciénegas Basins are notable examples in the Chihuahuan Desert [17].

The watersheds of the Rio Grande (Río Bravo del Norte) and the Rio Conchos are the only significant drainage systems in the Chihuahuan Desert [18]. The Sonoran Desert contains the Colorado delta, characterized by a flat fluvial plain formed from fine sediments of the Colorado River. The San Felipe Desert is also part of the Sonoran Desert, which consists largely of low hills formed from fault blocks. These hills run parallel to broad valleys in the middle, filled with increasing amounts of alluvial and eolian deposits as they extend towards the Gulf of California [19].

Wind erosion plays a crucial role in shaping surface features and impacting the biological productivity of ecosystems. Strong winds lift coarse particles (>100 µm) and carry them briefly (≈ 1 m) before they fall back to the ground, causing sandblasting. This process generates fine dust particles as smaller materials break free from larger aggregates [20]. Therefore, wind erosion is the primary mechanism responsible for the net loss of soil materials, while fluvial processes have a negligible impact. Intense wind erosion alters topography significantly, creating sand dunes or removing entire soil layers [21].

The Sonoran Desert's unique climate is strongly influenced by its topography, as the surrounding mountain ranges act as barriers that channel or block moisture-laden air masses. It is characterized by a bi-seasonal rainfall pattern with a mean annual range of 100-300 mm. From December to March, frontal storms coming from the North Pacific bring occasional gentle rain, while from July to September, the summer monsoon delivers violent thunderstorms. These distinct winter and summer rainfall patterns are known locally as "equipatas" (the Yaqui-Mayo word for rain) and "las aguas", respectively. Daytime temperatures can reach or exceed 40°C during May–September [22].

Khormali & Monger [23] pointed out that the heat absorbed by desert soils can raise the temperatures up to 80°C hotter than the surrounding air. This heat gradually permeates through the soil profile until it stabilizes around the approximate mean annual air temperature, typically at a depth of about 50 cm. However, during the night, desert temperatures are significantly cooler than daytime highs. Therefore, the Sonoran Desert rarely experiences frost in winter, unlike the Chihuahuan Desert, where minimum temperatures reach near or below freezing in the higher mountains. In the Chihuahuan Desert, the mean annual precipitation is about 235 mm (range 150-400 mm), with most of the precipitation falling in the summer (May to October), and an average annual temperature of 18.6°C (range 14-23°C) [6].

Deserts present extreme conditions for life, including drastic changes in temperature, high UV radiation, low moisture and nutrient availability, long periods of dryness, and osmotic stress [24 - 26]. All of these characteristics, present in the Sonoran and Chihuahuan Deserts, demand specific adaptations in organisms. Soil texture, depth, and structure, as well as certain aspects of soil development (*e.g.*, carbonates or organic matter accumulation), determine how the combined effects of precipitation, temperature, and evaporation shape soil water availability. This, in turn, is linked to plant productivity, composition, and response to and recovery from disturbance [26, 27].

Many deserts are characterized by high spatial heterogeneity in vegetation composition and structure, reflecting heterogeneity in hydrology and other factors influenced by variations in soil parent material, topographic position, and soil age [26, 27]. Vegetation patches can vary in shape, size, and species composition. They can be large areas ($\geq$100 m^2) with several species or can be made up of one or a few species (<15 m^2), in which case they are called islands of fertility [24, 27, 28]. The main characteristic of these patches is the high concentrations of organic matter, nutrients, and organisms, as well as greater soil aggregation, capture of suspended particles, wind speed reduction, resistance to erosion, and enhanced infiltration of water [24, 27 - 29]. These factors can increase plant richness, promote the island's productivity, and create new habitats for other organisms [28].

The Sonoran and Chihuahuan Deserts are widely recognized for their high plant diversity and endemism [7, 30]. Among the types of vegetation that are recognized, Rzedowski [7] identifies mainly grasslands and xeric shrublands. In large areas of the Sonoran grassland, there is a "mesquite-grassland", in which *Prosopis velutina* is the most abundant species. In the grasslands of Chihuahua, *Bouteloua gracilis, B. eriopoda,* and *B. curtipendula* are predominant, along with several species of *Aristida* and *Andropogon*. Additionally, in Chihuahua, *Hilaria mutica* also occupies large areas, while *Distichlis spicata* and *Eragrostis obtusiflora* can dominate and tolerate strong concentrations of salts [7].

The xeric shrublands are found in practically all topographic conditions and geological substrates of the Sonoran and Chihuahuan Deserts. In the northern regions of Sonora, particularly in hills and middle elevations, the shrubland dominated by *Cercidium microphyllum, Opuntia* spp., and *Carnegiea gigantea* develops on shallow soils. Along the Sonora coast, *Larrea* extends southward to the vicinity of Guaymas. In the Chihuahuan scrubland, *L. tridentata* and *Flourensia cernua* develop preferably on plains and lower parts of alluvial fans. There is a shrubland of *Fouquieria splendens* that mainly develops in the western part of the plateau, while the *Acacia vernicosa* scrub is characteristic of the shallow soils of the calcareous plains to the south [7].

The discontinuous cover of vascular plants creates a niche for other organisms that form microbiotic crusts, better known as biological crusts or biocrusts [29, 31, 32]. Biocrusts are widely distributed in deserts, forming a mantle that covers up to 70% of the soil, both in areas without vegetation and under plant canopies [29, 31 - 33]. Biocrusts are distinguishable from other types of crusts in that they are engineered by microbiota [29, 31]. They support a photosynthetic component (algae, cyanobacteria, or bryophytes) combined with a mat of fungi, archaea, and other bacteria, occurring at higher concentrations than the surrounding non-crusted soils [29, 32]. The appearance, biomass, and composition of the biocrusts vary depending on the climatic regime [33]. Additionally, soil type, rainfall, potential evapotranspiration, temperature, and biocrust age can also determine the type of organisms that develop [24, 29, 31].

Biocrusts are complex ecosystems that support multiple trophic levels that enrich the surrounding soils through the leaching of macronutrients [29]. Biocrusts can increase the amount of organic carbon in deserts up to three times or more than the surrounding soil [24, 33]. Additionally, biocrusts also perform other ecological services: 1) they are dust and water traps, collecting nutrients and regulating the hydrological cycle; 2) they stabilize topsoil, reducing erosion; and 3) they weather mineral rocks, facilitating local soil formation and releasing nutrients [24, 31, 33].

Another characteristic of drylands is the desert varnish, which is a reddish-brown to nearly black coating found on desert pavement. It consists of 2:1 silicate clay and oxyhydroxides of iron and manganese, which are believed to be precipitated by microbes [34]. Some types of desert varnish can form relatively quickly and protect microbial communities against harsh conditions like radiation, temperature fluctuations, and low moisture. Desert varnish is absent on limestone surfaces, except where silicified veins are exposed [35, 36].

Soils

Desert soils typically consist of bare rock, varnished stone pavements, and coarse-weathered mantles. They exhibit minimal weathering and leaching, resulting in coarser textures, shallow soil profiles, high concentrations of salts, evaporites, carbonates, and aeolian (wind-blown) dust accumulation. Lower erosion rates in arid lands promote the preservation of paleosols (ancient soils) and abrupt soil boundaries, especially in areas with low tectonic activity or ancient shields [37].

Over thousands to millions of years, desert soils have been formed through the erosion of major mountain ranges. This weathering process involves rain, wind, freeze-thaw cycles, and chemical processes. Debris from weathered rocks accumulates at mountain bases. It is sorted by rainwater, depositing finer particles like gravel, sand, and silt in lowlands, while larger fragments remain higher up in mountain canyons. Eventually, this accumulated debris can bury the mountains themselves [38].

Initially, the particles created through weathering are not suitable for supporting plant growth; they are mineral particles like those found in sand dunes. To become viable soil, these particles need to be stabilized and bound together into a structured form. This stabilization process takes many years and is facilitated by biocrusts. These crusts contribute to soil formation by synthesizing organic matter from atmospheric carbon dioxide, deriving energy from sunlight through photosynthesis, and fixing atmospheric nitrogen into cellular constituents [39].

The organic matter produced by these organisms stabilizes the soil, improves its structure, and acts as a sponge to retain water. Ultimately, this process creates conditions suitable for the establishment of higher plants in desert environments [39]. The most common pedogenic processes in deserts are listed below [23]:

Rock Weathering by salts

in situ rock disintegration in desert soils is less intense than in humid-temperate soils due to reduced root penetration and fewer freeze-thaw cycles. In contrast, carbonates and soluble minerals precipitate within rock crevices, aiding in rock breakdown [40]. Chemical weathering and mineral dissolution are limited by reduced soil water, though the formation of clay minerals like palygorskite and sepiolite is observed, especially in petrocalcic horizons [41].

Ferrugination

Metamorphism and biogenic mobilization of manganese and iron contribute to surface phenomena like desert varnish. Although ferrugination has been described in humid environments, in deserts, ferrugination occurs *in situ* as a part of a complex process involving periodic movement and accumulation of clay fractions, resulting in the development of a dense, red-brown, cloddy middle layer [23].

Oxidation/Reduction Processes

These processes are minimal compared to humid-temperate soils due to prolonged dry periods and good aeration. Redoximorphic features are more prominent in areas with groundwater discharge (beaches or floodplain edges) or sites with human irrigation [23].

Clay Illuviation

Clay illuviation is limited due to reduced wetting fronts, though clay coatings and argillans are common, leading to the formation of argillic horizons (Bt). Frequently, this argillic horizon co-exists with pedogenic carbonate as a Btk horizon [23, 42].

Shrink-Swell Characteristics

Vertisols in deserts present less frequent shrink-swell cycles compared to humid-temperate Vertisols. Consequently, cracks in deserts remain open for most of the year, closing only for a few days during the heavy storms' precipitation [42].

Precipitation of secondary minerals

Non-flushing of soluble material distinguishes desert soils from humid-temperate soils. In desert environments, dissolved cations and anions remain in the soil profile and precipitate from the drying soil solution. The formation of pedogenic carbonate is a clear example of this phenomenon. Additionally, desert soils accumulate other soluble salts, such as gypsum and silica [23].

Soil Organic Matter (SOM) and Dust Additions

Desert soils generally lack a robust source of SOM due to sparse vegetation, like woody shrubs and scattered grasses. This scarcity is reflected in global comparisons of SOM distribution patterns [39]. Dust additions to desert soils are complex and heterogeneous, influencing nutrient availability and water-holding capacity. Local sand particles, as well as distant fine sand and silt particles, contribute to soil profiles, often containing significant carbonate deposits [43]. Dust accumulation can be an important pedogenic factor since limiting weathering is unlikely to produce deep soils (>1 m) under granitic bedrocks. The "accretionary-inflationary" model represents a cumulative soil development where dust deposition and incorporation are responsible for the accumulation of most key secondary soil materials, promoting upward soil accretion and causing changes in soil volume [44].

Organic Matter Decomposition

Warm temperatures accelerate organic matter decomposition rates, but the dry conditions limit overall decomposition [45]. In these arid ecosystems, abiotic processes (photochemical oxidation or photodegradation) play a major role in facilitating the physical and chemical degradation of plant litter [46].

Diagnostic Horizons

Hot desert soils contain various diagnostic horizons, including cambic, argillic, natric, petrocalcic, gypsic, petrogypsic, salic, and duripan horizons. While none of these horizons are exclusive to hot desert soils, salic, gypsic, and petrogypsic are relatively rare in other soil types. Brief definitions of these diagnostic horizons are outlined below [23, 47]:

Cambic horizon (Bw, B, or Bq)

A subsurface layer that indicates soil formation through varying degrees of development, *i.e.*, higher oxide and/or clay contents compared to underlying layers, and often shows loss of carbonates or gypsum.

Argic horizon (Bt)

A subsurface horizon where silicate clays have accumulated either by translocation (illuviation) from upper horizons or formed in place. The evidence of clay translocation is the coating of clay on pore walls and surfaces of peds. It has a minimum horizon thickness requirement of 7.5 cm if loamy or clayey and 15 cm if sandy.

Natric horizon (Btn)

A dense subsurface horizon with a higher clay content than the overlying horizon(s). It contains elevated exchangeable sodium (sodium adsorption ratio $\geq$ 13 or exchangeable sodium percentage $\geq$ 15) and often exhibits columnar or prismatic structures.

Calcic horizon (Bk or Bkk)

A horizon with accumulated secondary calcium carbonate ($CaCO_3$), commonly found in subsurface layers. It may contain primary carbonates as well.

Petrocalcic horizon (Bkm or Bkkm)

A horizon cemented by $CaCO_3$, sometimes containing Magnesium carbonate ($MgCO_3$). It offers high resistance to penetration.

Gypsic horizon (By or Byy)

A horizon containing accumulations of secondary gypsum ($CaSO_4 \cdot 2H_2O$), either as a surface or subsurface layer.

Petrogypsic horizon (Bym or Byym)

A gypsum-cemented horizon that impedes root penetration, except through vertical fractures.

Salic horizon (Az, Bz, or Cz)

A horizon, either surficial (Az) or sub-surficial (Bz or Cz), which is rich in readily soluble salts and different from gypsum.

Petroduric horizon (Bqm)

Also known as duripan, it is a subsurface horizon cemented primarily by illuvial secondary silica (SiO_2). It often occurs below argic horizons in desert soils. Calcium carbonate can also be used as an additional cement.

Soils of the Sonoran and Chihuahuan Deserts

According to INEGI [48], Mexican deserts have a great variety of soils (Figs. **1** & **2**). We briefly discuss the different soil types present in Mexican deserts [47].

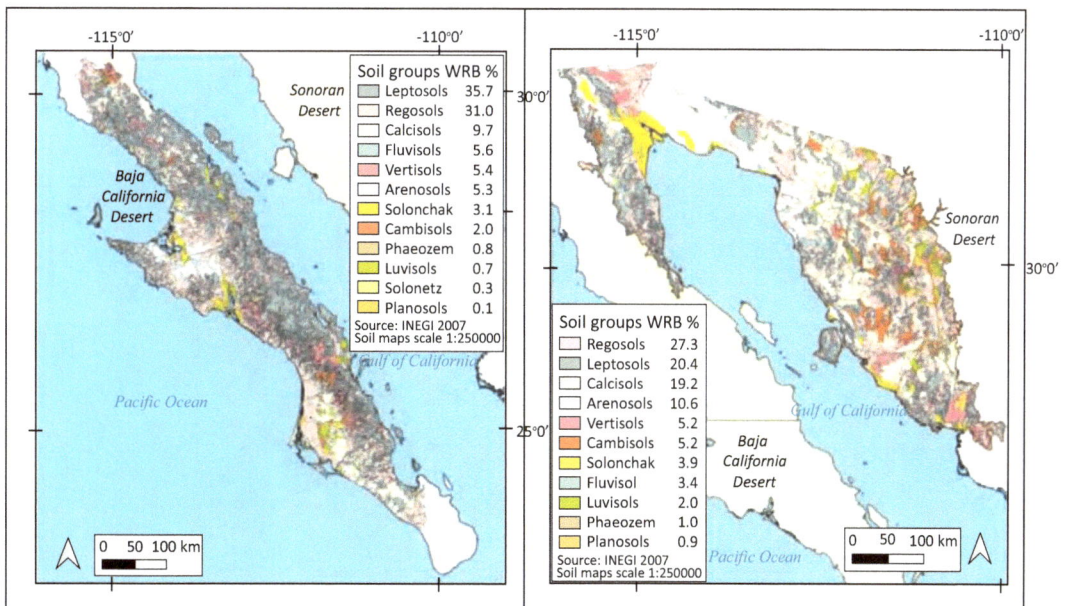

Fig. (1). Soils of the Baja California Desert (left) and Sonoran Desert (right). Source: Own elaboration.

Leptsols

These soils are either thin or abundant in coarse fragments, with significant limitations to root growth due to continuous rock starting ≤ 25 cm from the soil surface or with <20% fine earth averaged over a depth of 75 cm from the soil surface (*i.e.*, abundance of coarse fragments). Leptosols have a very small plant-available water capacity. These shallow soils are the most abundant in the Baja California (35.7%), the Chihuahuan Desert (31.6%), and the Sonoran Desert (20.4%). In regions with silicate rocks, shallow soils often occur where sedimentary cover is absent or thin and erosion exceeds weathering. Thus, shallow Leptosols are widespread across Baja California and the Sierra Madre Occidental, where both the Baja California and the Chihuahuan Deserts are located [22].

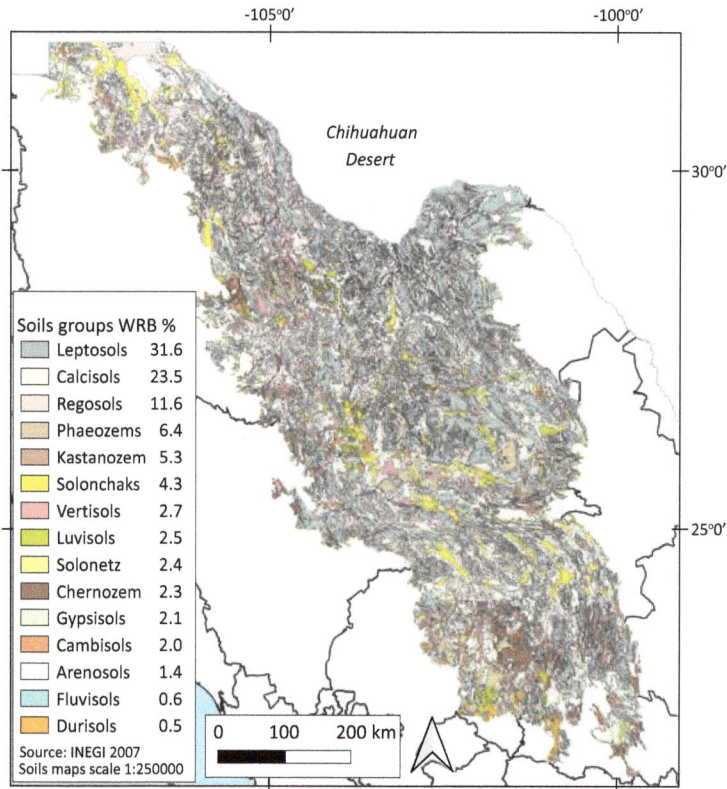

Fig. (2). Soils of the Chihuahuan Desert. Source: Own elaboration.

Regosols

Soils with little or no profile differentiation, often containing significant silt or clay contents. These young soils are the most common in the Sonoran Desert

(27.3%), the second most prevalent in the Baja California Desert, and the third in the Chihuahuan Desert (11.6%). Harsh conditions (poor chemical weathering due to lack of water) give place to weakly developed soils that conserve most of the original parent material. These are poor in organic matter and have a simple profile A/C [49].

Calcisols

Soils characterized by the accumulation of secondary carbonates, either soft (calcic horizon) or hard (petrocalcic horizon). These carbonates accumulate through rainwater (percolating only to a shallow depth or from ascending groundwater). Calcisols represent 23.5% of the Chihuahuan Desert, 9.7% in Baja California (9.7%) and 19.2% of the Sonoran Desert. Under these dry conditions, $CaCO_3$ forms easily due to water infiltration and evaporation, which promotes the translocation of the carbonates (pedogenic carbonates) from the soil surface. However, large areas of these deserts are covered with limestone and other sedimentary rocks containing $CaCO_3$. Additionally, Calcisols typically lack humus-enriched topsoil because of the low productivity of arid regions, which also limits the microbial decomposition of organic matter [19].

Vertisols

Soils with $\geq 30\%$ clay predominantly shrink-swell clay minerals (*e.g.*, smectite), leading to a constant churning (peloturbation), resulting in cracks that open and close periodically. Vertisols have vertic horizons characterized by slickensides (polished and grooved aggregate surfaces) and/or wedge-shaped aggregates tilted from the horizontal. These soils are hard when dry and sticky when wet and are, therefore, difficult to work, typically with high cation exchange capacity (CEC) and high base saturation. Approximately 5% of the Baja California and Sonoran Deserts are covered by Vertisols, while only 2% of the Chihuahuan Desert contains this type of soil. In the north of Mexico, these soils are present in sites where volcanic rocks and alluvial basins occur and whose weathering gives place to smectite clays [49].

Arenosols

Sandy or loamy soils are poor in organic matter. They cover large areas of recent sedimentation, often with dunes, and have little or no profile differentiation. They represent up to 10.6% of the Sonoran Desert, 5.3% of the Baja California Desert, and only 1.4% of the Chihuahuan Desert.

Solonchaks

Soils with high concentrations of soluble salts starting $\leq$ 50 cm from the soil surface. They possess high electrical conductivity and high osmotic forces. Solonchaks are accumulated by rainwater (percolating only to a shallow depth), ascending groundwater, or irrigation water. They are abundant in the Chihuahuan Desert (4.3%), followed by the Sonoran and Baja California Deserts (3.9 and 3.1%, respectively). High evaporation rates cause soluble salts to rise by capillarity, a process intensified by the mineral content of irrigation water, as in Baja California [50].

Cambisols

Moderately developed soils (little or no profile differentiation), with a subsoil that shows at least weak evidence of soil formation. This group of soil is more abundant in the Sonoran Desert (5.2%) than in the Baja California or Chihuahuan ones (2% each). These soils are in transition to a more developed soil; therefore, in arid regions, they are found where low precipitation limits clay illuviation or where local conditions do not promote the accumulation of carbonates, gypsum, or soluble salts [19].

Phaeozems

Soils characterized by a mollic horizon (high accumulation of organic matter) and a high base saturation at up to 1 m depth but with no secondary carbonates (unless very deep). Phaeozems represent 6.4% of the Chihuahuan Desert territory and up to 1% of the Sonoran and Baja California Deserts. The arid conditions of the Sonoran Desert hinder significant organic matter accumulation and soil development. Conversely, in the Chihuahuan Desert, with more humid climates, pedogenesis is more advanced, allowing the prevalence of Phaeozems [19].

Luvisols

Soils characterized by an argic horizon dominated by 2:1 clay minerals (high CEC) and a high base saturation at 50–100 cm depth. Luvisols are found in 2.5% of the Chihuahuan Desert, 2.0% of the Sonoran Desert, and 0.7% of the Baja California Desert. Clay illuviation plays a crucial role in the definition of this soil type, which is facilitated by rainfall that leaches clays through the soil profile. That is why it was traditionally believed that the Luvisols in the arid northern deserts of Mexico were remnants of a more humid climate. However, these Luvisols result from the neoformation of smectite and palygorskite rather than clay illuviation *per se* [51].

Solonetz

Soils that have a subsoil horizon with an accumulation of clay (and organic matter), a high content of exchangeable sodium, and a predominantly columnar structure (natric horizon). Solonetz are only found in the Chihuahuan and Baja California Deserts (2.4 and 0.3%, respectively). They are usually found in low flat lands or the periphery of closed valleys such [49] as those found in the Chihuahuan Desert.

Fluvisols

Soils are composed of fluviatile, lacustrine, or marine sediments whose original stratification is visible. It is common to find A horizons to be buried. This type of soil represents 5.6% of the Baja California Desert, 3.4% of the Sonoran Desert, and 0.5% of the Chihuahuan Desert. The young alluvial sediments are characterized by layered deposits with an irregular vertical distribution of clay and organic carbon. The dynamic nature of sedimentation in these areas prevents the complete development of pedogenetic processes. Consequently, the soils in these alluvial dynamic young deposits are classified as Fluvisols [19].

Planosols

Soils with stagnant water due to clay-richer subsoil. They have an abrupt textural difference (within slightly deeper zones, clay content increases twice or by $\geq 20\%$ more). Above and/or below the textural difference, there are reducing conditions with stagnic properties. Albic material is common above the textural difference. Planosols are reported as the least abundant soils, only in the Sonoran (0.9%) and Baja California Deserts (0.1%). Planosols form in flat areas that are periodically flooded, typically on substrates containing alluvial and colluvial clay deposits.

The next groups of soils are only reported for the Chihuahuan Desert: **Kastañozems**, 5.3%; **Chernozems**, 2.3%; **Gypsisols**, 2.1%; and **Durisols**, 0.5%. Kastañozems and Chernozems have well-developed humus horizons commonly found in subarid to dry climates, often with significant accumulations of secondary carbonates. Kastañozems have a mollic horizon. They typically develop on deep alluvial deposits where clay or calcium carbonates accumulate in the subsoil. However, where petrocalcic horizons exist, the development of Kastañozems is restricted. Chernozems are darker than Kastañozems and tend to form in slightly moister climates. Many Chernozems are developed from sediments rich in smectite, which can result in excessive compaction and cracking when dry [52].

Gypsisols

Gypsisols are characterized by the accumulation of secondary gypsum, either soft (gypsic horizon) or hard (petrogypsic horizon), deposited by rainwater or ascending groundwater. Many Gypsisols also contain secondary carbonates. In the Chihuahuan Desert, the Cuatro Ciénegas basin is known for its gypsum-rich soils, which include massive evaporite gypsum bedrock from Cretaceous-aged formations (Acatita Formation). The valleys also contain geologically recent gypseous soils formed alongside ancient salt lakes (Pleistocene), characterized by reprecipitated gypsum from eroded montane deposits. The famous gypsum dunes to the west derive from lacustrine gypsum through aeolian processes [53].

Durisols

Durisols are characterized by the accumulation of secondary silica (SiO_2), forming hard concretions (duric horizon) or a continuous hard layer (petroduric horizon or duripan). These soils are typical of arid and semiarid environments, ranging from very shallow to moderately deep and highly drained. Secondary silica cementation is commonly found within the upper meter of the soil profile. Durisols are reported at piedmont and colluvial deposits of the Sierra Madre Occidental [49].

CASE STUDY: SOIL ASSESSMENT AND RECONDITIONING FOR GREEN INFRASTRUCTURE IN HERMOSILLO, MEXICO

Cities in arid regions are particularly vulnerable to climate change, not only due to the expected widespread droughts but also because of the increase in temperature and heavy precipitation events [54]. Hermosillo, located in the Sonoran Desert of northwest Mexico, exemplifies these challenges with an annual precipitation of 250-350 mm and potential evaporation rates between 2400-2600 mm/year. Consequently, water scarcity is a great challenge for this fast-growing city, which is also affected by the seasonality of the North American Monsoon (NAM), which is responsible for severe storm events [55].

Green infrastructure (GI), defined as a strategically planned network of natural and semi-natural areas designed to provide ecosystem services, emerges as a crucial strategy for climate change adaptation. GI can be used as a component of a stormwater management system where soils and vegetation are used to infiltrate, evapotranspirate, and recycle stormwater runoff, allowing aquifer recharge, flooding reduction, and cooling [56].

However, implementing GI in urban areas often encounters challenges, particularly in soils that typically lack organic matter, contain low content of nutrients, have low soil microbial communities, exhibit compaction, contain mixed debris and sediment, pollution, and salt content. These degraded urban soils hinder the potential of GI to provide essential ecosystem services such as growing medium, flood regulation, water purification, climate regulation, and soil contaminant retention. Moreover, rapid soil changes due to urbanization processes can further diminish GI effectiveness over time [57]. Thus, before GI installation, it is not only necessary to make a soil inventory to identify those with the best suitable characteristics for GI, but also to plan a soil reconditioning to boost those suitable characteristics [58].

Based on these findings, a soil assessment was conducted in the urban soils of Hermosillo. The results revealed that most of the soils have predominantly coarse texture (CA and AC), composed of mixed debris with sediments that exhibited poor structure, compaction, high salt content, and pollutants. These conditions not only limit infiltration and water storage capacity but also increase susceptibility to erosion during dry and monsoon seasons. Therefore, before GI implementation, remediation efforts, including the incorporation of amendments, are necessary to enhance soil fertility, water retention, and pollutant retention capacities [58, 59].

One of the promising amendments for soil reconditioning is biochar, a biomaterial produced from biomass under thermochemical processes. Biochar is characterized by high porosity, a high specific surface area, and different physical, chemical, and biological properties that improve soil physical structure, hydraulic properties, and water retention capacity, thereby enhancing water availability for plants. Additionally, it may increase soil CEC, reduce organic and inorganic contaminant toxicity, and increase soil P content [57]. Consequently, biochar addition to desert urban soils could improve not only water retention but also soil fertility and ameliorate runoff pollution, providing a better substrate for green infrastructure establishment. As biochar is made from biomass residues, its production for soil amendment could also contribute to the circular economy of the city.

The reconditioning experiments conducted with Hermosillo's soils showed promising results with the addition of biochars derived from local biomass (*e.g.*, pruning wood). Biochar improved soil water retention, metal retention and reduced nutrient leaching, optimizing conditions for successful GI establishment [59].

CONCLUDING REMARKS

The desert soils of Mexico are shaped by unique environmental conditions, like intense solar radiation and a lack of water, which translate into minimal weathering and leaching, coarse textures, shallow profiles, and high concentrations of salts and carbonates. Under these conditions, key pedogenic processes include limited SOM accumulation, heterogeneous dust additions that influence nutrient availability, good aeration, persistent cracks due to less frequent shrink-swell process (promoted by the extended dry season), secondary mineral precipitation, limited clay illuviation, and common accumulations of pedogenic carbonates, silica, gypsum, and soluble salts -because of the reduced wetting fronts.

These pedogenic processes in Mexican Deserts mainly result in young and shallow soils (Leptosols and Regosols), which are the most common in the region. They are followed by soils with minimal profile differentiation, like Cambisols, Arenosols, and Fluvisols, which are influenced by the water and weathering dynamics of parent materials (igneous and sedimentary rocks, as well as alluvial, colluvial, and aeolian deposits). Dry conditions also contribute to the formation of soils with salt accumulation, including Solonchakz, Solonetz, Calcisols, Gypsisols, and Durisols.

Areas with alluvial basins are prompt to the inheritance or formation of smectite clays, which results in the presence of Vertisols, while localized rainfall patterns influence the illuviation process observed in Luvisols. In more humid and productive microenvironments across Mexican Deserts, there are localized patches with organic matter accumulation, forming Phaeozems. Finally, the relatively humid conditions specific to the Chihuahuan Desert support the presence of Chernozems and Kastañozems.

Due to these characteristics and harsh desert conditions (the delicate balance between low moisture and extreme temperatures, limited vegetation cover, low organic matter, high evaporation rates, and erosion), soils are extremely vulnerable to human-induced pressures like climate change. Therefore, urban areas located in deserts need urgent adaptation strategies to deal with water scarcity, urban heat islands, and flood management. As soils provide these crucial ecosystem services, it is essential to recondition the soils in desert cities before implementing any infrastructure to enhance ecosystem services.

AUTHORS' CONTRIBUTION

The authors confirm their contribution to the chapter as follows:

Conceptualization, methodology, investigation, data curation, writing-original draft preparation, writing-review and editing, and figure editing: Blanca González-Méndez; Methodology, formal analysis, investigation, data curation, writing-original draft preparation, writing-review and editing, and figure editing: Elizabeth Chávez-García.

QUESTIONS RELATED TO THE TEXT

1. How do seasonal climatic conditions influence the distribution of precipitation in the deserts of Mexico?
2. What are the geological and topographic factors that contribute to the formation and expansion of the Chihuahuan and Sonoran deserts?
3. Describe some adaptations that biocrusts have to survive and thrive in arid ecosystems.
4. What process facilitates the formation of deep soils in desert environments?
5. What factors limit the formation of argillic horizons in desert soils?
6. What are the challenges of desert soils in the face of environmental changes and anthropogenic pressures?

LIST OF ABBREVIATIONS

°	Degrees
°C	Degrees Celsius
Az, Bz, or Cz	Salic Horizons
Bk or Bkk	Calcic Horizons
Bkm or Bkkm	Petrocalcic Horizons
Bqm	Petroduric Horizon
Bt	Argillic Horizon
Btk	Argillic Horizon with Carbonates
Btn	Natric Horizon
Bw, B, or Bq	Cambic Horizons
By or Byy	Gypsic Horizons
Bym or Byym	Petrogypsic Horizons
$CaCO_3$	Calcium Carbonate
$CaSO_4 \cdot 2H_2O$	Gypsum
CEC	Cation Exchange Capacity
e.g.	For Example
GI	Green Infrastructure
i.e.	That is to Say
INEGI	Instituto Nacional de Estadística y Geografía
M	Meter
m²	Square Meters
$MgCO_3$	Magnesium Carbonate
mm	Millimeters
N	North

NAM	North American Monsoon
S	South
SOM	Soil Organic Matter
spp.	Species
USA	United States of America
UV	Ultraviolet Light

ACKNOWLEDGEMENTS

We thank Mauricio Peregrina-Llanes for the maps' elaboration of the Sonoran and Chihuahuan Deserts using INEGI maps. We also thank UNAM's Center for Academic Writing and Skills (CEHA) for reviewing this article and making pertinent recommendations.

REFERENCES

[1]   McGinnies WG, Goldman BJ, Paylore P. Deserts of the World: An Appraisal of Research Into Their Physical and Biological Environments. University of Arizona Press 1977.

[2]   Huerta-Martínez FM. Análisis de Gradientes en la Vegetación de El Huizache, San Luis Potosí, México. PhD dissertation. Colegio de Postgraduados 2002.

[3]   Van Devender TR, Brusca RC. Deep History of the Sonoran Desert.A natural history of the Sonoran Desert. University of California Press 2021.

[4]   Martínez-Yrízar A, Felger R, Búrquez A. Los ecosistemas de Sonora: un diverso capital natural.Divers Biológica Son. UNAM/CONABIO 2010; pp. 129-56.

[5]   Ezcurra E, Mellink E. Hot Deserts. John Wiley & Sons 2006.
       [http://dx.doi.org/10.1038/npg.els.0003178]

[6]   Schmidt RJ. Chihuahuan Climate. Second Symp. Resour. Chihuahuan Desert Reg. United States Mex., Alpine, Tex. Chihuahuan Desert Research Institute.

[7]   Rzedowski J.. Vegetación de México. 1a ed. Mexico City: Comisión Nacional para el Conocimiento y Uso de la Biodiversidad; 2006.

[8]   Sion BD, Harrison BJ, McDonald EV, Phillips FM, Axen GJ. Chronofunctions for new Mexico, USA soils show relationships among climate, dust input, and soil development. Quat Int 2022; 618: 35-51.
       [http://dx.doi.org/10.1016/j.quaint.2021.06.024]

[9]   Herbel CH, Gile LH, Fredrickson EL, Gibbens RP. Soil water and soils at soil water sites, Jornada Experimental Range. In: Gile LH, Ahrens RJ, Eds Suppl to Desert Proj soil Monogr, US Department of Agriculture, Soil Conservation Service. 1994; pp. 1-592.

[10]   Monger HC, Lynn WC. Clay mineralogy at the Desert Project and the Rincon Surface study area.In: Gile LH, Ahrens RJ, Eds. Suppl. to Desert Proj. soil Monogr. Vol. II, Soil Surv. Investig. Rep. No. 44, US Department of Agriculture, Soil Conservation Service; 1996; pp. 111-55.

[11]   Flores Vázquez JC, Rosas Barrera MD, Golubov J, Sánchez-Gallén I, Mandujano MC. Ecological Importance of bajadas in the Chihuahuan Desert. Plant Divers. Ecol. Chihuahuan Desert 2020; pp. 117-28.
       [http://dx.doi.org/10.1007/978-3-030-44963-6_8]

[12]   Baddock MC, Gill TE, Bullard JE, Acosta MD, Rivera Rivera NI. Geomorphology of the Chihuahuan Desert based on potential dust emissions. J Maps 2011; 7(1): 249-59.

[http://dx.doi.org/10.4113/jom.2011.1178]

[13]   McAuliffe JR. Landscape Evolution, Soil Formation, and Ecological Patterns and Processes in Sonoran Desert Bajadas. Ecol Monogr 1994; 64(2): 111-48.
        [http://dx.doi.org/10.2307/2937038]

[14]   Scarborough R. The geologic origin of the Sonoran Desert.In: Phillips SJ, Wentworth Comus P, Eds A natural history of the Sonoran Desert, University of California Press. 2012; pp. 71-86.

[15]   Morrison RB. Pliocene/Quaternary geology, geomorphology, and tectonics of Arizona.Soils and Quaternary Geology of the Southwestern United States. Geological SoMiety of America Special Paper 1985; 203: pp. 123-46.
        [http://dx.doi.org/10.1130/SPE203-p123]

[16]   Gile LH, Hawley JW, Grossman RB. Soils and geomorphology in the Basin and Range area of Southern New Mexico: Guidebook to the Desert Project. New Mexico: New Mexico Bureau of Mines and Mineral Resources 1981.

[17]   Souza V, Siefert JL, Escalante AE, Elser JJ, Eguiarte LE. The Cuatro Ciénegas Basin in Coahuila, Mexico: An Astrobiological Precambrian Park. Astrobiology 2012; 12(7): 641-7.
        [http://dx.doi.org/10.1089/ast.2011.0675] [PMID: 22920514]

[18]   Tamayo JL, West RC. The hydrology of Middle America.Wauchope, Handbook of Middle American Indians, Nat Environ Early Cult. Austin, Texas: University Texas Press 1964; pp. 84-121.

[19]   Krasilnikov P, Gutiérrez-Castorena MC, Ahrens RJ, Cruz-Gaistardo CO, Sedov S, Solleiro-Rebolledo E. The Soils of Mexico. London: Springer 2013.
        [http://dx.doi.org/10.1007/978-94-007-5660-1]

[20]   Bowker GE, Gillette DA, Bergametti G, Marticorena B, Heist DK. Fine-scale simulations of aeolian sediment dispersion in a small area in the northern Chihuahuan Desert. J Geophys Res Earth Surf 2008; 113.
        [http://dx.doi.org/10.1029/2007JF000748]

[21]   Gillette D, Monger HC. Eolian Processes on the Jornada Basin.Structure and Function of a Chihuahuan Desert Ecosystem: The Jornada Basin Long-Term Ecological Research Site. Oxford University Press 2006; pp. 189-210.
        [http://dx.doi.org/10.1093/oso/9780195117769.003.0013]

[22]   Tereshchenko I, Zolotokrylin AN, Titkova TB, Brito-Castillo L, Monzon CO. Seasonal variation of surface temperature-modulating factors in the Sonoran desert in northwestern Mexico. J Appl Meteorol Climatol 2012; 51(8): 1519-30.
        [http://dx.doi.org/10.1175/JAMC-D-11-0160.1]

[23]   Khormali F, Monger C. Hot desert soils—Global distribution and unique characteristics. Geoderma Reg 2020; 23: e00330.
        [http://dx.doi.org/10.1016/j.geodrs.2020.e00330]

[24]   Pointing SB, Belnap J. Microbial colonization and controls in dryland systems. Nat Rev Microbiol 2012; 10(8): 551-62.
        [http://dx.doi.org/10.1038/nrmicro2831] [PMID: 22772903]

[25]   Tchakerian V, Pease P. The Critical Zone in Desert Environments. Developments in Earth Surface Processes 2015; 19: 449-72.
        [http://dx.doi.org/10.1016/B978-0-444-63369-9.00014-8]

[26]   Duniway MC, Petrie MD, Peters DPC, Anderson JP, Crossland K, Herrick JE. Soil water dynamics at 15 locations distributed across a desert landscape: insights from a 27☐yr dataset. Ecosphere 2018; 9(7): e02335.
        [http://dx.doi.org/10.1002/ecs2.2335]

[27]   Weil RR, Brady N. The Nature and Properties of Soils. 15th ed., New Jersey: Prentice Hall 2017.

[28]   Montaño N, Ayala F, Bullock S, Briones O, Oliva F, Sánchez R, *et al.* Almacenes y flujos de carbono en ecosistemas áridos y semiáridos de México: síntesis y perpectivas. Carbon Stocks and Fluxes in Arid and Semiarid Ecosystems of Mexico: Synthesis and Prospects. Terra Latinoam 2016; 34: 39-59.

[29]   Steven B. The Biology of Arid Soils. New Haven, CT, USA: De Gruyter 2017.
[http://dx.doi.org/10.1515/9783110419047]

[30]   Conocimiento actual de la biodiversidad. Soberón J, Halffter G, Llorente-Bous J. Capital natural de MéxicoMexico City: Comisión Nacional para el Conocimiento y Uso de la Biodiversidad 2008; I.

[31]   Weber B, Burkhard B, Belnap J. Biological Soil Crusts: An Organizing Principle in Drylands. Utah, USA: Springer 2016.
[http://dx.doi.org/10.1007/978-3-319-30214-0]

[32]   Porras-Alfaro A, Herrera J, Natvig DO, Lipinski K, Sinsabaugh RL. Diversity and distribution of soil fungal communities in a semiarid grassland. Mycologia 2011; 103(1): 10-21.
[http://dx.doi.org/10.3852/09-297] [PMID: 20943560]

[33]   Belnap J, Lange O. Biological soil crust: Structure, function, and management. Utah, USA: Springer 2003.
[http://dx.doi.org/10.1007/978-3-642-56475-8]

[34]   Broecker WS, Liu T. Rock varnish: recorder of desert wetness? GSA Today 2001; 11(8): 4-10.
[http://dx.doi.org/10.1130/1052-5173(2001)011<0004:RVRODW>2.0.CO;2]

[35]   Garvie LAJ, Burt DM, Buseck PR. Nanometer-scale complexity, growth, and diagenesis in desert varnish. Geology 2008; 36(3): 215-8.
[http://dx.doi.org/10.1130/G24409A.1]

[36]   Gile LH, Grossman RB. The Desert Project soil monograph: Soils and landscapes of a desert region astride the Rio Grande Valley near Las Cruces, New Mexico. Washington; 1979.

[37]   Dregne HE. North American Deserts Deserts Arid Lands. Martinus 1984; pp. 145-56.
[http://dx.doi.org/10.1007/978-94-009-6080-0_8]

[38]   Dixon J. Canyonlands and Arches: Windows on Landscapes in the American Southwest.Geomorphol Landscapes World. Springer 2009; pp. 39-47.
[http://dx.doi.org/10.1007/978-90-481-3055-9_5]

[39]   Follett RF, Kimble JM, Lal R. The Potential of US Grazing Lands to Sequester Carbon and Mitigate the Greenhouse Effect. New York: CRC Press 2001.

[40]   Birkeland PW. Soil and Geomorphology. New York: Oxford University Press 1999.

[41]   Monger HC, Daugherty LA. Pressure Solution: Possible Mechanism for Silicate Grain Dissolution in a Petrocalcic Horizon. Soil Sci Soc Am J 1991; 55(6): 1625-9.
[http://dx.doi.org/10.2136/sssaj1991.03615995005500060021x]

[42]   Soil Taxonomy: A Basic System of Soil Classification for Making and Interpreting Soil Surveys. Natural Resources Conservation Service, United States Department of Agriculture 1999.

[43]   Field JP, Belnap J, Breshears DD, *et al.* The ecology of dust. Front Ecol Environ 2010; 8(8): 423-30.
[http://dx.doi.org/10.1890/090050]

[44]   McFadden LD. Strongly dust-influenced soils and what they tell us about landscape dynamics in vegetated aridlands of the southwestern United States. Spec Pap Geol Soc Am 2013; 500: 501-32.
[http://dx.doi.org/10.1130/2013.2500(15]

[45]   Pankova EI, Gerasimova MI. Desert soils: Properties, pedogenic processes, and classification. Arid Ecosyst 2012; 2: 69-77.

[46]   Gallo ME, Porras-Alfaro A, Odenbach KJ, Sinsabaugh RL. Photoacceleration of plant litter decomposition in an arid environment. Soil Biol Biochem 2009; 41(7): 1433-41.
[http://dx.doi.org/10.1016/j.soilbio.2009.03.025]

[47]   WRB FAO IWG. World Reference Base for Soil Resources. International soil classification system for naming soils and creating legends for soil maps. 4th ed. Vienna: International Union of Soil Sciences (IUSS); 2022; 4.

[48]   INEGI. The digital map of Mexico 2007. Available from: https://en.www.inegi.org.mx/temas/ mapadigital/

[49]   Herrera Pedroza R. Suelos de la Sierra Madre Occidental. In: Krasilnikov P, Jiménez Nava FJ, Reyna Trujillo T, García Calderón NE, Eds. Geografía de los suelos México, Facultad de Ciencias, UNAM; 2011; pp. 271-320.

[50]   Endo T, Yamamoto S, Larrinaga JA, Fujiyama H, Honna T. Status and Causes of Soil Salinization of Irrigated Agricultural Lands in Southern Baja California, Mexico. Appl Environ Soil Sci 2011; 2011: 1-12.
[http://dx.doi.org/10.1155/2011/873625]

[51]   Ducloux J, Petit S, Decarreau A, Delhoume JP. Clay Differentiation in Aridisols of Northern Mexico. Soil Sci Soc Am J 1995; 59(1): 269-76.
[http://dx.doi.org/10.2136/sssaj1995.03615995005900010043x]

[52]   Driessen P, Deckers J, Spaargaren O, Nachtergaele F. Lecture notes on the major soils of the world. Rome: FAO 2001.

[53]   Ochoterena H, Flores-Olvera H, Gómez-Hinostrosa C, Moore MJ. Gypsum and Plant Species: A Marvel of Cuatro Ciénegas and the Chihuahuan Desert.Plant Divers Ecol Chihuahuan Desert Emphas Cuatro Ciénegas Basin. Springer 2020; pp. 129-66.
[http://dx.doi.org/10.1007/978-3-030-44963-6_9]

[54]   IPCC. Climate Change 2014: Synthesis Report. Contribution of Working Groups I, II and III to the Fifth Assessment Report of the Intergovernmental Panel on Climate Change. Geneva, Switzerland; 2014.

[55]   Navarro Navarro LA, Moreno Vázquez JL. Cambios en el paisaje arbolado en Hermosillo: escasez de agua y plantas nativas. Region Soc 2016; 28(67): 28.
[http://dx.doi.org/10.22198/rys.2016.67.a194]

[56]   Shuster W, Dadio S. Soil investigation for Infiltration-based Green Infrastructure for Sewershed Management (Omaha, NE). Cincinnati, Ohio. 2014.

[57]   Mohanty SK, Valenca R, Berger AW, *et al.* Plenty of room for carbon on the ground: Potential applications of biochar for stormwater treatment. Sci Total Environ 2018; 625: 1644-58.
[http://dx.doi.org/10.1016/j.scitotenv.2018.01.037] [PMID: 29996460]

[58]   Environmental Protection Agency. Evaluation of urban soils: Suitability for green infrastructure or urban agriculture. Cleveland, Ohio. 2011.

[59]   Zuniga-Teran AA, González-Méndez B, Scarpitti C, *et al.* Green Belt Implementation in Arid Lands through Soil Reconditioning and Landscape Design: The Case of Hermosillo, Mexico. Land (Basel) 2022; 11(12): 2130.
[http://dx.doi.org/10.3390/land11122130]

CHAPTER 3

Water in the Arid and Semi-Arid Zones of Mexico

Elvia Manuela Gallegos-Neyra[1,*], **Francisco José Torner-Morales**[2], **América Patricia Garcia-Garcia**[1], **Cesar Alejandro Zamora-Barrios**[2], **Iván Andrés Arredondo-Fragoso**[2] and **Israel Valencia Quiroz**[3]

[1] *Research Laboratory on Emerging Pathogens, Interdisciplinary Research Unit in Health and Education Sciences (UIICSE), Faculty of Higher Studies Iztacala (FES)-Iztacala, Division of Research and Graduate Studies, National Autonomous University of Mexico, (UNAM), Tlalnepantla de Baz, Mexico State, Mexico*

[2] *Environmental Conservation and Improvement Project, Interdisciplinary Research Unit in Health and Education Sciences (UIICSE), Faculty of Higher Studies Iztacala (FES)-Iztacala, Division of Research and Graduate Studies, National Autonomous University of Mexico, (UNAM), Tlalnepantla de Baz, Mexico State, Mexico*

[3] *Phytochemistry Laboratory, UBIPRO, Superior Studies Faculty (FES)-Iztacala, National Autonomous University of Mexico (UNAM), Tlalnepantla de Baz, Mexico State, Mexico*

Abstract: This chapter analyzes the current situation and challenges faced by water management in the arid and semi-arid zones of Mexico. It examines the hydroclimatic characteristics of these regions, highlighting the importance of water for maintaining their unique ecosystems and supporting the rural populations that depend on it. Various water sources are studied, including groundwater aquifers, surface waters, rainwater harvesting, and treated wastewater. Each source is vital but presents distinct challenges for sustainable management, including the need for regulatory measures and innovative conservation techniques. Climate change and its effects, such as altered rainfall patterns, prolonged droughts, and rising temperatures, exacerbate water scarcity and threaten ecosystem stability, necessitating targeted adaptation and mitigation strategies tailored to local conditions. Innovative technologies and practices for water conservation and efficient use are discussed, such as advanced irrigation systems, water reuse, and real-time management through artificial intelligence and the Internet of Things. The chapter addresses the complexities of water use in agricultural, domestic, urban, and industrial sectors, with a focus on issues like overexploitation, pollution, climate change impacts, and conflicts over water use. It also underscores the significance of community-based approaches and the need for local engagement in achieving sustainable water management. Finally, the chapter provides an overview of strategies, policies, and legal frameworks aimed at promoting sustainable water

[*] **Corresponding author Elvia Manuela Gallegos-Neyra:** Research Laboratory on Emerging Pathogens, Interdisciplinary Research Unit in Health and Education Sciences (UIICSE), Faculty of Higher Studies Iztacala (FES)-Iztacala, Division of Research and Graduate Studies, National Autonomous University of Mexico, (UNAM), Tlalnepantla de Baz, Mexico State, 54090, Mexico; Tel: +525540394705.
E-mail: elvia.gallegos1@gmail.com

management, drawing lessons from successful case studies and emphasizing the importance of integrated resource management, community participation, and international cooperation to tackle the multiple water challenges faced by these regions of Mexico.

Keywords: Water management, arid zones, semi-arid zones, Mexico, hydroclimatic characteristics, groundwater aquifers, surface water, rainwater harvesting, climate change, water scarcity.

INTRODUCTION

Water is essential for humanity and fundamental for life on our planet. Access to water is a basic human right and a central goal of global initiatives, making it crucial for social and economic development. However, its scarcity can lead to conflicts, underscoring the need for sustainable and equitable water management [1, 2]. Since 2007, UNEP has implemented water sustainability policies with an ecosystem approach, promoting rainwater harvesting as an accessible solution in regions with low precipitation [2]. Designing efficient rainwater harvesting and reuse programs, including the recycling of wastewater for green areas or agricultural production, is essential [3, 4]. In Mexico, where 60% of the territory is arid or semi-arid, sustainable water management is vital for agriculture and livestock [5]. Effective policies and rainwater harvesting practices are essential to ensure water availability [2 - 4].

The arid and semi-arid zones of Mexico face significant challenges due to water scarcity and extreme climatic conditions, affecting approximately forty million people who depend on drought-resistant crops such as maize and agave [5 - 7]. Effective water management essential for improving water availability and agricultural productivity while addressing issues like aquifer overexploitation and pollution [8, 9]. Climate change in Mexico is altering rainfall and temperature patterns, increasing extreme events, and affecting the availability of water resources [4, 10]. Recent droughts have stressed water systems, impacting aquifer recharge and surface water availability for agriculture [11 - 18]. The agricultural sector in these regions is particularly susceptible to the impacts of climate variability and extreme events [4, 19, 20].

Addressing these challenges requires the implementation of advanced irrigation technologies and sustainable agricultural practices, promotion of ecosystem restoration, and fostering of international collaboration to develop climate-resilient water management policies [17 - 21].

Current Water Situation in Mexico

Mexico's geographical location, divided into tropical and temperate climatic zones by the Tropic of Cancer, significantly influences its biological, cultural, and economic diversity. Approximately 52% of the territory is classified as arid or semi-arid, impacting natural resources and economic activities. Mexico encompasses both temperate and tropical climates, with deserts in the north and a tropical climate in the south. The average annual precipitation is 760 mm, with 65% falling in the summer and 35% in the winter. In 2024, the population is estimated to be 135,421,451, which increases pressure on water resources. These resources are unevenly distributed, leaving some regions facing severe scarcity [4, 22 - 29].

The arid and semi-arid zones of Mexico, primarily located in the north and centre of the country, experience low precipitation and high evapotranspiration, with elevated temperatures and irregular rainfall patterns. In the Sonoran Desert, annual precipitation ranges from 50 to 300 mm, while the Mexican Plateau receives between 300 and 500 mm. The Chihuahuan Desert receives between 150 and 400 mm annually. Although the mountains of the Mexican Plateau can recharge aquifers, the plains of Sonora and Chihuahua have a lower capacity for groundwater storage [5, 24, 30, 31].

In Mexico, annual water availability is 446,777 hm^3, although not all of it is used. CONAGUA regulates water use through the Public Water Rights Registry (REPDA), granting annual concessions of 266,560 hm^3. Agriculture accounts for 76.3% of consumptive water use, while domestic use represents 14.6%. In 2022, the national average water use was 19.5%, considered low pressure, though it varies by region: the Southern Border has a usage rate of 1.7%, while the Valley of Mexico faces extremely high pressure at 128.6%, indicating an unsustainable situation [24, 32]. Mexico faces serious water challenges due to climate change, aquifer overexploitation, and competition for water. Climate change has increased the frequency of droughts and floods, and 105 of the country's 653 aquifers are overexploited. Competition among agriculture, industry, and domestic use exacerbates the issue. Excessive extraction has lowered water tables and led to saline intrusion in coastal areas, while agricultural and industrial activities pollute surface and groundwater [4, 12, 13, 20, 33 - 35].

Water Sources in Arid and Semi-arid Zones

Overexploitation in areas such as the Valley of Mexico and Bajío leads to the decline of water tables and land subsidence, necessitating regulatory and monitoring policies [5, 32, 33]. Rivers, lakes, and reservoirs, including the Cutzamala System, are essential resources, though their availability fluctuates with climate variations. These water sources face pollution from agricultural and industrial activities, highlighting the need for integrated management strategies [14, 36]. Rainwater harvesting and storage on roofs and paved surfaces are critical, especially in rural areas, to reduce reliance on overexploited aquifers. The collected water is stored in cisterns and underground tanks for domestic and irrigation use [30, 37]. Additionally, wastewater treatment and reuse are vital to addressing water scarcity. Treatment plants remove contaminants, making the water suitable for agricultural and industrial irrigation. In cities like Monterrey and Guadalajara, treated wastewater is crucial, relieving pressure on freshwater sources and enhancing water sustainability [38 - 40]. Various water sources are illustrated in Fig. **1**.

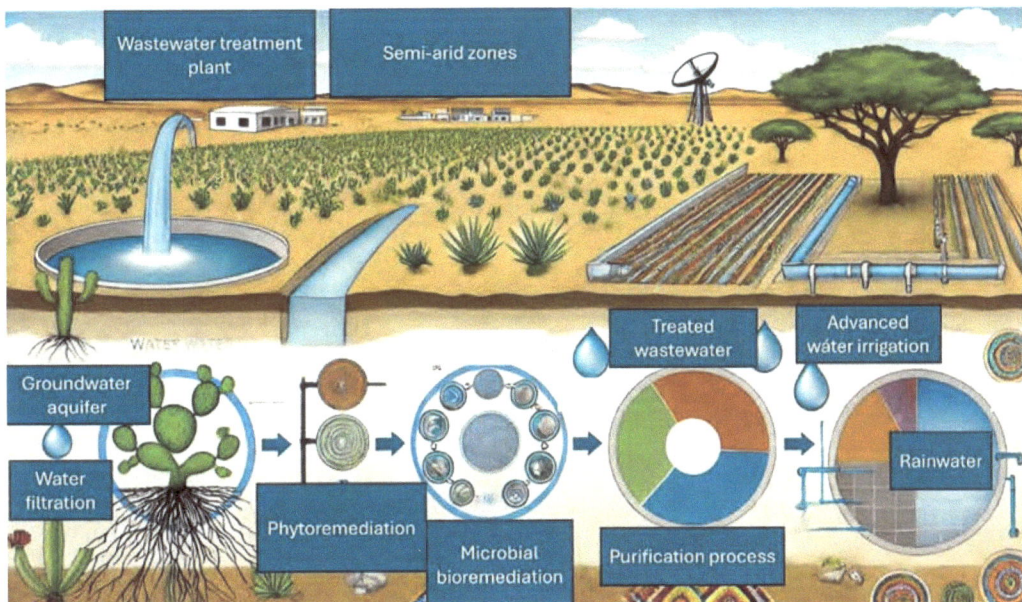

Fig. (1). Various water sources and conservation techniques in Mexico's arid and semi-arid zones, including groundwater aquifers, rainwater harvesting systems, treated wastewater, and advanced irrigation methods. Figure created with DALL-E.

Water Use And Management in Arid and Semi-Arid Zones of Mexico

Efficient water management in the arid and semi-arid zones of Mexico is essential for sustainability and economic development. Agriculture uses approximately 76% of the water in Mexico, and of this percentage, 67.8% is allocated to agricultural irrigation. Technologies such as drip irrigation and subsurface irrigation are crucial to improving water efficiency, increasing agricultural yields, and reducing evaporation and runoff [20, 21, 41, 42]. Urbanization and population growth have increased water demand in cities, and strategies such as water-saving devices, rainwater collection systems, and infrastructure modernization have been implemented to reduce losses. Additionally, in rural communities, improving access to drinking water and infrastructure is essential to overcome current limitations [2, 13, 18, 32, 33, 38].

Industrial use accounts for around 10% of water consumption in Mexico, and implementing recycling and reuse of technologies, such as treated wastewater for cooling systems, can significantly reduce the demand for freshwater [39]. Conservation and recycling strategies, such as the reuse of treated wastewater and rainwater harvesting, are essential for sustainable water management. These practices, along with public education, through awareness campaigns and specific programs, plays a crucial role in promoting efficient and responsible water use, as illustrated in Fig. (**2**) [19].

Fig. (2). A detailed illustration showing the impact of climate change on water resources in Mexico. The image includes three main scenes: 1) a dry riverbed with cracked earth, representing drought and reduced surface water availability, 2) a landscape featuring a reservoir with low water levels, indicating decreased aquifer recharge, and (3) a rural area with sparse rainfall, illustrating changes in rainfall patterns. Figure created with DALL-E.

Problems and Challenges in Water Management

Water management in Mexico, especially in arid and semi-arid zones, faces significant challenges, including water scarcity, aquifer overexploitation, pollution, climate change effects, and conflicts over usage. Approximately 40% of aquifers are overexploited, impacting agriculture and the supply of drinking water. The expansion of irrigated agriculture, population growth, and urbanization increase water use, resulting in scarcity and land subsidence [43 - 46]. Agricultural, industrial, and urban activities have polluted surface and groundwater bodies. Excessive use of fertilizers and pesticides, along with insufficient infrastructure for wastewater treatment, exacerbates the issue, affecting water quality and posing a risk to public health and the environment [47, 48].

Climate change exacerbates these challenges by reducing precipitation, increasing evaporation, and decreasing aquifer recharge, all of which intensify water scarcity. More frequent and intense droughts place significant pressure on water resources, impacting human consumption, agriculture, and industry [12, 49, 50]. Additionally, competition between sectors such as agriculture, industry, and domestic use generates tensions. The dispute over Colorado River water between the United States and Mexico, along with local conflicts between rural communities and industries, highlights the complexity of the issue. Comprehensive and sustainable solutions, along with sector-wide and intergovernmental cooperation, are essential to securing a sustainable water supply [12, 51, 52].

Strategies, Public Policies, and Legal Frameworks for Sustainable Water Management in Mexico

Sustainable water management in Mexico faces significant challenges stemming from increasing demand and the impacts of climate change. To address these issues, a range of strategies, public policies, and legal frameworks have been implemented to optimize water resource use, protect aquatic ecosystems, and ensure equitable access to water. Collaboration among the government, local communities, and international organizations is essential [32]. The National Water Program (PNH) 2020-2024 outlines guidelines to secure water quality and availability, enhance hydraulic infrastructure, and promote water-saving technologies [53 - 55]. The National Water Law (LAN) governs water management and empowers CONAGUA with the authority to oversee and allocate water resources, while other laws further support these regulations, focusing on broader environmental protection measures [56, 57].

Programs such as the Sustainable Watershed Management Program and PIAPAS focus on coordinating actions to promote equitable and sustainable water use, enhance drinking water and sanitation infrastructure, and reduce erosion [53, 54, 58, 59]. focus on investments in water infrastructure, the promotion of sustainable agricultural practices, and public education programs to encourage responsible water use [56, 57]. International cooperation, through agreements such as the 1944 Water Treaty and the 2030 Water Resources Group, has funded sustainable development projects promoting water efficiency in agriculture and wastewater treatment. Collaborations with the German Agency for International Cooperation (GIZ), the World Bank, and the FAO have advanced projects for wastewater reuse and the implementation of water-saving technologies [51, 60 - 62].

Case Studies and Successful Experiences in Water Management in Mexico

Effective water management in Mexico's arid and semi-arid zones is a critical challenge. Numerous inspiring examples highlight the effectiveness of advanced technologies, community participation, and robust policies in promoting sustainable water use. The Integrated Water Resources Management Project in the Mezquital Valley Basin, Hidalgo, utilizes efficient irrigation systems and sustainable agricultural practices, reducing water consumption and enhancing agricultural productivity with active participation from local farmers [53]. In the Mexicali Valley, Baja California, the Comprehensive Water Management Project optimizes water use in agriculture through drip irrigation systems and infrastructure improvements, achieving a 50% reduction in water use while significantly enhancing agricultural efficiency [15]. Additionally, conservation efforts like reforestation and rainwater harvesting in the Sierra Sur, Oaxaca, and ecological restoration in the Nazas River Basin, Durango, have bolstered community resilience by enhancing infiltration rates and supporting aquifer recharge [63, 64].

Biotechnological initiatives, such as using biofilms with algae and bacteria for wastewater treatment in Baja California and the industrial biofiltration project for wastewater in Querétaro, have greatly enhanced treated water quality, minimized environmental impact, and facilitated the reuse of treated water for agricultural purposes and aquifer recharge [65, 66]. These projects have demonstrated positive outcomes, including a 40% reduction in water consumption and a 30% increase in agricultural productivity in the Mezquital Valley along with increased water availability and reduced drought vulnerability in Oaxaca [63]. These examples underscore the importance of integrating technology, conservation, and community engagement to create effective water management solutions in arid and semi-arid regions.

Challenges and Future Perspectives of Water in the Arid and Semi-Arid Zones of Mexico

Sustainable water management in Mexico's arid and semi-arid zones faces major challenges from climate change, population growth, and urbanization. Addressing water scarcity requires the implementation of strategies grounded in technological innovation, conservation efforts, and active community participation. Integrating effective public policies, educational initiatives, and international cooperation is essential to securing a sustainable water future. For over 14 years, the Regional University Unit of Arid Zones at the Autonomous University of Chapingo (UACh) has promoted research and technology transfer in water use and management [9, 15, 67 - 70].

Future research should prioritize the development of sustainable and cost-effective water management technologies and creating public policies that promote water conservation and efficient use. International collaboration will be crucial to share knowledge and advanced technologies. Moreover, education and community awareness are fundamental, as educational programs and active community participation ensure solutions that are locally adapted and sustainable in the long term [32, 51, 52, 69, 71]. Table **1** provides examples of such initiatives.

Table 1. Emerging technologies for water sustainability and conservation in the arid and semi-arid zones of Mexico

Technology/Method	Description	Case Studies	References
Integrated Water Resources Management (IWRM)	Aims to maximize economic and social welfare while safeguarding ecosystem sustainability.	Basin of Mexico: Implementing Integrated Water Resources Management (IWRM) to balance water use among urban, agricultural, and ecological needs.	[72]
Information Technologies for Water Resource Management	The use of GIS and remote sensing helps identify groundwater sources, assess water quality, and monitor changes in water levels. These technologies significantly enhance water resource management and planning, enabling decision-making based on accurate data.	Lerma-Chapala River Basin (Estado de México): Enhancing water management through the use of GIS and remote sensing.	[73 - 75]

(Table 1) cont.....

Technology/Method	Description	Case Studies	References
Biotechnology and Artificial Intelligence in Water Management	Biotechnology plays a crucial role in bioremediation, while AI enhances decision-making, infrastructure efficiency, and real-time monitoring.	Coahuila Project: Predicting groundwater levels and precipitation variations. Sonora Project: Anticipating changes in water levels and rainfall patterns.	[76 - 84]
Advanced Biosensors for Soil Water Monitoring	Devices provide real-time data on soil moisture and water quality, improving water management in scarce regions.	Mexicali Valley (Baja California): Optimized water use and enhanced agricultural yields.	[85 - 87]
Water Conservation Technologies	Modernizing irrigation infrastructure and implementing conservation agriculture techniques to reduce water losses and improve efficiency in agricultural water use	Yaqui Valley, Sonora: Implemented modern irrigation infrastructure and conservation techniques to enhance water use efficiency.	[88]
Rainwater Harvesting	Installing water collection systems in homes and public buildings to provide additional resources for domestic and agricultural use.	Mixteca Valley, Oaxaca: Implemented rainwater collection and storage systems.	[89 - 91]
Fog and Dew Collection	Utilizing atmospheric humidity in regions with low rainfall to capture and store water droplets for community use.	San Cristóbal de las Casas, Chiapas: Installed fog collectors to enhance potable water availability.	[92, 93]
Water Storage and Collection Techniques	Modern storage systems and collection techniques, such as trenches and paved roofs, enhance water management in arid and semi-arid zones.	Northern Mexico: Utilizing underground cisterns and paved surface catchments to improve water storage.	[94]
Desalination and Brackish Water Treatment	Producing potable water through desalination and advanced treatment technologies in arid coastal regions.	Los Cabos (Baja California) Desalination Plant: Utilizes reverse osmosis and solar energy to supply potable water.	[95 - 97]
Innovations in Water Conservation	Using hydrogels and superabsorbent materials to enhance soil moisture retention and increase crop yields in agricultural soils.	Mezquital Valley, Hidalgo: Increased crop yield and reduced irrigation frequency using hydrogels. Sonoran Desert: Improved water retention and crop yield with superabsorbent polymers.	[98, 99]

CONCLUDING REMARKS

This chapter examines the challenges of water management in Mexico's arid and semi-arid zones, driven by climate change, population growth, and urbanization. It covers projections of water availability and demand, along with discussions on technological innovations, conservation projects, rainwater harvesting, desalination, and wastewater treatment. Furthermore, it emphasizes the importance of effective public policies and community education in achieving sustainable water management. Implementing innovative technologies and conservation practices can significantly enhance water availability and foster economic and social development in these regions.

It is crucial that governments, communities, and international organizations take decisive action to manage water sustainably. This includes implementing effective policies, adopting advanced technologies, and promoting conservation and efficient water use practices. The treatment and reuse of wastewater are vital to avoid detrimental effects on water bodies and soil, ensuring a continuous water supply, especially in developing countries. Technological innovations—such as advanced sensors, artificial intelligence, and biotechnology—combined with conservation initiatives like irrigation infrastructure modernization and rainwater harvesting, are key to enhancing water availability. Additionally, education and community engagement play a fundamental role in fostering sustainable water practices play a fundamental role in fostering sustainable water practices.

AUTHORS' CONTRIBUTION

Gallegos-Neyra EM is the lead author responsible for the chapter's conceptualization, information gathering, and primary manuscript writing.

Torner-Morales FJ contributed to the technical review of the content, focusing on scientific accuracy and integration of up-to-date references.

Garcia-Garcia AP contributed to the literature and resource gathering, ensuring the inclusion of key sources to support the chapter's content.

Zamora-Barrios CA conducted the critical review of the manuscript, ensuring the overall coherence and clarity of the text.

Arredondo-Fragoso IA assisted in the manuscript's spelling and grammar check, ensuring the linguistic quality of the chapter.

Valencia-Quiroz Israel supervised the final editing and coordinated the integration of all chapter sections, ensuring cohesion.

QUESTIONS RELATED TO THE TEXT

1. What are the primary water sources in the arid and semi-arid zones of Mexico, and what challenges are associated with each source?

2. How does climate change impact water availability in the arid and semi-arid zones of Mexico?

3. What are some advanced technologies and practices mentioned in the chapter that improve water management and conservation in these regions?

4. Describe the socio-economic and environmental challenges that water management faces in the arid and semi-arid zones of Mexico.

5. What strategies, public policies, and legal frameworks have been developed to promote sustainable water management in Mexico?

6. Analyze the importance of community participation and international collaboration in addressing water management challenges in the arid and semi-arid zones of Mexico.

CONSENT FOR PUBLICATION

All authors have read and approved the final version of this chapter and consent to its publication.

CONFLICT OF INTEREST

The authors declare that they have no conflict of interest.

LIST OF ABBREVIATIONS

CONAGUA	National Water Commission (Organization responsible for the management and regulation of water resources in Mexico.)
Cutzamala	Cutzamala System (Hydraulic system that supplies water to Mexico City and surrounding areas.)
FAO	Food and Agriculture Organization (A UN agency that fights hunger and promotes sustainable agriculture.)
FES-Iztacala, UNAM	Faculty of Higher Studies Iztacala, UNAM (A faculty of UNAM offering undergraduate, graduate, and research programs in health, Biology, psychology, and Social Sciences.)
GIZ	German Agency for International Cooperation (German agency for international cooperation in sustainable development.)

IMTA	Mexican Institute of Water Technology (Mexican institution focused on research and development in water technology.)
INEGI	National Institute of Statistics and Geography (Mexican institution responsible for national censuses and statistics.)
IPCC	Intergovernmental Panel on Climate Change (International organization that provides scientific assessments on climate change.)
IWRM	Integrated Water Resources Management (A coordinated process to manage water resources sustainably, balancing social, economic, and environmental needs.)
LAN	National Water Law (Legal framework in Mexico for water management and regulation.)
NASA	National Aeronautics and Space Administration (United States) (U.S. government agency responsible for space exploration and atmospheric research.)
OECD	Organization for Economic Cooperation and Development (An international organization promoting policies to improve global economic and social well-being.)
PIAPAS	Comprehensive Drinking Water, Sewerage, and Sanitation Program (Mexican government program aimed at improving access to drinking water and sanitation.)
PNH	National Water Program (Mexico's strategic plan for national water management.)
REPDA	Public Water Rights Registry (National registry that regulates and monitors water usage rights in Mexico.)
SEMARNAT	Secretariat of Environment and Natural Resources (A Mexican government agency responsible for protecting the environment and natural resources.)
UACh	Autonomous University of Chapingo (A Mexican institution specializing in agricultural and forestry education and research.)
UBIPRO	Unit of Biology, Technology, and Prototypes (A research unit at UNAM focused on applied biology, technology, and prototype development.)
UIICSE	Interdisciplinary Research Unit in Health and Education Sciences (A research unit at UNAM focused on interdisciplinary studies in health and education.)
UNAM	National Autonomous University of Mexico (Mexico's leading public institution for higher education and research.)
UNEP	United Nations Environment Program (A UN agency responsible for coordinating environmental activities and promoting sustainable development.)
UNESCO	United Nations Educational, Scientific and Cultural Organization (A UN agency that promotes peace and security through education, science, and culture.)
WWAP	World Water Assessment Programme (A UN initiative that evaluates global water resources and provides recommendations.)

REFERENCES

[1] United Nations Environment Programme (UNEP). Freshwater strategic priorities 2022-2025. UNEP. 2023. Available from: https://www.unep.org/resources/publication/freshwater-strategic-prioriti-s-2022-2025

[2] United Nations Educational Scientific and Cultural Organization (UNESCO). UN World Water Development Report 2021: Valuing water. Paris: UNESCO. 2021. Available from:

https://www.unwater.org/publications/un-world-water-development-report-2021

[3]   Food and Agriculture Organization (FAO). Irrigation methods and water management in arid regions. Rome: FAO. 2020. Available from: https://www.fao.org/4/W3094E/w3094e05.htm

[4]   Intergovernmental Panel on Climate Change (IPCC). Climate change and its impact on agriculture. Geneva: IPCC. 2021. Available from: https://www.ipcc.ch/report/ar6/wg1/

[5]   Instituto Nacional de Estadística y Geografía (INEGI). National statistics on arid and semi-arid regions in Mexico. INEGI. 2021. Available from: https://www.inegi.org.mx/

[6]   Aguilar CN, Meléndez- Renteria NP, Ramirez-Guzman KN, Eds.. Exploration and valorization of natural resources from arid zones. In: Rodríguez-González L, Martínez-Medina G A, Méndez-Carmona J Y, Saldaña-Mendoza S A, Rodriguez J, Aguilar CN, Ramírez-Guzmán N, Eds. Arid zone description and Its biodiversity. 1st Edition. New York: Apple Academic Press. [Internet]. 2024; pp 1-24.
[http://dx.doi.org/10.1201/9781032672236]

[7]   López-Lambraño AA, Martínez-Acosta L, Gámez-Balmaceda E, Medrano-Barboza JP, Remolina López JF, López-Ramos A. Supply and demand analysis of water resources. Case study: irrigation water demand in a semi-arid zone in Mexico. Agriculture 2020; 10(8): 333.
[http://dx.doi.org/10.3390/agriculture10080333]

[8]   Kinzelbach W, Brunner P, von Boetticher A, Kgotlhang L, Milzow C. Sustainable water management in arid and semi-arid regions.Groundwater modelling in arid and semi-arid areas International hydrology series. Cambridge: Cambridge University Press 2010; pp. 119-30.
[http://dx.doi.org/10.1017/CBO9780511760280.009]
[http://dx.doi.org/10.1017/CBO9780511760280.009]

[9]   Hernández-Cruz A, Sandoval-Solís S, Mendoza-Espinosa LG. An overview of modeling efforts of water resources in Mexico: Challenges and opportunities. Environ Sci Policy 2022; 136: 510-9.
[http://dx.doi.org/10.1016/j.envsci.2022.07.005] [http://dx.doi.org/10.1016/j.envsci.2022.07.005]

[10]   Silva JA, Monroy Becerril DM, Martínez Díaz E. Effects of climate change on water resources in Mexico. Manag Environ Qual 2023; 34(2): 408-27.
[http://dx.doi.org/10.1108/MEQ-03-2022-0081] [http://dx.doi.org/10.1108/MEQ-03-2022-0081]

[11]   National Aeronautics and Space Administration (NASA). Earth observatory. Drought parches Mexico. NASA. 2023. Available from: https://earthobservatory.nasa.gov/images/152908/drought-parche--mexico

[12]   Circle of Blue. Water depletion: A pivotal concern in Mexico's 2024 election. Circle of Blue. 2024. Available from: https://www.circleofblue.org/2024/hotspots/water-depletion-a-pivotal-conce-n-in-mexicos-2024-election/

[13]   Americas Quarterly. Mexico's water crisis is spilling over into politics. Americas Quarterly. 2024. Available from: https://www.americasquarterly.org/article/mexicos-water-crisis-is-spilling-ove--into-politics/

[14]   Comisión Nacional del Agua (CONAGUA). Almacenamiento en presas del Sistema Cutzamala. CONAGUA. 2024. Available from: https://gobierno.cdmx.gob.mx/acciones/cuidado-del-agua/

[15]   Instituto Nacional de Ecología y Cambio Climático (INECC). Efectos del cambio climático en México. Ciudad de México: INECC. 2023 [cited: 2024 Jun 28]. Available from: https://www.gob.mx/inecc/acciones-y-programas/efectos-del-cambio-climatico

[16]   Pérez CP, Amado ÁJ, Segovia OE, Conesa GC, Alarcón CJ. La degradación ambiental y sus efectos en la contaminación de las aguas superficiales en la cuenca del río Conchos (Chihuahua-México). Cuad Geogr 2019; 58(1): 47-67.
[http://dx.doi.org/10.30827/cuadgeo.v58i1.6636]

[17]   Ali SA, Armanous AM, Eds. Groundwater in arid and semi-arid areas: monitoring, assessment, modelling, and management. 1st ed., Springer 2023.

[http://dx.doi.org/10.1007/978-3-031-43348-1]

[18]   Secretaría de Medio Ambiente y Recursos Naturales (SEMARNAT). Informe de la situación del medio ambiente en México 2019. SEMARNAT. 2020. Available from: https://apps1.semarnat.gob.mx:8443/dgeia/informe18/index.html

[19]   Nacional del Agua C. (CONAGUA) Informe sobre la sequía en México: retos y estrategias. Ciudad de México: CONAGUA 2022. https://www.gob.mx/conagua/articulos/atencion-a-la-sequia-en-mexico

[20]   Secretaría de Agricultura y Desarrollo Rural (SADER). Impacto de la variabilidad climática en la agricultura mexicana. Ciudad de México: SADER. 2023. Available from: https://www.gob.mx/agricultura/es/articulos/el-cambio-climatico-afecta-al-campo-como-enfrentarlo

[21]   Dhak D, Chiavola A, Mishra A, Dhak P. Editorial: Innovations and challenges in green and sustainable water purification and waste water management. Front Chem 2023; 11: 1235757. [http://dx.doi.org/10.3389/fchem.2023.1235757] [PMID: 37426332]

[22]   Countrymeters. Mexico population (2024) live. Countrymeters. 2024. Available from: https://countrymeters.info/es/Mexico

[23]   Worldometer. Mexico Population (2024). Worldometer. 2024. Available from: https://www.worldometers.info/world-population/mexico-population/

[24]   Nacional del Agua C. (CONAGUA) Estadísticas del agua en México. Ciudad de México: CONAGUA 2020. https://files.conagua.gob.mx/conagua/publicaciones/Publicaciones/EAM%202021.pdf

[25]   Latin America & Caribbean Geographic (LAC Geo). The Latin American region: a mosaic of geography, biodiversity, climate, and culture. LAC Geo. 2023. Available from: https://lacgeo.com/latin-american-region

[26]   WorldAtlas. Mexico's most famous geographical features. WorldAtlas. 2024. Available from: https://www.worldatlas.com/articles/mexico-s-most-famous-geographical-features.html

[27]   Instituto Nacional de Estadística y Geografía (INEGI). Censo de población y vivienda 2020. INEGI. 2021. Available from: https://www.inegi.org.mx/programas/ccpv/2020/

[28]   Comisión Nacional del Agua (CONAGUA). Informe de resultados del proyecto de predicción hidrológica en Sonora. CONAGUA. 2021. Available from: https://sinav30.conagua.gob.mx:8080/

[29]   García E. Modificaciones al sistema de clasificación climática de Köppen. 5.ª ed. Ciudad de México: Instituto de Geografía, UNAM; 2004. Available from: http://doi.org/www.publicaciones.igg.unam.mx/index.php/ig/catalog/view/83/82/251-1

[30]   Ríos-Sánchez KI, Chamizo-Checa S, Galindo-Castillo E, *et al.* The groundwater management in the Mexico megacity peri-urban interface. Sustainability (Basel) 2024; 16(11): 4801. [http://dx.doi.org/10.3390/su16114801.] [http://dx.doi.org/10.3390/su16114801]

[31]   Universidad Nacional Autónoma de México (UNAM). Sustainable groundwater management in Mexico. UNAM. 2021. Available from: https://www.revista.unam.mx/vol.14/num10/art37/

[32]   Food and Agriculture Organization (FAO). Rainwater harvesting: a lifeline for human well-being. FAO. 2021. Available from: https://www.bebuffered.com/downloads/UNEP-SEI_Rainwater_Harvesting_Lifeline_090310b.pdf

[33]   Diario Oficial de la Federación (DOF). Norma oficial mexicana NOM-001-SEMARNAT-2018. 2019. Available from: https://www.dof.gob.mx/nota_detalle.php?codigo=5510140&fecha=05/01/2018#gsc.tab=0

[34]   Pedrozo Acuña A. La sobreexplotación de acuíferos: camino seguro hacia la quiebra hídrica. Perspectivas IMTA 2021; 2(48): 48. [http://dx.doi.org/10.24850/b-imta-perspectivas-2021-48.] [http://dx.doi.org/10.24850/b-imta-perspectivas-2021-48]

[35]   United Nations Educational Scientific and Cultural Organization (UNESCO). UNESCO-Water. UN

World Water Development Report 2023: Partnerships and cooperation for water. UNESCO. 2023. Available from: https://www.unesco.org/reports/wwdr/2023/en

[36] Secretaría de Medio Ambiente y Recursos Naturales (SEMARNAT). Informe de la situación del medio ambiente en México 2020. SEMARNAT. 2021. Available from: https://apps1.semarnat. gob.mx:8443/dgeia/informe18/index.html

[37] Tariq J, Shahrour I, Comair F. Smart rainwater harvesting for sustainable potable water supply in arid and semi-arid areas. Sustainability 2022; 14.15: 9271.
[http://dx.doi.org/10.3390/su14159271]

[38] Organization for Economic Co-operation and Development (OECD). Water resources allocation: sharing risks and opportunities. OECD. 2020. Available from: https://www.oecd.org/en/publications/water-resources-allocation_9789264229631-en.html

[39] Aguilar-Aguilar A, de León-Martínez LD, Forgionny A, Acelas Soto NY, Mendoza SR, Zárate-Guzmán AI. A systematic review on the current situation of emerging pollutants in Mexico: A perspective on policies, regulation, detection, and elimination in water and wastewater. Sci Total Environ 2023; 905: 167426.
[http://dx.doi.org/10.1016/j.scitotenv.2023.167426] [PMID: 37774864]

[40] World Water Assessment Programme (WWAP). The United Nations World water development report 2021: Valuing water. WWAP. 2021. Available from: https://www.unesco.org/reports/wwdr/2021/en

[41] Palermo SA, Maiolo M, Brusco AC, *et al.* Smart technologies for water resource management: an overview. Sensors (Basel) 2022; 19;22(16): 6225.
[http://dx.doi.org/10.3390/s22166225] [PMID: 36015982]

[42] Wu Y, Wan S, Palaiahnakote S. Editorial: Intelligent computing in farmland water conservancy for smart agriculture. Front Plant Sci 2023; 14: 1236010.
[http://dx.doi.org/10.3389/fpls.2023.1236010] [PMID: 37546278]

[43] Esteller MV, Diaz-Delgado C. Environmental effects of aquifer overexploitation: a case study in the highlands of Mexico. Environ Manage 2002; 29(2): 266-78.
[http://dx.doi.org/10.1007/s00267-001-0024-0] [PMID: 11815828]

[44] Organization for Economic Co-operation and Development (OECD). Water governance in Mexico: overcoming fragmentation for better management. OECD. 2023. Available from: https://www.oecd.org/en/topics/water-governance.html

[45] Cruz-Ayala MB, Megdal SB. An overview of managed aquifer recharge in Mexico and its legal framework. Water 2020; 12(2): 474. [http://doi,org/10.3390/w12020474
[http://dx.doi.org/10.3390/w12020474]

[46] Sustainable Earth Reviews. State level water security indices in Mexico. Sustainable Earth Reviews. 2023. Available from: https://sustainableearthreviews.biomedcentral.com/articles/10.1186/s42055-020-00031-4

[47] Ochoa-Noriega CA, Aznar-Sánchez JA, Velasco-Muñoz JF, Álvarez-Bejar A. The use of water in agriculture in Mexico and its sustainable management: a bibliometric review. Agronomy (Basel) 2020; 10(12): 1957.
[http://dx.doi.org/10.3390/agronomy10121957] [http://dx.doi.org/10.3390/agronomy10121957]

[48] Cruz Ayala M-B, Tortajada C. Managed aquifer recharge in Mexico: proposals for an improved legal framework and public policies. Water Int 2023; 48(1): 165-83. https://ideas.repec.org/a/taf/rwinxx/v48y2023i1p165-183.html
[http://dx.doi.org/10.1080/02508060.2022.2132668]

[49] Intergovernmental Panel on Climate Change (IPCC). Climate change 2022: Impacts, adaptation and vulnerability. Contribution of Working Group II to the Sixth Assessment Report of the Intergovernmental Panel on Climate Change. IPCC. 2022. Available from: https://www.ipcc.ch/report/ar6/wg2/

[50]   Martinez-Austria PF. Climate change and water resources in Mexico.In: Raynal-Villasenor J, Ed Water resources of Mexico. Cham: Springer 2020; p. 9. [http://dx.doi.org/10.1007/978-3-030-40686-8_9]

[51]   Ingram H. Transboundary groundwater on the U.S.–Mexico border: is the glass half full, half empty, or even on the table? Nat Resour J 2000; 40(2): 185-8. [http://www.jstor.org/stable/24888634

[52]   Wilson Center. Bilateral water management: water sharing between the US and Mexico along the border. Wilson Center. 2023. Available from: https://www.wilsoncenter.org/article/bilateral-wate--management-water-sharing-between-us-and-mexico-along-border

[53]   Secretaría de Medio Ambiente y Recursos Naturales (SEMARNAT). Programa nacional hídrico. SEMARNAT. 2022. Available from: https://www.gob.mx/conagua/articulos/consulta-para-el--el-programa-nacional-hidrico-2019-2024-190499

[54]   Comisión Nacional del Agua (CONAGUA). Plan nacional de desarrollo y programa nacional hídrico. CONAGUA. 2023. Available from: https://www.gob.mx/conagua/articulos/consulta-para-el--el-programa-nacional-hidrico-2019-2024-190499

[55]   Secretaría de Medio Ambiente y Recursos Naturales (SEMARNAT). Programa nacional hídrico. SEMARNAT. 2020. Available from: https://www.gob.mx/conagua/articulos/consulta-para-el--el-programa-nacional-hidrico-2019-2024-190499

[56]   Secretaría de Medio Ambiente y Recursos Naturales (SEMARNAT). Ley de aguas nacionales y su reglamento. SEMARNAT. 2023. Available from: https://biblioteca.semarnat.gob.mx/janium/Documentos/211855.pdf

[57]   Diario Oficial de la Federación (DOF). Ley de aguas nacionales. 2022. Available from: https://www.diputados.gob.mx/LeyesBiblio/pdf/LAN.pdf

[58]   Dávila Pórcel RA, Covarrubias Pérez GC. Integrated water resources management and the Mexican prospects. Environ Earth Sci 2017; 76(11): 390. [http://dx.doi.org/ 10.1007/s12665-017-6633-6] [http://dx.doi.org/10.1007/s12665-017-6633-6]

[59]   Secretaría de Medio Ambiente y Recursos Naturales (SEMARNAT). Programa "agua para todos". SEMARNAT. 2022. Available from: https://www.gob.mx/semarnat/documentos/presentacion-d--programa-nacional-de-reservas-de-agua?state=published

[60]   World Economic Forum. Ensuring sustainable water management for all by 2030. Sustainable Water Management. 2023. Available from: https://www.weforum.org/impact/sustainable-wate--management/

[61]   Deutsche Gesellschaft für Internationale Zusammenarbeit (GIZ). Sustainable water management in Mexico. GIZ. 2023. Available from: https://www.giz.de/en/worldwide/33041.html

[62]   Food and Agriculture Organization (FAO). Cooperación internacional en la gestión del agua. FAO. 2021. Available from: https://www.fao.org/water/es/

[63]   Galvez V, Rojas R, Bennison G, Prats C, Claro E. Collaborate or perish: water resources management under contentious water use in a semiarid basin. International Journal of River Basin Management 2020; 18(4): 421-37. [http://dx.doi.org/10.1080/15715124.2019.163408]

[64]   Barrios AJ, Franquesa SJ. 2023. https://upcommons.upc.edu/bitstream/handle/2117/408554/BarriosFranquesa2023_Archidoct.pdf?sequence=3&isAllowed=y

[65]   Villada-Canela M, Muñoz-Pizza DM, García-Searcy V, Camacho-López R, Daesslé LW, Mendoza-Espinosa L. Public participation for integrated groundwater management: The case of Maneadero Valley, Baja California, Mexico. Water 2021; 13(17): 2326. [http://dx.doi.org/10.3390/w13172326] [http://dx.doi.org/10.3390/w13172326]

[66]   Pedrozo-Acuña A. Hacia una agricultura eficiente y sustentable. Perspectivas IMTA 2020; p. 21. [http://dx.doi.org/10.24850/b-imta-perspectivas-2020-21]

[67]   Food and Agriculture Organization (FAO). Innovaciones tecnológicas y conservación del agua. FAO. 2021. Available from: https://openknowledge.fao.org/server/api/core/bitstreams/f24094d0-17b- -4221-8414-439c0ae27313/content

[68]   World Bank. Managing Water Resources in River Basins. World Bank [the Internet]. 2019. Available from: https://www.worldbank.org/en/topic/waterresourcesmanagement

[69]   United Nations Educational Scientific and Cultural Organization (UNESCO). UNESCO-Water. UN World water development report 2023. Future directions for water research and policy. UNESCO. 2023. Available from: https://www.unwater.org/publications/un-world-water-development-re- ort-2023

[70]   Lluch-Cota SE, Velázquez Zapata JA, Nieto Delgado C. Agricultura, agua y cambio climático en zonas áridas de México. RECURSOS NATURALES Y SOCIEDAD 2022; 8(2): 35-48. [http://dx.doi.org/10.18846/renaysoc.2022.08.08.02.0004]

[71]   Benítez Ávila IS, Camacho Lomelí R, Talledos Sánchez E, Santacruz de León G. Gestión comunitaria del agua para riego en dos comunidades campesinasen la región de los Valles Centrales de Oaxaca. Rev Geogr Agric 2022; (68): 23-42. [http://dx.doi.org/10.5154/r.rga.2022.08.08.02.0004]

[72]   Grison C, Koop S, Eisenreich S, *et al.* Integrated water resources management in cities in the world: global challenges. Water Resour Manage 2023; 37(6-7): 2787-803. [http://dx.doi.org/10.1007/s11269-023-03475-3]

[73]   Chang FJ, Chang LC, Chen JF. Artificial intelligence techniques in hydrology and water resources management. Water 2023; 15(10): 1846. [http://dx.doi.org/10.3390/w15101846]

[74]   Tan J, Zou XY. Water-related technological innovations and water use efficiency: international Evidence Emerg Mark Finance Trade. 2023;59 (15):4138-57. [http://dx.doi.org/10.1080/1540496X.2023.2181663]

[75]   Comisión Nacional del Agua (CONAGUA). Uso de sistemas de información geográfica y teledetección en la gestión de la cuenca del río Lerma-Chapala. CONAGUA. 2020. Available from: https://files.conagua.gob.mx/conagua/generico/PNH/PHR_2021-2024_RHA_VIII_LSP.pdf

[76]   Egbemhenghe AU, Ojeyemi T, Iwuozor KO, *et al.* Revolutionizing water treatment, conservation, and management: Harnessing the power of AI-driven ChatGPT solutions. Environmental Challenges 2023; 13: 100782. [http://dx.doi.org/ 10.1016/j.envc.2023.100782]

[77]   Alkhafaji MA, Ramadan GM, Jaffer Z, Jasim L. Revolutionizing agriculture: the impact of AI and IoT. E3S Web Conf 2024; 491: 01010. [http://dx.doi.org/10.1051/e3sconf/202449101010]

[78]   Hernandez M, Diaz A, Ramirez P. Comprehensive assessment of groundwater quality in Mexico and application of new water classification scheme based on machine learning (evaluación integral de la calidad de las aguas subterráneas en México y aplicación de un nuevo esquema de clasificación del agua basado en el aprendizaje automático). Rev Mex Ing Quim 2023; 22(2): 1723-34.

[79]   Alharbi S, Felemban A, Abdelrahim A, Al-Dakhil M. Agricultural and technology-based strategies to improve water-use efficiency in arid and semiarid areas. Water 2024; 16(13): 1842. [http://dx.doi.org/10.3390/w16131842]

[80]   Smart Water. The future of water conservation: innovations and technology. Smart Water. 2023. Available from: https://smartwateronline.com/news/the-future-of-water-conservation-innovatio- s-and-technology

[81]   Estrada PF, Zavala HJ, Martínez AA, Raga G, Gay GC, Eds. State and perspectives of climate change in Mexico: a starting point. 2023. [ISBN: 978-607-30-8172-6]. [https://doi.org//cambioclimatico. unam.mx/wp-content/uploads/2023/12/State-and-Perspectives-of-Climate-Change-in-Mexico-a-

Starting-Point.pdf

[82]   World Wide Fund for Nature (WWF). Innovative water conservation technologies. Gland: WWF. 2023. Available from: https://www.worldwildlife.org/magazine/issues/fall-2017/articles/water-for-all

[83]   Kumar V, Kedam N, Sharma KV, Mehta DJ, Caloiero T. Advanced machine learning techniques to improve hydrological prediction: a comparative analysis of streamflow prediction models. Water 2023; 15(14): 2572.
[http://dx.doi.org/10.3390/w15142572]

[84]   Instituto Mexicano de Tecnología del Agua (IMTA). Uso de inteligencia artificial para la predicción de cambios hidrológicos en Coahuila. IMTA. 2019. Available from: https://www.gob.mx/imta/archivo/documentos?idiom=es

[85]   González AM, Ortega RA, Eds. La gestión hídrica en México: casos de estudio y propuestas de políticas públicas. México: El Colegio de la Frontera Norte 2020. [ISBN: 9786074793628].
[http://dx.doi.org/libreria.colef.mx/detalle.aspx?id=7753]

[86]   Gavrilaş S, Ursachi CŞ, Perţa CS, Munteanu FD. Recent trends in biosensors for environmental quality monitoring. Sensors (Basel) 1513; 15;22(4): 1513.
[http://dx.doi.org/10.3390/s22041513]

[87]   Guzmán-Ramirez EM, López-Hernandez JG, Guerrero-Guzman A, *et al.* Use of an automatized system to avoid the deterioration of soil in the cultivation of asparagus in the Mexicali Valley of the Baja California, Mexico. Partners Universal International Innovation Journal 2023; 1(3): 292-302.
[PUIIJ].

[88]   Ray RL, Sishodia R, Olutimehin T. Rainwater harvesting for sustainable water resource management under climate change.Climate Risk and Sustainable Water Management. Cambridge: Cambridge University Press 2022; p. 21.
[http://dx.doi.org/10.1017/9781108787291.021]

[89]   Secretaría de Medio Ambiente y Recursos Naturales (SEMARNAT). Experiencias y proyectos en Oaxaca. 2019 [cited: 2024 Jul 4]. Available from: https://www.gob.mx/semarnat/acciones---programas/experiencias-y-proyectos-en-oaxaca

[90]   Zamani N, Maleki M, Eslamian F. Fog water harvesting investigation as a water supply resource in arid and semi-arid areas. Water Pract Technol 2021; 1: 43-52.
[http://dx.doi.org/10.22034/WPJ.2021.272244.1029]

[91]   Comisión Nacional del Agua (CONAGUA). Programa nacional para captación de agua de lluvia y enotecnias en zonas rurales (PROCAPTAR). 2017 [cited: 2024 Jul 4]. Available from: https://www.gob.mx/conagua/acciones-y-programas/programa-nacional-para-captac-on-de-agua-de-lluvia-y-ecotecnias-en-zonas-rurales-procaptar

[92]   Holwerda F, Gotsch S. La niebla y la ecohidrología del bosque mesófilo de montaña en México. Coordinación editorial, CCA, UNAM. Centro de Ciencias de la Atmósfera. Boletín Infoatmosfera. 2012;(06):2-6. [cited: 2024 Jul 4]. Available from: https://www.atmosfera.unam.mx/boletines/Infoatmosfera_06.pdf

[93]   Pascual AJ, Naranjo MF, Payano R, Medrano PO. Tecnología para la recolección de agua de niebla.In: IV Simposio Internacional Tecnohistoria. Chiapas, México 2011.
[http://dx.doi.org/10.13140/RG.2.1.4806.7048]

[94]   Navarro FA, Delgado MR. Tecnologías para la gestión sostenible del agua. 1ª ed. Ciudad de México: CLAVE Ed. 2021. [https://www.aguanet.com.mx/archivos/Tecnologias_para_la_gestion_sostenible_del_agua.pdf

[95]   Darre NC, Toor GS. Desalination of water: a review. Curr Pollut Rep 2018; 4(2): 104-11.
[http://dx.doi.org/10.1007/s40726-018-0085-9]

[96]   Obotey Ezugbe E, Rathilal S. Membrane technologies in wastewater treatment: a review. Membranes (Basel) 2020; 10(5): 89.

[http://dx.doi.org/10.3390/membranes10050089] [PMID: 32365810]

[97] Ahmed FE, Khalil A, Hilal N. Emerging desalination technologies: Current status, challenges and future trends. Desalination 2021; 517: 115183.
[http://dx.doi.org/10.1016/j.desal.2021.115183]

[98] Albalasmeh AA, Mohawesh O, Gharaibeh MA, Alghamdi AG, Alajlouni MA, Alqudah AM. Effect of hydrogel on corn growth, water use efficiency, and soil properties in a semi-arid region. J Saudi Soc Agric Sci 2022; 21(8): 518-24.
[http://dx.doi.org/10.1016/j.jssas.2022.03.001]

[99] Centro de Investigación en Alimentación y Desarrollo (CIAD). Uso del hidrogel para la agricultura. CIAD. 2022 [cited: 2024 Jul 4]. Available from: https://www.ciad.mx/uso-del-hidrogel-par--la-agricultura/

[100] Mora-Ravelo SG, Alarcon A, Rocandio RM, Vanoye EV. Bioremediation of wastewater for reutilization in agricultural systems: a review. Appl Ecol Environ Res 2017; 15(1): 33-50.
[http://dx.doi.org/10.15666/aeer/1501_033050]

[101] Yang F, Cen R, Feng W, Liu J, Qu Z, Miao Q. Effects of super-absorbent polymer on soil remediation and crop growth in arid and semi-arid areas. Sustainability (Basel) 2020; 12(18): 7825.
[http://dx.doi.org/10.3390/su12187825]

[102] Sosa PG, Hermosillo RD, Jurado GP, Pomposo A, Bustamante M, García PJ. Efectividad de un polímero retenedor de humedad para elevar la supervivencia en plantaciones de mezquite (*Prosopis glandulosa* Torr.) en zonas áridas. Eur Sci J 2021; 17(7): 55.
[http://dx.doi.org/10.19044/esj.2021.v17n7p55]

[103] Khan I, Iqbal B, Khan AA, *et al.* Inamullah, Rehman A, Fayyaz A, Shakoor A, Farooq TH, Wang LX. The interactive impact of straw mulch and biochar application positively enhanced the growth indexes of maize (*Zea mays* L.) crop. Agronomy (Basel) 2022; 12(10): 2584.
[http://dx.doi.org/10.3390/agronomy12102584]

[104] Barus J, Ernawati R, Nila W, Yulia P, Nandari DS, Slameto S. Improvement in soil properties and soil water content due to the application of rice husk biochar and straw compost in tropical upland. Int J Recycl Org Waste Agric 2023; 12: 85-95.
[http://dx.doi.org/10.30486/ijrowa.2022.1942099.1355citation{index=0}&#8203]

[105] Pedroza-Parga EH, Pedroza-Sandoval A, Velásquez-Valle MA, Sánchez-Cohen I, Trejo-Calzada R, Samaniego-Gaxiola JA. Efecto de la cobertura del suelo sobre el crecimiento y productividad del zacate buffel (*Cenchrus ciliaris* L.) en suelos degradados de zonas áridas. Rev Mex Cienc Pecu 2022; 13(4): 866-78.
[http://dx.doi.org/10.22319/rmcp.v13i4.5963]

[106] Babiker AE, Elnasikh MH, Abd Elbasit MAM, Abuali AI, Abu ZM, Liu G. Potential of low-cost subsurface irrigation system in maize (*Zea mays* L.) production in high water scarcity regions. Agric Eng Int CIGR J 2021; 23(3): 42-51.
[http://dx.doi.org/10.35633/inmatec1-1.2022.31.1.44]

[107] Secretaría de Protección al Ambiente de Baja California. Programa ambiental estratégico región vitivinícola del Valle de Guadalupe. 2016. [cited:2024 Jul 13]. Available from: https://bajacalifornia.gob.mx/Documentos/sest/desarrollo_sustentable/Programas_Proyectos/Estrategic o-Ambiental-Valle-GPE/PROGRAMA%20AMBIENTAL%20ESTRATEGICO_REGION% 20VITIVINICOLA%20DEL%20VALLE%20DE%20GUADALUPE.%202016.pdf

[108] Lucero-Vega G, Troyo-Diéguez E, Murillo-Amador B, *et al.* Diseño de un sistema de riego subterráneo para abatir la evaporación en suelo desnudo comparado con dos métodos convencionales. Agrociencia 2017; 51(5): 487-505.
[http://dx.doi.org/www.redalyc.org/articulo.oa?id=30252307002]

[109] Gulomov SB, Sherov AG. Study on drip irrigation system as the best solution for irrigated agriculture. IOP Conf Ser Earth Environ Sci 2020; 614(1): 012144.

[http://dx.doi.org/10.1088/1755-1315/614/1/012144]

[110] Chen X, Zhao C, Yun P, *et al.* Climate-resilient crops: Lessons from xerophytes. Plant J 2024; 117(6): 1815-35.
[http://dx.doi.org/10.1111/tpj.16549] [PMID: 37967090]

[111] Lluch CS, Velázquez ZJ, Nieto DC. Agricultura, agua y cambio climático en zonas áridas de México. Rev Digit Divulg Científica 2022; 8(2): 35-47.
[http://dx.doi.org/10.18846/renaysoc.2022.08.08.02.0004]

[112] Instituto Nacional de Investigaciones Forestales Agrícolas y Pecuarias (INIFAP) Informe anual de autoevaluación del INIFAP, correspondiente al año 2019 Primera Reunión Ordinaria 2020 de la H. Junta de Gobierno del INIFAP 2019; pp. 1-63.

[113] Flores-Gallardo H, Sifuentes-Ibarra E, Flores-Magdaleno H, Ojeda-Bustamante W, Ramos-García CR. Técnicas de conservación del agua en riego por gravedad a nivel parcelario. Rev Mex Cienc Agric 2018; 5(2): 241-52.
[http://dx.doi.org/ 10.29312/remexca.v14i8.3202]

[114] Acciona. La EDAR de Atotonilco (México), la mayor planta de tratamiento de aguas residuales del mundo cumple un año desde su puesta en marcha. 2018. Available from: https://www.acciona-mx.com/actualidad/noticias/edar-atotonilco-cumple/

CHAPTER 4

'OMICS' Studies on Rhizosphere-Microorganism Interactions in Soils

Edgar Antonio Estrella-Parra[1,*], José G. Avila-Acevedo[1], Adriana Montserrat Espinosa González[1], Ana M. García-Bores[1], Jessica Hernández-Pineda[2], Nallely Alvarez-Santos[1], José Cruz Rivera-Cabrera[3] and Erick Nolasco Ontiveros[1]

[1] *Phytochemistry Laboratory, UBIPRO, Superior Studies Faculty (FES)-Iztacala, National Autonomous University of Mexico (UNAM), Tlalnepantla de Baz, Mexico State, Mexico*

[2] *Department of Infectious Diseases and Immunology, INPer, SSA, Mexico City, Mexico*

[3] *Liquid Chromatography Laboratory, Department of Pharmacology, Military School of Medicine, CDA, Palomas S/N, Lomas de San Isidro, Mexico City, Mexico*

Abstract: Soil is an ecosystem in which millions of microorganisms live and interact with plant roots. It has phytoremediation properties, sequestering pollutants such as heavy metals (cadmium, lead, and sulfur, among others), microplastics, and a great diversity of products of anthropogenic origin. Recently, the indiscriminate discharge of pharmaceuticals into public sewage systems has become a major concern, resulting in a public health problem due to the multi-resistance of clinically important bacteria and fungi to these pharmaceuticals. Similarly, the constant use of soil for agriculture, as well as the application of pesticides to combat economically important pests, has damaged both the native soil microbiome and impoverished both the biotic and abiotic properties of the soil. This issue is further exacerbated by the detrimental effects of global climate change. This has led to the search for methods to detoxify soils and reduce the deleterious effects of pollutants. Thus, omics tools, such as metabolomics, metagenomics, proteomics, genomics, and transcriptomics, detect the presence of these pollutants and develop detoxification strategies. For example, in soils exposed to copper (Cu), the earthworm *Eisenia fetida* induces metabolites such as pyruvic acid. In China, the restoration of black soils is possible due to the metabolomic profiling of 287 detected metabolites, which permitted the identification of specific biomarker metabolites that serve for the restoration of degraded soil. Thus, omics tools have become indispensable for the monitoring, diagnosis, and remediation of soils with a high rate of alteration due to anthropogenic activities.

Keywords: Agriculture soils, Contaminated soils, Omics tools, Rhizosphere-microorganism interactions.

* **Corresponding author Edgar Antonio Estrella-Parra:** Phytochemistry Laboratory, UBIPRO, Superior Studies Faculty (FES)-Iztacala, National Autonomous University of Mexico (UNAM), Tlalnepantla de Baz, Mexico State, Mexico; Tel: +525556231137; E-mail: estreparr@iztacala.unam.mx

Israel Valencia Quiroz (Ed.)

INTRODUCTION

Soil is a vital resource for plant sustenance and provides a habitat for microorganisms in their roots [1]. Often referred to as the soul of the earth, soil is home to trillions of living organisms [2]. Wu (2023) defines 'soil health' as "a state in which the quantities of harmful substances and soil-borne phytopathogens contained in soil do not exceed the threshold of safety to local environments and agroforestry production (sic)". It has been estimated that there are 2.6 x10^{29} bacteria in soil ecosystems; moreover, per gram of soil may contain 10,000 prokaryotic species [3]. However, anthropogenic pollution has significantly increased the levels of pollutants in marine and terrestrial ecosystems [4].

Currently, research on degraded soils is a multidisciplinary and multidimensional field that evaluates both their degradation and recovery processes [5]. Soil biota plays an essential role in the decomposition of organic matter, soil nutrient cycling, and bioturbation and inhibits the transmission of soil diseases [6]. Thus, determining the exposure of soil microbiota to contaminants is crucial for restoring soil degradation [5]. According to the United Nations Environment Program (UNEP), soil health deterioration negatively impacts food security and public health [7]. During 1970 and 2015, approximately 35% of all wetlands, marine coasts, and land were lost due to human activities and natural degradation [8]. Plant roots release organic and inorganic compounds that are involved in respiration, secretion, and nutrient absorption in the soil, forming a plant-soil exchange process [9]. Thus, the rhizosphere is the region of soil influenced by plant roots and contains a diverse set of microorganisms, including fungi, bacteria, algae, and protozoa [9]. For this reason, the microorganisms present in the rhizosphere are of great importance due to their ecological multifunctionality, including their ability to remove pollutants in soil, adapt to the environment, and ameliorate the effects of global warming [8]. In addition, microorganisms can colonize various natural and anthropogenic environments through their resilient metabolism [10]. Plant-microbe interactions affect different physiological processes in plants, regulate plant growth, and maintain ecosystem stability, development, and maintenance [8]. Plants and microorganisms can act as metal hyperaccumulators, sequestering metals from the soil, acting as bioremediators through genes, and detoxifying metal transporters, activities that can occur through plant-microbiota relationships [11]. To understand the process of phytoremediation, it is necessary to understand the development of bacteria in the rhizosphere and the mechanisms of adaptability and growth, even in contaminated soil [1]. Moreover, plant root metabolites provide carbon and energy for microorganisms and regulate the microbial environment, altering microbial forms and communities [8]. Similarly, the study of volatile organic compounds (VOCs) emitted by organisms in soil is in its early stages, but it is necessary to elucidate the metabolomic profile of the soil [12].

The use of "OMICS" tools makes it possible to obtain complete microbiome profiles directly from samples [10]. Advances in omics tools, such as metatranscriptomics, metagenomics, metaproteomics, and metabolomics, allow us to understand the interactions between metabolites and proteins and their relationship with the environment [13]. These tools also facilitate the generation of a large amount of data of different order, such as metabolites, proteins, and genes, that are regulated by distinct signaling pathways interacting with each other [14]. Metagenomics allows the recovery of genetic material from environmental samples, allowing researchers to understand the dynamics of the remediation process in an ecosystem mediated by microorganisms [4]. Furthermore, studying the soil metabolome expands knowledge and understanding of the interaction and functional significance of microorganisms through the interaction of metabolites, which encompasses various biological hierarchies [12]. For example, combining the 16S rRNA unit with the metabolomic analysis of agricultural soil microorganisms allows the evaluation of the toxicity of natural and synthetic triketone-type pesticides used in crop cultivation [15].

This chapter's objective is to understand omics tools for studying soil-microorganism interactions in contaminated soils.

Trash Landfills, Microplastics, and Antibiotics: Soil Pollution

The soil is a complex microecological environment in which microbial activity and root exudate secretion are derived [5]. Biotic and abiotic agents interact ecologically in soil ecosystems to form complex and stable microbial communities [3]. Among the most critical organisms in the soil, bacteria are the most influential group to be used as biotransformation products in soil due to their versatility in metabolism, making them ideal for use in biotransformation processes within the soil [10]. Thus, microorganisms possess several enzymes that are harmful to the contaminants in their environment. These enzymes help to degrade and transform hazardous substances into less dangerous forms [7], including contaminants in landfills [16].

However, the overexposure of microorganisms to contaminants reduces their capacity to sequester contaminants in the soil [17]. Removing contaminants such as heavy metals from soil is necessary to avoid adverse health effects [2]. Because of industrialization and population growth, the levels of substances such as hydrocarbons, heavy metals, and aromatic hydrocarbons have drastically increased [18]. Heavy metals negatively affect economically important plants in terms of physiological conditions, reducing their germination rate, generating oxidative stress, and altering their photosynthesis. Heavy metals such as arsenic, cadmium, chromium, cobalt, copper, mercury, nickel, and lead are thus released into the soil*via*anthropogenic activities [19]. Organisms that act as biomarkers of contaminants in the soil, such as earthworms (*Eisenia fetida*), report specific metabolites associated with energy metabolism, such as pyruvic acid, when Cu is present in the soil [20]. Alarmingly, it is estimated that by 2050, 3,600 million tons of solid waste will be generated [16].

Solid waste is managed through various methods, such as incineration, landfilling, and composting [21]. One of the problems of landfills is the high generation of sludge, which loses water and fluids with time. In the early stages of sludge formation, hydrolytic and methanogenic bacteria dominate the microbial community [22]. In India, municipal solid waste (MSW) generation increased dramatically from 6 to 48 tons in a short period from 1994 to 1997, with plastics becoming an emerging concern. It is estimated that 4.9 million tons of plastics are now present in landfills or natural environments around the world [23]. Landfill gas generation poses a risk to human health. However, studies have shown that interaction between plants and microorganisms in soil can help mitigate these harmful effects [24].

Another pollutant, due to its use in people's daily lives, is plastic. In recent times, the accumulation of microplastics in soil has increased because of their resistance and long life span in ecosystems [6]. Furthermore, microplastics alter the enzymatic activity and physicochemical properties of soil, affecting soil microbial diversity [6]. Millions of tons of plastic and its excessive use have resulted in the contamination of marine ecosystems, landfills, and water bodies, among others [13]. For this reason, plastics are classified according to their size. Microplastics and nanoplastics are 5 mm-1μm and 1 μm-1nm particle sizes, respectively [42]. Alarmingly, in 2019, it was estimated that there were 368 million tons of plastic in the world, of which only a small percentage is recycled (6-23%, approximated) [6]. This apotheotic generation of plastic pollution does not diminish its production, with projections for 2050 expected to exceed its production of 33 billion metric tons globally [25]. Moreover, in 2021, the European continent generated 390 million tons of plastic [42].

Similarly, it is estimated that there are 4–23 times more microplastics on continental soil than in marine environments [6]. Among the most widely used plastics, polyethylene accounts for 36% of all plastics on the market [26]. The degradation of plastics is an ongoing issue. Recent research has shown that a critical step for their total degradation lies in the C-C and C-H bonds of polyethylene, which renders them more durable and less biodegradable towards the enzyme activity of microorganisms [26]. In this way, polystyrene-degrading enzymes, such as PEase, have been identified through transcriptomic, proteomic, and metagenomic analyses [26]. Moreover, earthworms indicate that nanoplastics, even at low concentrations, pose a toxicological risk to soil-dwelling wildlife [25].

Another problem is the degradation of antibiotics, which can only be achieved with bacterial activity in the soil [27]. The European Union has launched the use of treated water for agricultural activities in Africa, but this water contains a mixture of contaminants such as residues of pharmaceuticals, cosmetic products, and micro-contaminants [28]. For example, tetracycline is a drug found in wastewater that interferes with soil microbial communities and produces antibiotic-resistance genes [29]. Sulfonamide-type drugs are a contaminant in the soil that creates genes resistant to drugs used in clinical therapy as antibiotics [27]. Sulfadiazine did not affect alpha bacterial communities in the soil. Still, it did affect the remaining microbial communities present. Hence, the *Verrumicrobiata phylum* increased after sulfadiazine treatment, including sulfadiazine, induces lower biomass, shorter roots, a lower number of pigments useful for photosynthesis, and a smaller plant size [27]. Moreover, a high amount of organic carbon in soil interacts with genes associated with drug resistance, such as the ackA and pta genes, which contribute to multidrug resistance, including resistance to vancomycin [30].

Due to the urgency of studying soil-rhizome interactions in contaminated soils, the U.S. government, through its Environmental Protection Agency (EPA), has established the omics tool to determine the nature, environment, and intrinsic functions of microbial communities present in highly contaminated systems [18]. It has even been proposed to use the term holo-omics, which, unlike multiomics, refers to all the omic domains (genomics, transcriptomics, proteomics, and metabolomics) of two interacting individuals, *i.e.*, host-guest, for example, microbes in a human being or other mammals [31]. The omics, metagenomics, metatranscriptomics, and metaproteomics areas are tools employed to identify specific genes involved in the environmental remediation of xenobiotic substances [2]. In this way, metagenomics can be used to decipher microbial communities and identify unexplored genes [13]. Therefore, a metagenomics study in a landfill in India identified bacteria such as *Thermobifida fusca* and *Pseudomonas mendocina*, which degrade PET. Additionally, the study highlighted enzymes involved in the degradation of plastics, such as oxidoreductase, cutinase, lipase, and other enzymes [23].

Agricultural-type Soils: Food Challenges and Overexploitation

By 2050, the world's population is projected to reach 9.2 billion, increasing food demand by 60-70% and necessitating the development of more efficient agricultural practices [32]. The rapid increase in pollutants in crop soil —even in small quantities—has serious detrimental effects on individuals and the environment [2]. In modern agriculture, synthetic pesticides are used to combat pests [33], although pesticides used in agricultural soils affect soil microbial communities [15]. The use of pesticides on cultivars has adverse effects on humans and the environment [34]. The annual use of approximately 3.5 million tons of pesticides reduces biodiversity in oceans and on continents [34]. Fortunately, microorganisms of various taxa, such as fungi, bacteria, and algae, can degrade pesticides through their metabolic processes [33]. Moreover, bioorganic fertilizers promote efficiency in the phytoremediation of heavy metals and saline soils, improving the properties of the rhizosphere, particularly the microbiome [35]. Research has shown that the toxicity induced by the natural triketone-type pesticide hydroxy-leptorpermone differs from the synthetic pesticide analog that induces the metabolite CMBA. This suggests that bacterial populations can select, tolerate, and degrade by different metabolic pathways [15]. Compounds such as trichloroethylene —a compound used in industry—are pollutants found in wastewater; therefore, their elimination is necessary. It has been demonstrated that the fungus *Yarrowia lipolytica* can effectively eliminate trichloroethylene [36]. Furthermore, the Bacillus genus has been shown to promote plant growth even under high salinity conditions, making various bacterial strains of Bacillus beneficial for agriculture, especially in saline soils [32]. It has also been reported that in domesticated rice cultivars, the symbiotic relationship with mycorrhizal fungi is modified by the pressure caused by climate change [37]. Nitrosomonadaceae and Nitrospiraceae bacteria are more abundant in wild rice rhizospheres, leading to increased levels of signaling compounds such as strigolactones [37]. In China, the restoration of black soils —known as mineral soils induced by exhaustive agricultural activities—is possible due to the metabolomic profiling of 287 detected metabolites, such as benzenoid, policies, alkaloids, and others, and the identification of specific biomarker metabolites that serve for the restoration of degraded soil [5].

In the alfalfa cultivar (M. sativa), bryobacter such as *Blastococcus*, *Modestobacter*, *Actinophytocola*, *Bacillus*, and *Streptomyces*, as well as mycorrhizal fungi such as *Leohuymicola*, *Funneliformis*, and *Certatobasidiaceae*, promote plant nutrients as well as soil stability, increase rhizomicrobiota and their activity, and restore the microbiota community [35]. In cultivars of cucumber, maize, and ryegrass, bacteria such as Proteobacteria, Actinomycetes, and Firmicutes are predominant in cultivar soils [9]. Likewise, herbaceous vegetation, such as *Rumex acetosa*, shows high oxidation capacity in the presence of methane, implying a colonization effect of microorganisms that are tolerant to this oxidation in the rhizosphere [21].

Thus, in addition to pesticides, contaminants such as arsenic and cadmium have been reported in agricultural soils, posing food and health risks. Elevated levels of arsenic and cadmium are known to affect cultivar genes, with insects such as collembolans and microbes being bio-indicators of these heavy metals [38]. In addition, cadmium inhibits the development of soybean seeds; however, if the roots of soybean seeds are inoculated with the bacterium *Oceanobacillus picturae*, they improve their growth rate, biomass, root length, and even promote the recovery of the physiological conditions of the soil, such as pH. Thus, using metabolomic and transcriptomic analysis, 157 genes associated with the synthesis of aspartic acid, cysteine, and even flavonoid synthesis have been identified [39]. Metabolites such as dimethylglycine, betaine, and scyllo-inositol are considered biomarker metabolites in earthworms found in chromium-contaminated soil [40].

Another problem associated with soils used for agricultural purposes is plastic film contamination. In agricultural soil, the plastic films used are a source of microplastic pollution. In China, 1.4 million tons of plastic were used in agriculture in 2017 [6]. In cultivar plants, microplastics, along with soil, alter the distribution of oxygen in the soil and the balance of aerobic and anaerobic organisms [6]. Microplastics such as polyethylene inhibit the growth of *Brassica parachinensis*, causing a 50% decrease in biomass and a 48.63% decrease in decay. Metabolomic analysis has shown that microplastics interfere with the synthesis of amino acids and oxidative stress and affect the pathways associated with the growth of the root [41]. Moreover, in Europe, irrigating commercial plants with wastewater has demonstrated changes in the microbial community in soil, such as the presence of acid bacteria, delta proteobacteria, and Chloroflexi, particularly in lettuce crops [28].

In this manner, multi-microbial analysis allows us to determine the roles of the composition of metabolites and their functioning in degraded soils through the use of biomarkers, allowing us to establish a strategy for soil remediation [5]. Ecological metabolomics studies the ecological and ecophysiological functions of organisms and ecosystems based on their metabolic composition [12], allowing the creation of studies and remediation strategies for different bacterial communities.

Study of Soils and/or Microbiota in Contaminated Soils Through OMICS Tools

In general, we will summarize findings that have been published by different authors on the subject. According to Chen (2023) [21], the methods for soil-rhizosphere metabolome analysis are as follows:

DNA Extraction

In general, a portion of the plant's roots is taken, eliminating the adhering soil. The rhizospheres are then placed on ice for transport to the laboratory. Once in the laboratory, the root is refrigerated at -80 °C for DNA extraction. Subsequently, DNA is isolated using an extraction kit such as the Mobio Powersoil DNA isolation kit.

Metabolomics Analyses

For a non-specific metabolomics study, an LC-MS is used, and the experimental spectra are compared with those of a database such as HMDB. Subsequently, the KEGG program is used to search for metabolomics pathways. Thus, sample preparation is initiated. For this purpose, 50 mg of the solid sample is transferred into 100 µL of solvent, and 400 µL of acetonitrile: methanol (1:1) is added and stirred by vortexing for 30 s. The sample is then centrifuged at 4°C at 13,000 g for 15 m. The supernatant is removed, dried with nitrogen, and re-suspended in 120 µL (acetonitrile:water 1:1). The sample is then subjected to ultra-high frequency at 5°C for 5 m and centrifuged under the above conditions, after which it is inserted into the vial to enter the LC-MS. The chromatographic and spectrometric conditions are adjusted according to the sample.

Other methods commonly used for soil-rhizosphere metabolome analysis are in agreement with those put forward by Liu (2024) [35]. First, 50 mg of freeze-dried and extradited soil plants is taken. Then, a 400 µL methanol: water (v/v=4:1) solution with 0.02 mg of 1-L-2-chlorophenylalanin is used as the internal standard. The mixture is allowed to settle at -20 °C and treated by a high-throughput tissue crusher at 50 Hz for 6 min, followed by vortex for 30 s and ultrasound at 40 kHz for 30 min at 5 °C. The samples are placed at -20 °C for 30 min to precipitate proteins. After centrifugation of 13000 g at 4 °C for 15 min, the supernatants are carefully transferred to sample vials for LC-MS/MS analysis. After acquiring the metabolome mass spectra, bioinformatic analysis is performed. The MS raw data is imported into Progenesis QI (version 2.3) for peak detection and alignment. The preprocessing results generated a data matrix consisting of the retention time, mass-to-charge ratio (m/z), and peak intensity. Metabolic features detected at least 80% in any set of samples are retained, whereas those with a relative standard deviation of QC > 30% are discarded. After filtering, minimum metabolite values are imputed for specific samples in which the metabolite levels fell below the lower limit of quantitation, and each metabolic feature is normalized by the sum. Subsequent statistical analysis is performed on log10-transformed data to identify significant differences in metabolite levels between comparable groups (Liu *et al.*, 2024). Other bioinformatics platforms, such as MSDial, MZmine, and KNIME, can also be used for bioinformatics analysis.

Bacterial Metagenomics

According to Jurado (2020) [10], we performed the bacterial metagenome as described.

In the case of landfill sludge or soil, samples are taken from specific processes, such as composting areas. Note that the sample must have at least three replicates. A commercial DNeasy power oil DNA isolation kit (Qiagen) was used for microbial DNA extraction. Subsequently, bacterial DNA was sequenced and amplified by PCR using the Bakt341F and Bakt 805R primers. The raw data obtained were separated based on their specific barcodes. The quality of the FASTQ files was verified using the FastQC software. The Cutadapt 1.3 software was used to remove sequences with less than 300 base pairs. Sequences were taxonomically assigned using QIIME software. The OTU was assigned to the microbial taxon using UCLUST algorithms (confidence levels less than 0.005% are excluded). Finally, we discarded rRNAs >97% similarity and >380 bp in length, in which the 16s portion was located in the v3-v4 hypervariable region.

Exposure of Soil to Nano-plastic Contaminants by Earthworm Analysis

A multiomics study analyzing the effects of nanoplastic contaminants on soil was reported by Tang (2023) [25]. The earthworms were crushed and then extracted with methanol:water solution (4:0.85; v/v) and then with chloroform:water (4:2; v:v). The extracted samples were then added to deuterated water with a buffered phosphate solution (0.2 M, pH 7.4). Next, they were subjected to metabolomic analysis by two-dimensional NMR (1H-1H and 1H-13C). The spectra were analyzed using the PLS-DAD (partial least squares-discriminant analysis) method.

For proteomic analysis, a lysis buffer (SDS; 4% SDS, 100 mM trs-HCL, 1 nM dithiothreitol, pH 7.6) was used. Subsequently, a protein assay kit (Bio-Rad, USA) was used to measure the protein content. Following this, the proteins were then digested with trypsin to obtain peptides, which were extracted in C18 columns. The peptides were then recovered and concentrated by reconstitution in 40 µL of 0.1% formic acid. Proteins were separated by polyacrylamide gel electrophoresis (PAGE). Liquid chromatography coupled with mass spectrometry (LC-MS) was carried out for protein analysis. MASCOT software was used for bioinformatics analysis.

For transcriptomics, RNA was first extracted using TRIzol reagent according to the manufacturer's manual of indications. Subsequently, its purity and integrity were calculated to build its library. To calculate the gene expression of each transcript, fragments per kilobase were calculated. RNA-seq (RSEM) is then used to quantify genes and isoforms. The transcripts were then analyzed using the Go and KEGG platforms. The statistical analysis was performed using the Bonferroni test ($P<0.05$).

For the soil metabolome analysis, soil samples were collected from the surface layer (0-20 cm) and stored at 4°C until use. Subsequently, its physicochemical properties, such as sand, organic matter, organic carbon, cation exchange, pH, moisture, and salts, were evaluated as necessary.

Later on, the soil was macerated in methanol, and the residual solvent was removed using a rotary evaporator. The dried samples were then stored at approximately -20ºC. This soil extract was injected (10 µL) into an LC-MS system under specific chromatographic and spectrometric conditions. For bioinformatics analysis, the Metlin platform can be used to evaluate retention time, aligned ions, and detected peaks. The experiments were performed in triplicate.

For example, Mughal (2024) reported on the metabolomic analysis between the rhizosphere and soil of a maize-soybean cultivar in the mountains of southeastern China. In this study, soybean and corn germplasm were used, planting two corn plants and two soybean plants with a spacing of 20 cm between them. The soil was collected at different stages of development, *i.e.*, flowering, fruiting, and the fourth and seventh stages of leaf development. For soil sampling, 25 g of soil was taken from a depth of 10-25 cm. The collected soil was preserved in liquid nitrogen at -80 °C. Metabolomics analysis was performed using a UPLC---TOF/MS with an Agilent eclipse Plus-C18 column. The soil samples preserved in liquid nitrogen were thawed. Subsequently, 50 mg of dry soil was taken, and 10 mL of 80% methanol was added to a tube. They were then extracted by ultrasonic stimulation and centrifuged at 10,000 xg for 5 minutes. Finally, the supernatant was filtered through a filtration screen with a mesh size of 0.22 µm. Regarding the chromatographic conditions, they used a gradient-type mobile phase, using water and acetonitrile acidified with 0.1% formic acid. A flow rate of 300 µL per minute was used. A volume of 1 µL was injected into the system. The spectrometric conditions were scanned in positive mode, using nitrogen as a carrier gas, with the injection chamber maintained at 350 °C. One spectrum per second was acquired, with a capillary voltage of 3800 V. The quadrupole scanned masses from 50 to 1000 u (atomic mass units). For data processing, Profinder B.06.00 software was used under the criteria of peak identification, peak filtering, and peak alignment. The accuracy of the results was also verified individually. Subsequently, these data were exported to a data matrix in Simca-P 14.1 software (Umetrics AB, Sweden) for multivariate statistical analysis. Principal component analysis (PCA) was obtained, along with orthogonal partial least squares-discriminant analysis (OPLS-DA). Finally, a Student's t-test was performed for values below p 0.05. To conclude, the data were imported into MetaboAnalyst 3.0 software. In this program, the peak area was analyzed, and normalization and standardization processes, such as sum normalization, Pareto scaling, and log transformation, were performed. These analyses allow elucidating the interconnection between all metabolites, using heat maps and cluster analysis [43].

In general, we propose studying the use of different omics tools on contaminated soil topics, as described in Fig. (**1**).

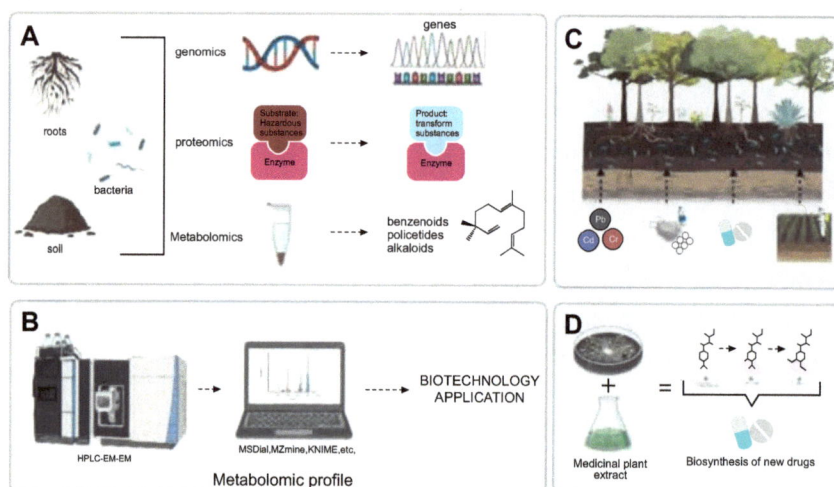

Fig. (1). Considerations for an omics-type study in soil-plant interactions. A) The soil-root-microorganism sample is taken, according to the required analysis, and regularly stored in low-temperature conditions until it is analyzed. B). After its extraction, which can be through solvents of the organism type, we proceed to obtain its metabolome, proteome, and genome through standardized processes, and then a sample (from 1 μL) is injected into a chemical-analytical technique such as mass spectrometry or nuclear magnetic resonance to obtain electronic archives and analyze them through computer processing. C) Subsequently, and already analyzed by algorithms created or taken from a bioinformatics platform, the obtained biological profile is obtained, such as the detection of heavy metals, traces of contaminants such as drugs, secondary metabolites, proteins of interest, and/or relevant genes. D). Moreover, although it is not the goal of this chapter, if drug development is desired, this type of analysis allows a preliminary study to be carried out for its subsequent development. Figure made with Biorender.

CONCLUDING REMARKS

Understanding biological intra- and interspecific relationships is indispensable for finding solutions to any problem in an ecosystem. Thus, comprehending the biological interactions within the rhizosphere, as well as the contamination caused by anthropogenic activities in soils of economic and food interest, is crucial. The application of omics tools is indispensable for addressing these issues. Therefore, utilizing omics tools in biological interactions can be considered a gold standard technique for diagnosing and remediating specific soil conditions.

AUTHORS' CONTRIBUTION

José G. Avila-Acevedo developed the part on soil contamination in agriculture.

Adriana Montserrat Espinosa González developed the part on contamination by pharmaceuticals.

Ana M. García-Bores developed the part on contamination by microplastics and landfills.

Hernández-Pineda J. developed the methodological part of the genetics of microorganisms.

Nallely Alvarez-Santos created and developed the figure.

José Cruz Rivera-Cabrera contributed to the omics tools part because of their 'expertise'.

Nolasco Ontiveros Erick developed the methodological part of contaminants by microplastics.

Edgar Antonio Estrella-Parra developed the methodological part of metabolomics and compiled the wording of each topic developed by the authors.

QUESTIONS RELATED TO THE TEXT

1. What roles do rhizosphere microorganisms play in soil ecosystems, particularly in the context of pollutant bioremediation?
2. Describe the various omics tools mentioned in the chapter and explain how each contributes to the study of soil-microorganism interactions.
3. How do heavy metals and other pollutants affect soil health and microbial activity, and what are some of the bioremediation strategies discussed in the chapter?
4. What are the major challenges and potential solutions for managing

microplastic contamination in agricultural soils, as described in the chapter?

5. How can multi-omics approaches be applied to understand and improve the resilience of agricultural soils to contaminants such as pesticides and antibiotics?

6. Explain the significance of bioinformatics in omics studies and how it aids in the analysis and interpretation of complex data from soil samples.

LIST OF ABBREVIATIONS

C-C	Carbon-Carbon
C-H	Carbon-Hydrogen
CMBA	2-Chloro-4-Methylsulfonyl Benzoic Acid
EPA	Environmental Protection Agency
HMDB	Human Metabolome Database
KEGG	Kyoto Encyclopedia of Genes and Genomes
LC-MS	Liquid Chromatography Coupled Mass Spectrometry
LC-MS/MS	Liquid Chromatography Coupled Mass Spectrometry Tandem
MSW	Municipal Solid Waste
OPLS-DA	Orthogonal Partial Least Squares-Discriminant Analysis
PAGE	Polyacrylamide Gel Electrophoresis
PEase	Polyethylene Degrading Enzyme
PLS-DAD	Partial Least Squares-Discriminant Analysis
rRNA	Ribosomal RNA
SDS	Sodium Dodecyl Sulfate
u	Atomic Mass Units
UPLC-Q-TOF/MS	Ultra-High Performance Liquid Chromatography with Quadrupole Time-of-Flight Mass Spectrometry
VOCs	Volatile Organic Compounds

REFERENCES

[1]   Bhanse P, Kumar M, Singh L, Awasthi MK, Qureshi A. Role of plant growth-promoting rhizobacteria in boosting the phytoremediation of stressed soils: Opportunities, challenges, and prospects. Chemosphere 2022; 303(Pt 1): 134954.
[http://dx.doi.org/10.1016/j.chemosphere.2022.134954] [PMID: 35595111]

[2]   Gautam K, Sharma P, Dwivedi S, *et al.* A review on control and abatement of soil pollution by heavy metals: Emphasis on artificial intelligence in recovery of contaminated soil. Environ Res 2023; 225: 115592.
[http://dx.doi.org/10.1016/j.envres.2023.115592] [PMID: 36863654]

[3]   Wu D, Wang W, Yao Y, Li H, Wang Q, Niu B. Microbial interactions within beneficial consortia promote soil health. Sci Total Environ 2023; 900: 165801.

[http://dx.doi.org/10.1016/j.scitotenv.2023.165801] [PMID: 37499809]

[4]   Atif Khurshid W, Nahid A, Nafiaah N, *et al.* Bioprospecting culturable and unculturable microbial consortia through metagenomics for bioremediation. CLCE 2022; 2: 100017.
[http://dx.doi.org/10.1016/j.clce.2022.100017]

[5]   Yang J, He J, Jia L, Gu H. Integrating metagenomics and metabolomics to study the response of microbiota in black soil degradation. Sci Total Environ 2023; 899: 165486.
[http://dx.doi.org/10.1016/j.scitotenv.2023.165486] [PMID: 37442461]

[6]   Li Y, Liu Q, Junaid M, Chen G, Wang J. Distribution, sources, transportation and biodegradation of microplastics in the soil environment. Trends Analyt Chem 2023; 164: 117106.
[http://dx.doi.org/10.1016/j.trac.2023.117106]

[7]   Maqsood Q, Sumrin A, Waseem R, Hussain M, Imtiaz M, Hussain N. Bioengineered microbial strains for detoxification of toxic environmental pollutants. Environ Res 2023; 227: 115665.
[http://dx.doi.org/10.1016/j.envres.2023.115665] [PMID: 36907340]

[8]   Wang Y, Bai J, Zhang L, *et al.* Advances in studies on the plant rhizosphere microorganisms in wetlands: A visualization analysis based on CiteSpace. Chemosphere 2023; 317: 137860.
[http://dx.doi.org/10.1016/j.chemosphere.2023.137860] [PMID: 36649898]

[9]   Lv L, Huang H, Lv J, *et al.* Unique dissolved organic matter molecules and microbial communities in rhizosphere of three typical crop soils and their significant associations based on FT-ICR-MS and high-throughput sequencing analysis. Sci Total Environ 2024; 919: 170904.
[http://dx.doi.org/10.1016/j.scitotenv.2024.170904] [PMID: 38354799]

[10]   Jurado MM, Camelo-Castillo AJ, Suárez-Estrella F, *et al.* Integral approach using bacterial microbiome to stabilize municipal solid waste. J Environ Manage 2020; 265: 110528.
[http://dx.doi.org/10.1016/j.jenvman.2020.110528] [PMID: 32421558]

[11]   Sofie T, Tori L, Jaco V. The Bacterial and Fungal Microbiota of Hyperaccumulator Plants: Small Organisms, Large Influence.Advances in Botanical Research. Academic Press 2017; pp. 43-86.
[http://dx.doi.org/10.1016/bs.abr.2016.12.003]

[12]   Brown RW, Reay MK, Centler F, *et al.* Soil metabolomics - current challenges and future perspectives. Soil Biol Biochem 2024; 193: 109382.
[http://dx.doi.org/10.1016/j.soilbio.2024.109382]

[13]   Shilpa NB, Basak N, Meena SS. Exploring the plastic degrading ability of microbial communities through metagenomic approach. Mater Today Proc 2022; 57(4): 1924-32.
[http://dx.doi.org/10.1016/j.matpr.2022.02.308]

[14]   Samuel J, Gunasekaran R, Awantika R, *et al.* Paradigm of integrative OMICS of microbial technology towards biorefinery prospects. ISBAB 2024; 58: 103226. [doi: 10.1016/j.bcab.2024.103226].

[15]   Romdhane S, Devers-Lamrani M, Beguet J, *et al.* Assessment of the ecotoxicological impact of natural and synthetic β-triketone herbicides on the diversity and activity of the soil bacterial community using omic approaches. Sci Total Environ 2019; 651(Pt 1): 241-9.
[http://dx.doi.org/10.1016/j.scitotenv.2018.09.159] [PMID: 30236841]

[16]   Qian Y, Hu P, Lang-Yona N, Xu M, Guo C, Gu JD. Global landfill leachate characteristics: Occurrences and abundances of environmental contaminants and the microbiome. J Hazard Mater 2024; 461: 132446.
[http://dx.doi.org/10.1016/j.jhazmat.2023.132446] [PMID: 37729713]

[17]   Smułek W, Zdarta A, Grzywaczyk A, *et al.* Evaluation of the physico-chemical properties of hydrocarbons-exposed bacterial biomass. Colloids Surf B Biointerfaces 2020; 196: 111310.
[http://dx.doi.org/10.1016/j.colsurfb.2020.111310] [PMID: 32911293]

[18]   Sharma P, Singh SP, Iqbal HMN, Tong YW. Omics approaches in bioremediation of environmental contaminants: An integrated approach for environmental safety and sustainability. Environ Res 2022; 211: 113102.

[http://dx.doi.org/10.1016/j.envres.2022.113102] [PMID: 35300964]

[19]   Dinakaran E, Keisham DD, Hemanth KJ, *et al.* Agronomic, breeding, and biotechnological interventions to mitigate heavy metal toxicity problems in agriculture. J Agric Res (Lahore) 2022; 10: 100374. [http://dx.doi.org/10.1016/j.jafr.2022.100374].

[20]   Zhang Y, Huang C, Zhao J, *et al.* Insights into tolerance mechanisms of earthworms (Eisenia fetida) in copper-contaminated soils by integrating multi-omics analyses. Environ Res 2024; 252(Pt 2): 118910. [http://dx.doi.org/10.1016/j.envres.2024.118910] [PMID: 38604487]

[21]   Chen S, Fu W, Cai L, *et al.* Metabolic diversity shapes vegetation-enhanced methane oxidation in landfill covers: Multi-omics study of rhizosphere microorganisms. Waste Manag 2023; 172: 151-61. [http://dx.doi.org/10.1016/j.wasman.2023.10.021] [PMID: 37918308]

[22]   Xing Y, An Y, Lin L, *et al.* Microbiological mechanisms of sludge property variations under long-term landfill: From micro-omics perspective. Chem Eng J 2024; 486: 150275. [doi: 10.1016/j.cej.2024.150275]. [http://dx.doi.org/10.1016/j.cej.2024.150275]

[23]   Kumar R, Pandit P, Kumar D, *et al.* Landfill microbiome harbour plastic degrading genes: A metagenomic study of solid waste dumping site of Gujarat, India. Sci Total Environ 2021; 779: 146184. [http://dx.doi.org/10.1016/j.scitotenv.2021.146184] [PMID: 33752005]

[24]   Shangjie C, Yongqiong W, Fuqing X, *et al.* Synergistic effects of vegetation and microorganisms on enhancing of biodegradation of landfill gas. Environ Res 2023; 227: 115804. [http://dx.doi.org/10.1016/j.envres.2023.115804] [PMID: 37003556]

[25]   Tang R, Zhu D, Luo Y, *et al.* Nanoplastics induce molecular toxicity in earthworm: Integrated multi-omics, morphological, and intestinal microorganism analyses. J Hazard Mater 2023; 442: 130034. [http://dx.doi.org/10.1016/j.jhazmat.2022.130034] [PMID: 36206716]

[26]   Jin J, Arciszewski J, Auclair K, Jia Z. Enzymatic polyethylene biorecycling: Confronting challenges and shaping the future. J Hazard Mater 2023; 460: 132449. [http://dx.doi.org/10.1016/j.jhazmat.2023.132449] [PMID: 37690195]

[27]   Wang JX, Li P, Chen CZ, Liu L, Li ZH. Biodegradation of sulfadiazine by ryegrass (Lolium perenne L.) in a soil system: Analysis of detoxification mechanisms, transcriptome, and bacterial communities. J Hazard Mater 2024; 462: 132811. [http://dx.doi.org/10.1016/j.jhazmat.2023.132811] [PMID: 37866149]

[28]   Gallego S, Montemurro N, Béguet J, *et al.* Ecotoxicological risk assessment of wastewater irrigation on soil microorganisms: Fate and impact of wastewater-borne micropollutants in lettuce-soil system. Ecotoxicol Environ Saf 2021; 223: 112595. [http://dx.doi.org/10.1016/j.ecoenv.2021.112595] [PMID: 34390984]

[29]   Chen X, Zhu Y, Chen J, Yan S, Xie S. Multi-omic profiling of a novel activated sludge strain Sphingobacterium sp. WM1 reveals the mechanism of tetracycline biodegradation and its merits of potential application. Water Res 2023; 243: 120397. [http://dx.doi.org/10.1016/j.watres.2023.120397] [PMID: 37499542]

[30]   Zhang D, Li H, Yang Q, Xu Y. Microbial-mediated conversion of soil organic carbon co-regulates the evolution of antibiotic resistance. J Hazard Mater 2024; 471: 134404. [http://dx.doi.org/10.1016/j.jhazmat.2024.134404] [PMID: 38688217]

[31]   Nyholm L, Koziol A, Marcos S, *et al.* Holo-Omics: Integrated host-microbiota multi-omics for basic and applied biological research. iScience 2020; 23(8): 101414. [http://dx.doi.org/10.1016/j.isci.2020.101414] [PMID: 32777774]

[32]   Valencia-Marin MF, Chávez-Avila S, Guzmán-Guzmán P, *et al.* Survival strategies of Bacillus spp. in saline soils: Key factors to promote plant growth and health. Biotechnol Adv 2024; 70: 108303. [http://dx.doi.org/10.1016/j.biotechadv.2023.108303] [PMID: 38128850]

[33] Sarker A, Nandi R, Kim J-E, Islam T. Remediation of chemical pesticides from contaminated sites through potential microorganisms and their functional enzymes: Prospects and challenges. Environmental Technology & Innovation 2021; 23: 101777.
[http://dx.doi.org/10.1016/j.eti.2021.101777]

[34] Goh MS, Lam SD, Yang Y, *et al.* Omics technologies used in pesticide residue detection and mitigation in crop. J Hazard Mater 2021; 420: 126624.
[http://dx.doi.org/10.1016/j.jhazmat.2021.126624] [PMID: 34329083]

[35] Liu T, Wang Q, Li Y, *et al.* Bio-organic fertilizer facilitated phytoremediation of heavy metal(loid)s-contaminated saline soil by mediating the plant-soil-rhizomicrobiota interactions. Sci Total Environ 2024; 922: 171278.
[http://dx.doi.org/10.1016/j.scitotenv.2024.171278] [PMID: 38417528]

[36] Yuan M, Chen G, Xiao Y, Qu Y, Ren Y. The mechanisms of yeast extracellular metabolites in stimulating microbial degradation of trichloroethylene: Physiological characteristics and omics analysis. Environ Res 2024; 255: 119193.
[http://dx.doi.org/10.1016/j.envres.2024.119193] [PMID: 38777296]

[37] Tian L, Wang J, Chen H, Li W, Tran L-SP, Tian C. Integrative multi-omics approaches reveal that Asian cultivated rice domestication influences its symbiotic relationship with arbuscular mycorrhizal fungi. Pedosphere 2024; 34(2): 315-27.
[http://dx.doi.org/10.1016/j.pedsph.2023.09.007]

[38] Ren XY, Zheng YL, Liu ZL, Duan GL, Zhu D, Ding LJ. Exploring ecological effects of arsenic and cadmium combined exposure on cropland soil: from multilevel organisms to soil functioning by multi-omics coupled with high-throughput quantitative PCR. J Hazard Mater 2024; 466: 133567.
[http://dx.doi.org/10.1016/j.jhazmat.2024.133567] [PMID: 38271874]

[39] Yang S, Han X, Li J, *et al.* Oceanobacillus picturae alleviates cadmium stress and promotes growth in soybean seedlings. J Hazard Mater 2024; 472: 134568.
[http://dx.doi.org/10.1016/j.jhazmat.2024.134568] [PMID: 38749246]

[40] Tang R, Li X, Mo Y, *et al.* Toxic responses of metabolites, organelles and gut microorganisms of Eisenia fetida in a soil with chromium contamination. Environ Pollut 2019; 251: 910-20.
[http://dx.doi.org/10.1016/j.envpol.2019.05.069] [PMID: 31234257]

[41] Li X, Cheng X, Wu J, Cai Z, Wang Z, Zhou J. Multi-omics reveals different impact patterns of conventional and biodegradable microplastics on the crop rhizosphere in a biofertilizer environment. J Hazard Mater 2024; 467: 133709.
[http://dx.doi.org/10.1016/j.jhazmat.2024.133709] [PMID: 38330650]

[42] Yu W, Leilei X, Wulf A, *et al.* Micro- and nanoplastics in soil ecosystems: Analytical methods, fate, and effects. TrAC 2023; 169: 117309.
[http://dx.doi.org/10.1016/j.trac.2023.117309]

[43] Mughal N, Long X, Deng J, *et al.* Metabolomics analysis of rhizospheric soil: New evidence supporting the ecological advantages of soybean maize strip intercropping system. Appl Soil Ecol 2024; 202: 105564. [doi; https://doi.org/10.1016/j.apsoil.2024.105564].
[http://dx.doi.org/10.1016/j.apsoil.2024.105564]

CHAPTER 5

Bioengineering Techniques for Soil Erosion Prevention

Iván E. Barrera[1,2,*] and **Paris O. Gonzalez[2]**

[1] *Postgraduate in Biological Sciences, Postgraduate Studies Unit, National Autonomous University of Mexico (UNAM), Coyoacan, Mexico City, Mexico*

[2] *Phytochemistry Laboratory, UBIPRO, Superior Studies Faculty (FES)-Iztacala, National Autonomous University of Mexico (UNAM), Tlalnepantla de Baz, Mexico State, Mexico*

Abstract: This chapter addresses the impact of soil erosion caused by both natural processes and human activity. Soil erosion can create serious environmental damage, such as the reduction of farmable, productive land and increased sedimentation in water. To combat this, bioengineering methods are a way to harness sustainable and environmentally friendly approaches. These methods involve using living vegetation, wood, or other natural materials to manage soil, particularly on slopes, banks of rivers, or retaining walls. Some bioengineering methods discussed are bush mattresses, wattle fences, and jute netting. Bush mattresses involve laying down branches and vegetation across eroding areas to slow down the movement of water and capture sediment. Wattle fences use stakes with woven vegetation to trap soil and prevent erosion. Meanwhile, jute netting is a breathable and biodegradable weave-like fabric that can be laid over the soil to protect it from erosion until vegetation establishes. The effectiveness of these methods relies on careful planning, site-specific conditions, and maintenance of the body of work. While bioengineering methods can support fairly intense workloads and may require a huge number of materials, they offer ecological benefits to prevent soil degradation in the long term. Here, we emphasize that these methods can have a remarkable impact if the appropriate method is selected and routinely maintained. By implementing sustainable practices for soil erosion and environmental management, the goal of sustainable land management can be achieved, contributing to the protection of the environment from the adverse impacts of soil erosion.

Keywords: Bioengineering, Construction methods, Deep rooting, Fascines, Living branches, Natural resources, Prevention, Revegetation, Slope stabilization, Soil erosion, Soil reinforcement, Techniques, Terraces.

---
* **Corresponding author Iván E. Barrera:** Postgraduate in Biological Sciences, Postgraduate Studies Unit, National Autonomous University of Mexico (UNAM), Coyoacan, Mexico City, Mexico;
E-mail: ivan_140493@hotmail.com

Israel Valencia Quiroz (Ed.)

INTRODUCTION

Soil Erosion

The term 'erosion' encompasses different concepts depending on its focus. In general, soil erosion refers to any alteration in the soil or underlying bedrock caused by an external vector. These vectors are also used as classification points to determine the type of erosion being discussed, which is mainly divided into two groups: biotic (caused by living organisms) and abiotic (due to inanimate factors). These groups are subdivided into different types. Biotic vectors include plant organisms (phytogenic erosion), animals (zoogenic erosion), and human activity (anthropogenic erosion), while abiotic factors are divided by the type of phenomenon they generate, whether chemical (corrosion) or physical (mechanical erosion) [1].

Erosion occurs naturally through geomorphic factors such as water, snow, ice, wind, and both plant and animal organisms [2 - 5]. However, climate change and human activity are factors of particular concern as they accelerate soil deterioration. In economically and socially important activities such as agriculture, which cause the elimination of vegetation and soil exposure, it is necessary to carry out adequate planning based on the type of soil to be exploited to ensure that it can be maintained with the least amount of possible alterations in its original qualities. Techniques must be employed either to prevent accelerated deterioration as much as possible or to improve the soil's resilience, *i.e.*, the natural recovery capacity of soils [1, 6 - 9].

If erosion conditions persist, the soil may undergo changes in its properties, such as the strength of the material, infiltration capacity, and plant productivity; these are characteristics that the soil generates in its natural formation processes [10, 11]. For example, the loss of nutrients and organic matter in soils due to erosion are important factors because they can indicate the loss of agricultural productivity, mainly when cultivation practices are not planned. The potential impact of such activities on the soil must be considered, particularly when selecting crops that are less harmful to each specific soil type [6, 12 - 14].

Worldwide, attempts have been made to estimate soil loss due to erosion [15 - 17]. For example, in 2015, the average soil loss in the European Union was estimated at 2.46 mg ha^{-1} yr^{-1} (organic matter in megagrams per hectare per year). However, in some areas of Africa, these values exceed 20 mg ha^{-1} yr^{-1}, and this problem increases every year for foretold reasons [18, 19]. Therefore, soil erosion is a major global concern because of its threat to the food industry and the environmental loss it can cause [20 - 23].

Bioengineering as a Soil Erosion Prevention Strategy

Various strategies have been developed to prevent soil deterioration. Bioengineering uses biological knowledge to generate possible solutions to current problems [24, 25]. In the case of soil erosion, there are techniques to secure unstable slopes and embankments that can be affected by erosion phenomena. These methods make use of whole plants or their parts, rocks, wood structures, and construction materials to build structures that prevent the progressive deterioration of soils and allow their use in a more sustainable and safe manner [26 - 29].

Soil Erosion Prevention Techniques on Slopes and Embankments

Bush Mattress

This type of construction is carried out using live branches to protect and stabilize slopes, reduce the speed of rainwater, control erosion on embankments and slopes, and improve riprap and sediment accumulation to prevent it from being washed away (Fig. **1**).

The construction consists of rectangular sections made with branches. The size of these branches varies depending on the area that needs to be covered. They can be small, 60 cm long by 6 mm wide, or for larger areas, they can be 2 to 3 m long and 2.5 cm in diameter. The branches are tied together to form a cylindrical fascine between 15 and 30 cm long and are used to build a mattress, with the buds facing the same side. The branches are placed on the slope using wooden stakes (can also be live) of 1 meter in length. These stakes are driven into the ground at a depth of 60 - 70 cm and serve as supports to place the strips of branches. Using this structure, it is also possible to plant directly on the fascines by filling them with soil.

This method is effective immediately after placement, as plants with dense rooting are used for its construction. It is important to select plants that can grow adequately according to the type of soil to be protected and can also be used as a base for growing new plants. However, its main disadvantage is the amount of material needed to build it and the occasional maintenance required by the shrubs that are formed.

Installation Process

1. Clean the area and remove any debris from the area to be protected.
2. Dig a trench 10–30 cm deep at the base of the slope.
3. Place cuttings flat on the slope, crisscrossed by pushing the roots against the

trench with the tips inclined in a direction parallel to the slope.

4. Drive the pegs between the branches at half their length, with a distance of 1 m between them.
5. The branches should be wired to the pegs as tightly as possible.
6. Thrust the pegs further to tighten the wire and press the branches into the ground.
7. Place the fascines along the trench, over the bottom of the branches, and cover the bottom with soil.
8. Fill any gaps in and between branches with soil to ensure proper rooting.
9. Provide periodic maintenance to ensure mattress and slope bonding.

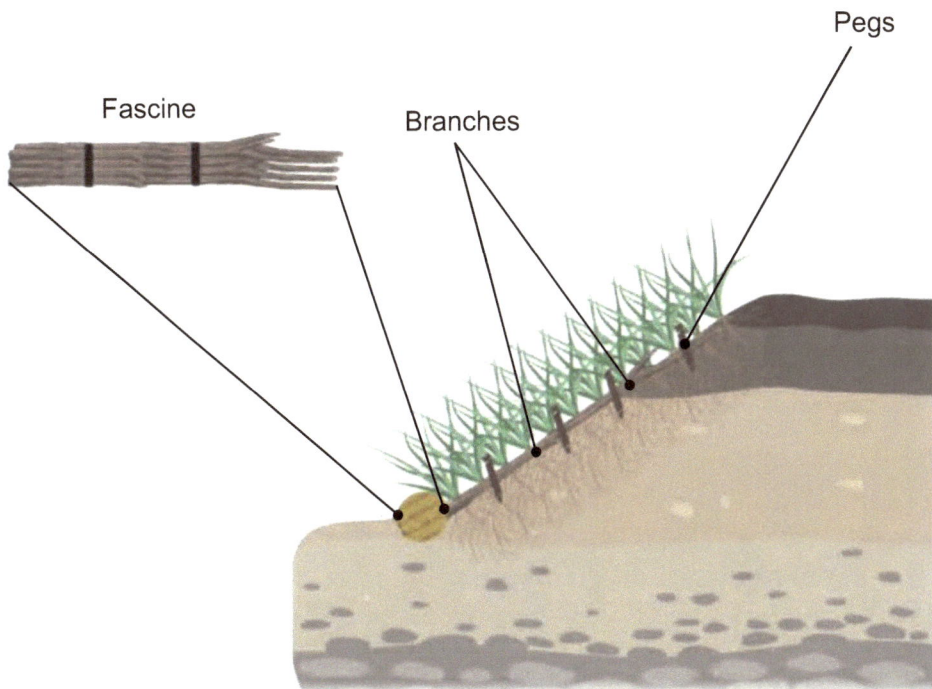

Fig. (1). Bush Mattress in the slope scheme. Figure made with Biorender.

Wattle Fences

Fences made from living plant matter such as branches, stakes, and reeds act as small walls that allow the capture of debris that comes off the slope and reinforce it through the rooting process of the fence itself (Fig. 2). These fences also help to modify the slope to generate planting surfaces.

Pegs that are 1 m long are buried in the ground at a depth of 60 cm with a spacing of 1 m. Smaller pegs of living material are placed between them at a depth of 30 cm. Strong branches woven around the previously placed pegs form the fence. The tips of the branches woven into the stakes are placed in the soil at a depth of 10 cm. The structure is placed on the inner surface of the slope at distances of approximately 40 cm. The spaces generated between the fence and slope are filled with excavated soil to stabilize the surface and create planting spaces.

This method allows the stabilization of the land and prevents the soil from moving or washing away when establishing terraces. It also serves as a base for plantings. In contrast to these advantages, constant maintenance of the fence is required while the stakes begin to grow, as it is necessary to continue weaving the branches around them as they grow, so it requires constant use of flexible branches. It is not recommended in areas where rocks may fall on the slope, as the fence may not provide sufficient resistance in such conditions.

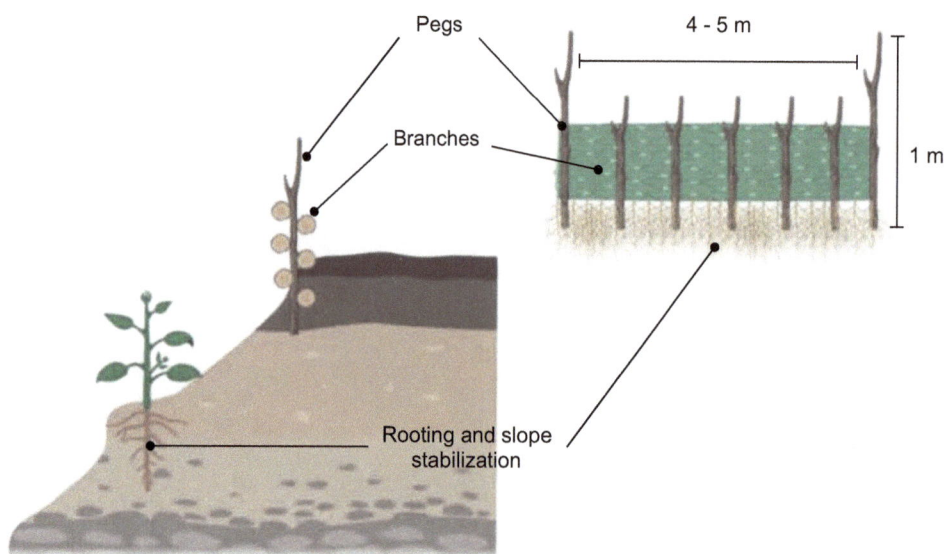

Fig. (2). Wattle fence in slope scheme. Figure made with Biorender.

Installation Process

1. Clean the area where the fence is to be installed.
2. Mark the sections where the stakes are to be placed with 4–5-meter spacings.
3. Dig the holes where the pegs will be driven at 1 m intervals on the marked lines.
4. Place the larger pegs in the holes and insert the smaller ones between them.

5. Dig 15 cm deep trenches between the pegs.
6. Place the cuttings in the trench, firmly backfill with soil, and weave them between the pegs until the fence is complete.
7. Fill the space between the fence and slope with soil to form terraces for planting.
8. Weave the cuttings while the branches and cuttings are growing.

Log Brush Barrier

This technique is used to reduce the velocity of water runoff during flooding, stabilize slopes, and control soil surfaces (Fig. **3**). By using logs with diameters larger than those of the branches used previously (approximately 25 cm) and lengths of up to 6 m, a barrier can be formed that enhances soil capture and provide greater resistance in the event of large rocks falling on the slope where it is placed.

Unbranched logs are fixed to the ground with thick stakes approximately 10 cm in diameter and driven 30 cm deep, forming level lines on the slope where they are placed. If there are spaces between the logs and the slope, they should be filled with soil or rocks to avoid the filtration of fine materials.

This method promotes proper sedimentation of the soil while preventing soil erosion, but it requires a higher amount of plant matter and a significant construction effort, depending on the size of the logs to be used.

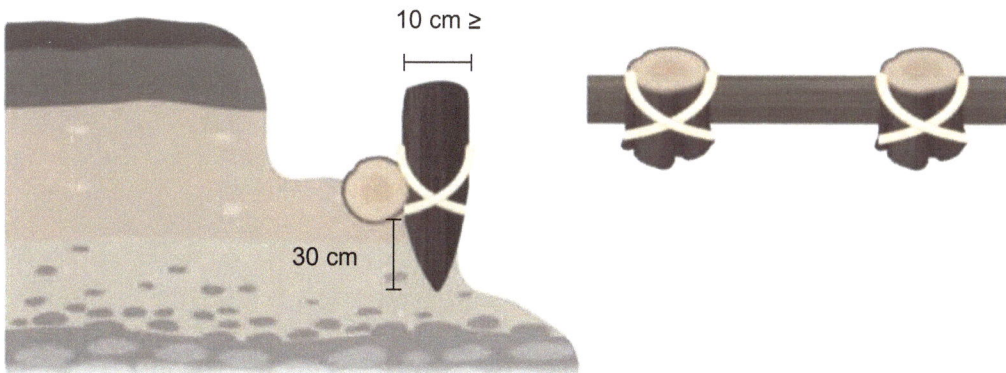

Fig. (3). Log brush barrier in slope scheme. Figure made with Biorender.

Installation Process

1. Clean the areas where the barrier will be placed and select the places where the logs will be set.
2. Place the pegs at the selected points.

3. Place the logs and fix them with the help of pegs using wire, ensuring contact between the logs and the ground on the slope.
4. Fill the space between the slope and the log to form terraces.
5. If there are gaps or spaces between the log and slope, these should be filled with soil or rocks, as well as the ends of each barrier, to provide greater stability.

Fascines (Bush Wattles)

This type of construction is simpler than the previous methods, consisting only of cylindrical fascines of live branches placed directly on the surface of the slope (Fig. **4**). These are anchored with the help of 1.5 m diameter pegs with a separation of 30 cm between them.

The fascines are composed of live branches 1 m long and 3 cm in diameter, and the quantity can vary from 4 to 30 branches, depending on the conditions of the slope being protected. These can be used to reinforce slopes and improve drainage and soil infiltration. They are placed perpendicularly on slopes, trenches, ravines, and banks of water bodies. The spacing between them depends on the inclination of the slope; if it is less than 30°, a separation of 4 m is adequate, but if it is greater, a separation of approximately 2 m is needed.

The construction is simple, it doesn't require much effort and is useful for areas with wet slopes. However, it requires many flexible branches, it's not resistant to falling rocks and protects mainly the superficial layers of soil.

Installation Process

1. Clean the site and flatten it to remove any lumps or depressions.
2. Mark the sections to be covered by the fascines using lines that follow the contour of the slope with suitable spacing depending on the angle of the slope.
3. Dig trenches 10 cm deep and 30 cm wide along the lines starting from the bottom of the slope and moving upward.
4. The live branches are tied together using a rope or cable to form the fascines.
5. Place the fascines along the excavated trenches.
6. Secure the fascines with pegs directly above and below them, leaving an approximately 7 cm overhang.
7. Backfill the trenches with soil covering the sides of the fascines while leaving the top uncovered.

Fences

These structures consist of walls made of different types of materials, such as concrete, rocks, and wood, but live plant materials can also be used for their construction (Fig. **5**). Their purpose is to reduce the velocity of runoff water and contain sediments, thus stabilizing soil in riverbeds and natural drainages.

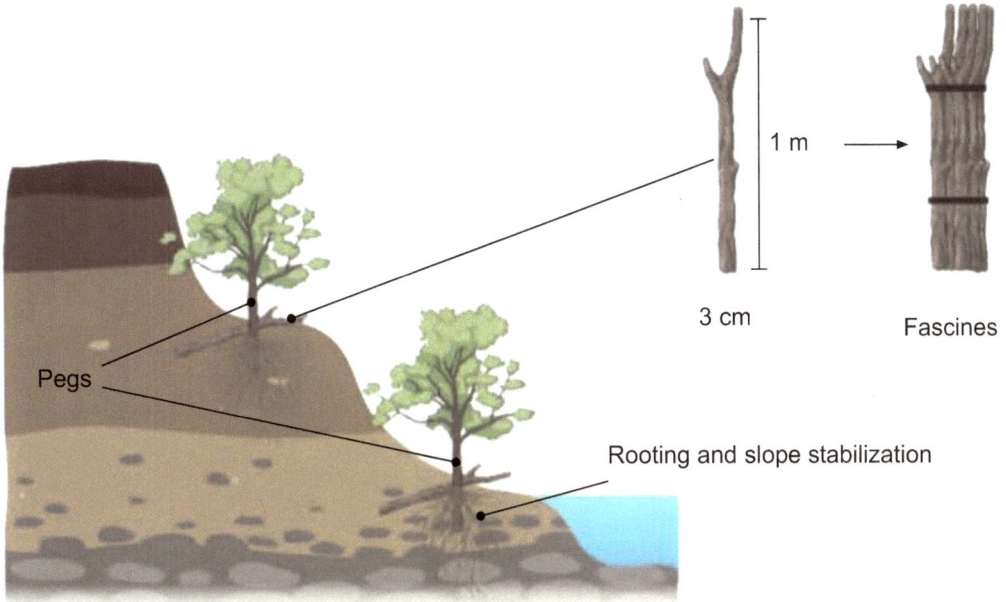

Fig. (4). Fascines with pegs in the slope scheme. Figure made with Biorender.

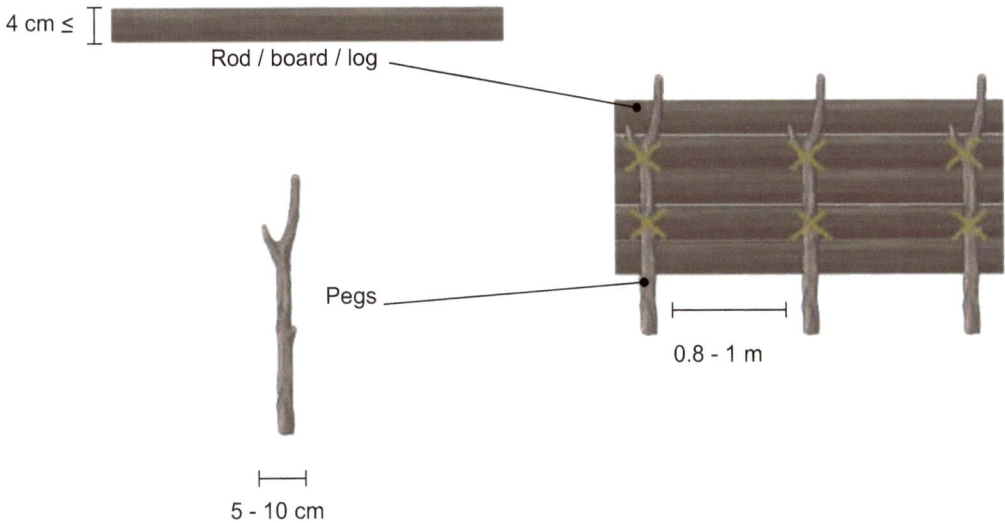

Fig. (5). General fence scheme. Figure made with Biorender.

To anchor these structures, pegs between 5 and 10 cm in diameter are driven at a depth of 40 cm directly into the ground vertically with spaces between 80 cm and 1 m. Rods of at least 4 cm in diameter are fixed horizontally to the stakes with wire tightly to avoid any gaps between them. The length of the fence must ensure a slope against erosion, and it should be placed in 5 m sections to prevent the structure from collapsing completely in case of any inconvenience.

Installation Process

1. Identify critical locations of abrupt slope changes in natural drainage.
2. Clean the area where the structure will be placed.
3. Excavate the bottom to level it.
4. Position pegs along the width of the drainage area.
5. Lay the first rod, board, or log horizontally and tie it securely to the pegs.
6. Continue placing horizontal structures until the desired height is reached, considering that for structures greater than 1 m in height, reinforcement of anchor pegs is required.
7. In the case of slopes, distribute the fences at intervals of 4 – 5 m downwards, depending on the slope angle.

Live Cribwall

These structures are made from logs or bamboo canes that function as scaffolding, which when filled with rocks and soil, function as a base for placing vegetation to help increase stability on the slope where it is placed (Fig. **6**).

The logs (of approximately 2 m) that make up the structure are placed in a crisscrossed manner to form the bases in which the living plant material will be included to increase the stability of the slope, bank, or embankment due to the deep rooting, so it is important to consider that the cuttings must have adequate size to be able to cross the structure and reach the wall of the slope, and it is worth mentioning that for its correct operation, it is required that the amount of soil to be stabilized is small.

The structure generates great stability for the slope in the long term, initially due to the same construction that functions as a support and later due to the additional vegetation it includes, but it should be noted that the effort and materials needed to place it are greater in comparison with previously discussed techniques, since it is required to excavate the surface where it will be placed, and nails or screws might be needed to assemble the structure properly.

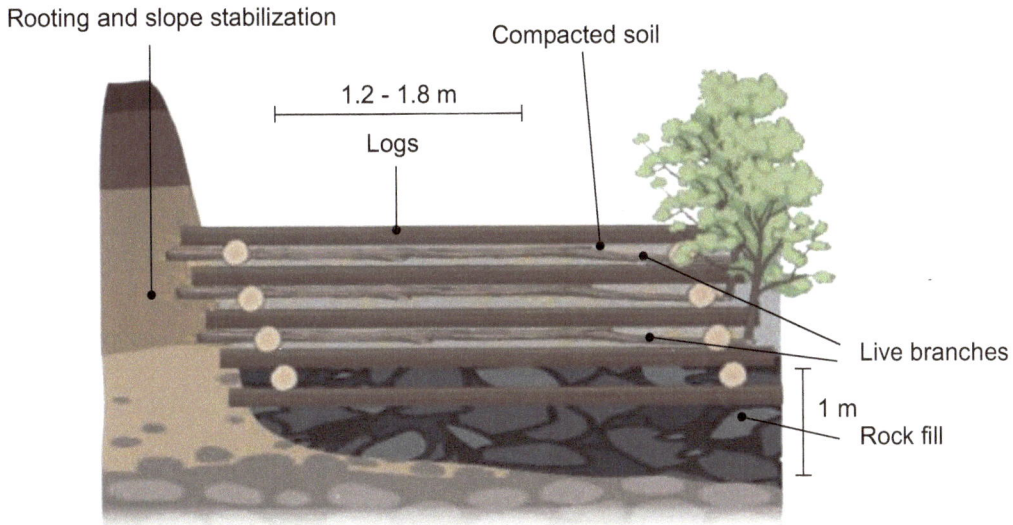

Rooting and slope stabilization

Compacted soil

1.2 - 1.8 m

Logs

Live branches

1 m

Rock fill

Fig. (6). Live cribwall scheme. Figure made with Biorender.

Installation Process

1. Clean the area where the structure will be placed.
2. Dig a trench 1 – 2 m wide and 1 m deep at the base of the slope or embankment.
3. Backfill the trench with rocks.
4. Place two rows of logs end to end on the rocks, marking the front and back of the wall.
5. Place smaller logs at the ends of the larger ones perpendicularly from front to back to form a box-like structure, leaving 15 cm free at each end. If necessary, you can join them using nails or screws.
6. Place live branches between the logs from the front to the back, ensuring that they exit at the front and reach the slope behind the structure.
7. Cover the cuttings with soil and rocks to fill the space between them and press down.
8. Repeat the above steps until the required height has been reached, moving 15 cm toward the slope at each level.

Jute Netting

There are techniques based on the use of a fibrous material of plant origin that serves as a substrate and anchoring point for the growth of plants on slopes with a very steep angle of inclination (35° - 80°) (Fig. **7**).

In this case, the fibers of the plant commonly known as jute (*Corchorus capsularis*) are used to form woven nets or mats that serve as a base for propagating different plant species, such as grasses, which can prevent landslides on steep slopes. These nets completely cover the slope they are intended to protect and are designed to prevent erosion and landslides on roads and highways.

Because it can be constructed with vegetable fiber, the cost of its production and installation is considerably low, but it compensates for the time required to weave the jute nets. Fortunately, it is possible to acquire it commercially.

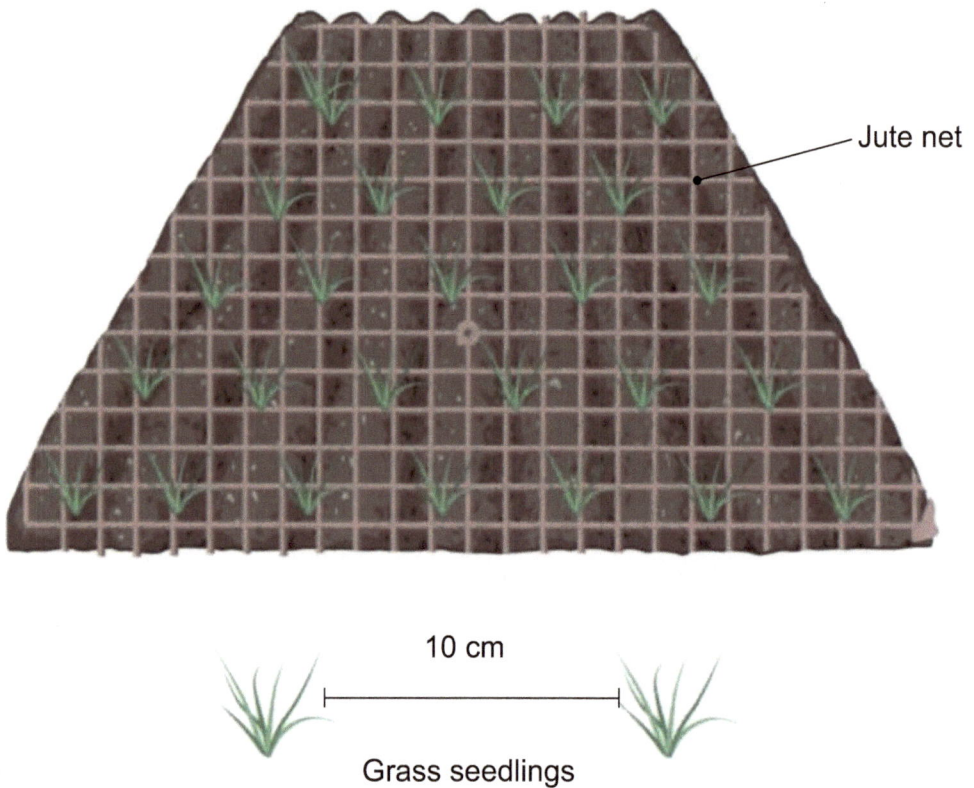

Fig. (7). Jute netting on a slope with grass seedlings. Figure made with Biorender.

Installation Process

1. Flatten the slope surface.
2. Place fertile soil on the flattened surface.
3. Cover with mulch.
4. Lay the net in straight vertical lines from the top, securing it to the ground using pegs with a spacing of 30 cm.

5. Unroll the net down the slope and secure it with live pegs at 0.5 – 1 m intervals.
6. Completely cover the surface of the slope by performing the above steps.
7. Sow the grass diagonally, using seeds or seedlings, through the fabric with a 10 cm spacing until the entire area is covered.
8. Provide constant maintenance and care until the grass grows properly.

Jute-mat Log

A variation of the above technique is to use jute rolls or "logs" to reinforce stream banks (Fig. **8**). These logs are constructed using jute or coconut fiber mats filled with hay and branches to give them the proper weight and density to protect the slope from stream erosion.

Their simple design makes them practical for easy field applications, and they can be adapted depending on the needs of the stream bank to be protected. Their size can vary from 30 cm to approximately 3 m, and if necessary, they can be stacked to cover a larger area. Besides being flexible, they can be easily adapted to the shape of natural drains.

Installation Process

1. Cut enough mat to cover the required length by 60 cm.
2. Spread out the blanket and cover it with a layer of straw, leaving 30 cm free at each end.
3. Place live cuttings of local deciduous vegetation at one end of the mat.
4. Fold the ends of the blanket inward over the straw.
5. Roll the blanket toward the cuttings and place them on the outermost layer of the log.
6. The log is secured by tying it in several sections to keep its shape.
7. Clean the area where the log will be placed, eliminating bulges and hollows.
8. Place the log starting from the downstream end and overlap the next one at a distance of 0.5 m. This should be done when the water level is average and the cuttings face the bank. It is preferable to use one larger log rather than several smaller ones for greater strength and stability.
9. Secure the log with wire approximately every 0.5 m, surrounding the log and securing it firmly to the bank with pegs.
10. Drive live pegs along the log to secure them and increase vegetative cover.

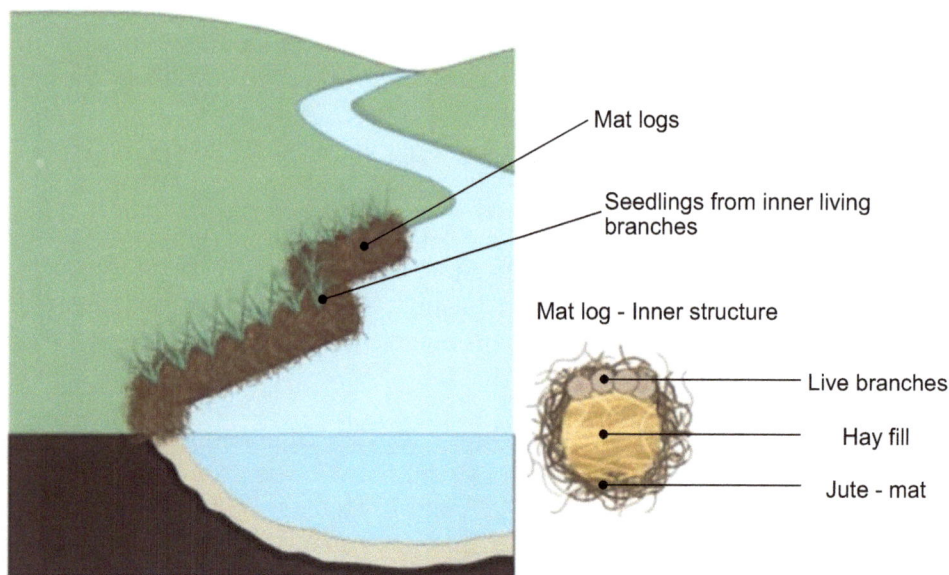

Fig. (8). A jute-mat log for the river bank scheme. Figure made with Biorender.

Brush Layering

This technique is similar to live cribwall but simpler, as it does not require the construction of rigid boxes to contain the live plant material (Fig. **9**). It is useful for road construction and securing embankments.

In this method, live branches or cuttings placed perpendicularly to the slope to be reinforced are used and secured by covering them with soil. Other materials, such as coconut or jute fiber trunks, can also be used. However, if a considerable height needs to be covered, soil is the best option to reduce costs.

This method prevents erosion of the slope by covering the surface of the slope and generating terraces reinforced with live branches, which will take root over time, giving greater firmness to the bank formed. Therefore, it is necessary to have branches of sufficient length (approximately 1.5 - 5 m) that can reach the slope through the bank and have a diameter of approximately 5 cm.

Installation Process

1. Clean the base of the slope before building the terraces
2. Excavate at the base to form 0.5 - 1 m long horizontal banks following the contour of the slope.
3. Tilt the branches slightly forward to form a groove at the base.

4. Place the live branches in the trench at the base of the bench by crisscrossing them, leaving about a quarter of them protruding, using approximately 25 branches per meter.
5. Cover the branches with compacted soil 5 – 10 cm high to secure them.
6. Excavate the upper section of the slope to form the next bench, where the branches of the next terrace will be placed.
7. Alternatively, cover the soil layer, material such as straw with seeds can be placed between the soil layers.
8. Construct the necessary terraces consecutively until the slope is completely covered.

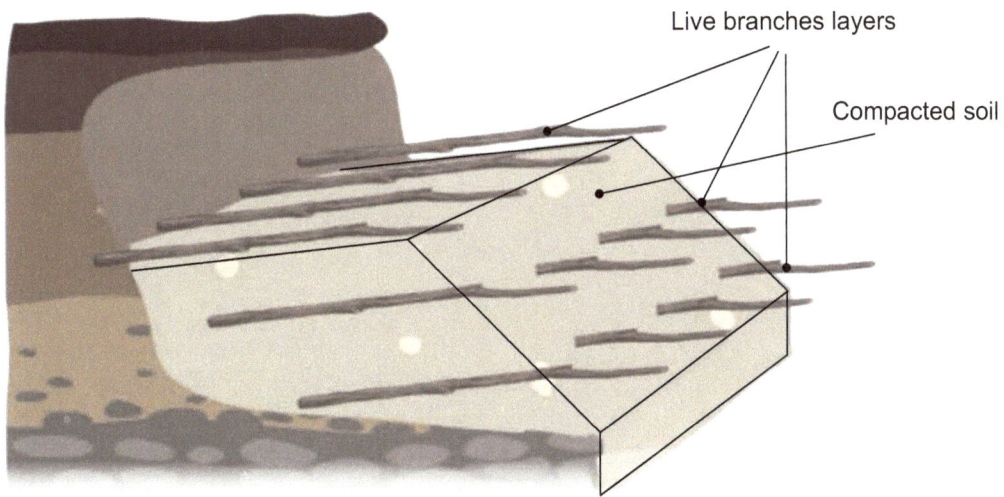

Fig. (9). Brush layering on the slope scheme. Figure made with Biorender.

Live Reinforced Earth Walls

There are ways to combine techniques to adapt them to the different needs of the terrain. In this case, live-reinforced earth walls combine fences and brush layering (Fig. **10**).

By combining these methods, it is possible to protect overhanging cuts or cavities. Using logs or reeds, a fence containing the soil is formed. Layers of live branches are interspersed with soil beds to form a reinforced wall.

In this way, a solid structure is formed with the help of the fence and the live branches of brush layering. Rooting helps generate the stability of the contained soil that fills the cavity and slope, covering the exposed area and avoiding both erosion and the fall of the overhanging cut.

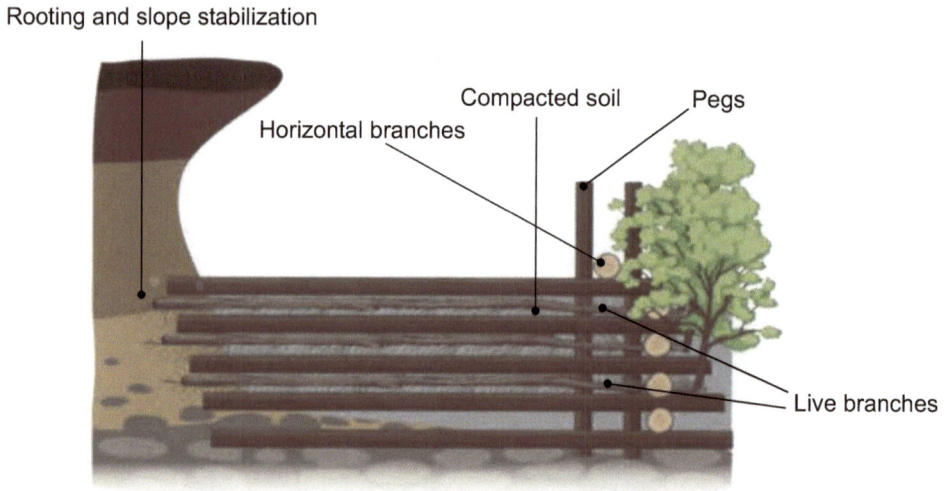

Fig. (10). Live-reinforced earth wall on an overhanging cut. Figure made with Biorender.

Installation Process

1. Clean the installation area.
2. Mark the sections where the pegs will be placed. Dig the holes where the pegs will be driven at 1 m intervals on the marked lines.
3. Place a rod, board, or log horizontally at the base of the fence, which will serve as the initial support point, and attach it to the pegs with rope or wire.
4. Fill the space formed at the bottom of the structure with soil.
5. Place the first row of branches on the soil bed, ensuring that they reach the slope wall, leaving a quarter free at the front end.
6. Cover the branches with compacted soil 5–10 cm high to secure them.
7. Place the next pegs and the corresponding horizontal branches to form the next terrace, reducing their distance from the wall slightly to shape the slope.
8. Continue adding levels until the overhanging cut or cavity is completely covered.

Considerations

The remarkable number of different bioengineering techniques available allows us to have a wide variety of options available to protect different types of surfaces, slopes, banks, or embankments from the erosion to which they may be exposed. However, careful consideration is needed to select the best option for each particular case [30].

The first consideration is the type of surface to be protected against erosion. While most bioengineering techniques for the prevention of soil erosion are applied to slopes and embankments, they are also applied to natural drains, such as rivers, to reduce the speed of the current, promote adequate sedimentation, and prevent water erosion that would occur naturally in these soils. Therefore, it is necessary to evaluate the particular needs of the area to be treated to choose the optimal method [31].

The availability of both living and inert materials, as well as tools or machinery in large-scale projects, to carry out the construction and installation of the different structures is also a critical factor in selecting a particular technique. It is essential to ensure that all necessary resources are available to implement the chosen technique effectively or, if not, to adapt it to the current conditions. This includes considering the capital available for investment [32].

Many of the bioengineering techniques are based on the formation of stable surfaces that allow the cultivation of plants with the capacity to generate dense roots. These roots are the elements that promote the securing of the soil. Therefore, it is important to select plant species that have the capacity to generate roots easily [33 - 36].

Ideally, the chosen species should be native to the area where the intervention is intended to be performed to avoid the entry of exotic or invasive species that may cause long-term damage instead of benefit. It is also necessary to consider the type of soil being protected to ensure that the selected vegetation will not have the opposite of the desired effect [37, 38]. Some plants suitable for these techniques are listed below (Table **1**):

Table 1. Plants useful for bioengineering techniques for soil erosion prevention [26, 39, 40].

Plant	Scientific Name	Resource Used
Willows	*Salix spp.*	Branches and cuttings
Poplars	*Populus spp.*	Branches and cuttings
Dogwoods	*Cornus spp.*	Branches and cuttings
Jute	*Corchorus capsularis*	Fiber
Coconut tree	*Cocos nucifera*	Fiber
Chestnut	*Castanea sativa*	Stakes
Banana	*Platanus spp.*	Branches and cuttings
Eleagnus	*Elaeagnus spp.*	Branches and cuttings
Spruce	*Pseudotsuga menziesii*	Logs and lumber
Black maple	*Acer nigrum*	Logs and lumber

(Table 1) cont.....

Plant	Scientific Name	Resource Used
Juniper	*Juniperus spp.*	Logs and lumber
Cedar	*Cedrus spp.*	Logs and lumber
Pine	*Pinus spp.*	Logs and lumber
Larch	*Larix spp.*	Logs and lumber
Beech	*Fagus sylvatica*	Logs and lumber

CONCLUDING REMARKS

The appropriate use of different bioengineering techniques is regarded as a highly efficient option for the prevention of soil erosion. By utilizing living plant materials and inert elements of the natural environment, these techniques help prevent the gradual deterioration of soil under different conditions. Additionally, they offer additional benefits, such as the formation of terraces that can be used for crops, the construction of banks and living walls to prevent flooding or outflows, and the promotion of the maintenance of native flora in localities.

AUTHORS' CONTRIBUTION

The authors confirm their contribution to the chapter as follows:

Documental research, writing, and edition: Iván E. Barrera; Design of figures and schemes: Paris O. Gonzalez

QUESTIONS RELATED TO THE TEXT

1. What are the main causes of soil erosion, as discussed in this chapter, and how do human activities exacerbate this process?
2. Describe the bioengineering technique known as the 'bush mattress.' What are its main components, and how is it installed?
3. What are the advantages and disadvantages of using wattle fences for soil erosion prevention on slopes?
4. Explain the 'log brush barrier' technique. What are its benefits, and in what situations is it most effective?
5. Identify the primary considerations for selecting the appropriate bioengineering technique for soil erosion prevention, as outlined in the chapter.
6. Discuss the role of plant selection in bioengineering techniques for soil erosion prevention. What are some recommended plant species mentioned in the chapter?

ACKNOWLEDGEMENT

This chapter is part of the productivity of Iván E. Barrera during his studies to obtain a Master in Sciences degree in the Posgrado en Ciencias Biológicas UNAM, and thanks the support of the Consejo Nacional de Humanidades, Ciencias y Tecnologías (CONAHCYT) through the scholarship granted whit CVU: 1249142.

REFERENCES

[1] Zachar D, Ed. Ed. Soil erosion. Amsterdam: Elsevier scientific publishing Company 1982.
[http://dx.doi.org/10.18172/cig.1094]

[2] Sadeghi SH, Raeisi MB, Hazbavi Z. Influence of freeze-only and freezing-thawing cycles on splash erosion. Int Soil Water Conserv Res 2018; 6(4): 275-9.
[http://dx.doi.org/10.1016/j.iswcr.2018.07.004]

[3] Marin-Diaz B, Govers LL, van der Wal D, Olff H, Bouma T. The importance of marshes providing soil stabilization to resist fast-low erosion in case of dike breach Ecological aplications 2022; 32: e2622.
[http://dx.doi.org/10.1002/eap.2622]

[4] Tuo D, Lu Q, Wu B, *et al.* Effects of wind-water erosion and topographic factor on soil properties in the Loess Hilly region of China. Plants 2023; 12(13): 2568.
[http://dx.doi.org/10.3390/plants12132568] [PMID: 37447129]

[5] Liu W, Chen Z, Rong L, *et al.* Soil erosion thickness and seasonal variations together drive soil nitrogen dynamics at the aerly stage of vegetation restoration in the dry-hot valley. Microorganisms 2024; 12(8): 1546.
[http://dx.doi.org/10.3390/microorganisms12081546] [PMID: 39203388]

[6] Brunel N, Seguel O. Effects of erosion on soil properties. Agro Sur 2011; 39(1): 1-12.
[http://dx.doi.org/10.4206/agrosur.2011.v39n1-01]

[7] Jiao J, Han L, Jia Y, Lei D, Wang Nm Li L. Seed morphology characteristics in relation to seed loss by water erosion in the Loess Plateau SpringerPlus 2013; 2)1: 9.
[http://dx.doi.org/10.1186/2193-1801-2-S1-S9]

[8] Li J, Zhou Y, Li Q, Yi S, Peng L. Exploring the effects of land use changes on the landscape pattern and soil erosion of western Hubei province from 2000 to 2020. Int J Environ Res Public Health 2022; 19(3): 1571.
[http://dx.doi.org/10.3390/ijerph19031571] [PMID: 35162595]

[9] Zhou C, Xia H, Yang T, Zhang Z, Zheng G. Grassland degradation affected vegetation carbon density but not soil carbon density. BMC Plant Biol 2024; 24(1): 719.
[http://dx.doi.org/10.1186/s12870-024-05409-6] [PMID: 39069617]

[10] Hasim G, Coughlan K, Syers J, *et al.* On-site nutrient depletion: an effect and a cause of soil eroision.Soil erosion at multiple scales: principles and methods for assessing causes and impacts. New York: IBSRAM 1998; pp. 207-21.

[11] Toy TJ, Foster GR, Renard KG, Eds. Soil Erosion: Processes, Prediction Measurement, and Control. New York: John Wiley & Sons, Inc. 2002.

[12] Camas R, Turrent A, Cortes JI, *et al.* Soil erosion, runoff and nitrogen and phosphorus loss on slopes under different management systems in Chiapas, Mexico. Rev Mex Cienc Agric 2012; 3(2): 231-43.
[http://dx.doi.org/10.29312/remexca.v3i2.1459]

[13] Steinhoff-Knopp B, Kuhn TK, Burkhard B. The impact of soil erosion on soil-related ecosystem services: development and testing a scenario-based assessment approach. Environ Monit Assess 2021;

193(S1) (Suppl. 1): 274.
[http://dx.doi.org/10.1007/s10661-020-08814-0] [PMID: 33988744]

[14]   Suresh B, Vinay D. Soil Bioengineerind to deal with soil erosion and landslides in developing nations. Technoarete Transactions on Recent Research in Applied Microbiology and Biotechnology 2022; 1(2): 7-13.

[15]   Borrelli P, Robinson DA, Fleischer LR, *et al.* An assessment of the global impact of 21st century land use change on soil erosion. Nat Commun 2017; 8(1): 2013.
[http://dx.doi.org/10.1038/s41467-017-02142-7] [PMID: 29222506]

[16]   Borrelli P, Alewell C, Alvarez P, *et al.* Soil erosion modelling: A global review and statistical analysis. Sci Total Environ 2021; 780: 146494.
[http://dx.doi.org/10.1016/j.scitotenv.2021.146494] [PMID: 33773346]

[17]   Saadon A, Abdullah J, Mohd Yassin I, Muhammad NS, Ariffin J. Nonlinear multi independent variables in quantifying river bank erosion using Neural Network AutoRegressive eXogenous (NNARX) model. Heliyon 2024; 10(4): e26252.
[http://dx.doi.org/10.1016/j.heliyon.2024.e26252] [PMID: 38404813]

[18]   Panagos P, Borrelli P, Poesen J, *et al.* The new assessment of soil loss by water erosion in Europe. Environ Sci Policy 2015; 54: 438-47.
[http://dx.doi.org/10.1016/j.envsci.2015.08.012]

[19]   Omuto C, Vargas R. Soil Loss Atlas of Malawi. Rome: Food an Agriculture Organization of the United Nations 2019.

[20]   Aviles D, Wesström I, Joel A. Effect of vegetation removal on soil erosion and bank stability in agricultural drainage ditches. Land (Basel) 2020; 9(11): 441.
[http://dx.doi.org/10.3390/land9110441]

[21]   Qiu L, Zhang Q, Zhu H, *et al.* Erosion reduces soil microbial diversity, network complexity and multifunctionality. ISME J 2021; 15(8): 2474-89.
[http://dx.doi.org/10.1038/s41396-021-00913-1] [PMID: 33712698]

[22]   Tesfahunegn GB, Ayuk ET, Adiku SGK. Farmers' perception on soil erosion in Ghana: Implication for developing sustainable soil management strategy. PLoS One 2021; 16(3): e0242444.
[http://dx.doi.org/10.1371/journal.pone.0242444] [PMID: 33651832]

[23]   Bezak N, Borrelli P, Mikoš M, Jemec Auflič M, Panagos P. Towards multi-model soil erosion modelling: An evaluation of the erosion potential method (EPM) for global soil erosion assessments. Catena 2024; 234: 107596.
[http://dx.doi.org/10.1016/j.catena.2023.107596]

[24]   Stachew E, Houette T, Gruber P. Root systems research for bioinspired resilient design: A concept framework for foundation and coastal engineering Frontiers in robotics and AI 202(8): 548444.
[http://dx.doi.org/10.3389/frobt.2021.548444]

[25]   Mmbando GS, Ngongolo K. The recent genetic modification techniques for improve soil conservation, nutrient uptake and utilization. GM Crops & foodn 2024; 15(1): 233-247.
[http://dx.doi.org/10.1080/21645698.2024.2377408]

[26]   Georgi N, Stathakopoulos JE. Bioengineering techniques for soil erosion protection and slope stabilization. Proceeding of the 46th Congress of the European Regional Science Association; 2006 Ago 30 – Sep 03; Volos, Greece Researchgate.

[27]   Martínez J, Negrete F, Benavides J, Eds., *et al.* Prácticas para prevención y control de erosion en suelos de ladera. Colombia: Gráficas del Caribe 2012.
[http://dx.doi.org/10.21930/agrosavia.manual.2012.2]

[28]   Shrestha AB. GC E, Adhikary RP, Rai SK Eds. Resource manual on flash flood risk management - Module 3: Structural measures. Kathmandu: ICIMOD. 2012.

[29] Girona-García A, Cretella C, Fernández C, Robichaud PR, Vieira DCS, Keizer JJ. How much does it cost to mitigate soil erosion after wildfires? J Environ Manage 2023; 334: 117478.
[http://dx.doi.org/10.1016/j.jenvman.2023.117478] [PMID: 36796191]

[30] Carrasco GO, Vargas MD, Vargas EJ. How bioengineering influences slope stabilization and water erosion mitigation? Perfiles ingenieria 2023.
[http://dx.doi.org/10-.31381/perfilesingenieria.v19i20.6313]

[31] Wei H, Zhao W, Wang H. Effects of vegetation restoration on soil erosion on the Loess Plateau: A case study in the Ansai watershed. Int J Environ Res Public Health 2021; 18(12): 6266.
[http://dx.doi.org/10.3390/ijerph18126266] [PMID: 34200518]

[32] Istanbuly MN, Krása J, Jabbarian Amiri B. How socio-economic drive4rs explain landscape soil erosion regulation services in Polish catchemnts. Int J Environ Res Public Health 2022; 19(4): 2372.
[http://dx.doi.org/10.3390/ijerph19042372] [PMID: 35206558]

[33] Ford H, Garbutt A, Ladd C, Malarkey J, Skov MW. Soil stabilization linked to plant diversity and environmental context in coastal wetlands Journal of vegetation Science 2016. 27_ 259-268.
[http://dx.doi.org/10.1111/jvs.12367]

[34] Dorairaj D, Osman N. Present practives and emerging opportunities in bioengineering for slope stabilization in Malaysia: An overview. PeerJ 2020; •••
[http://dx.doi.org/10.7717/peerj.10477] [PMID: 33520435]

[35] Chen S, Yao F, Mi G, Wang L, Wu H, Wang Y. Crop rotation increases root biomass and promotes the correlation of soil dissolved carbon with the microbial community in the rhizosphere. Front Bioeng Biotechnol 2022; 10: 1081647.
[http://dx.doi.org/10.3389/fbioe.2022.1081647] [PMID: 36561045]

[36] Cao T, Zhang H, Chen T, et al. Research on the mechanism of plant root protection for soil slope stability. PLoS One 2023; 18(11): e0293661.
[http://dx.doi.org/10.1371/journal.pone.0293661] [PMID: 38011254]

[37] Ghestem M, Cao K, Ma W, et al. A framework for identifying plant species to be used as 'ecological engineers' for fixing soil on unstable slopes. PLoS One 2014; 9(8): e95876.
[http://dx.doi.org/10.1371/journal.pone.0095876] [PMID: 25105571]

[38] Dorairaj D, Suradi MF, Mansor NS, Osman N. Root architecture, rooting profiles and physiological responses of potential slope plants grown on acidic soil. PeerJ 2020; 8: e9595.
[http://dx.doi.org/10.7717/peerj.9595] [PMID: 32904129]

[39] Eubanks E, Meadows D. Soil Bioengineering Techniques.A Soil Bioengineering Guide for Streambank and Laskeshore Stabilization Minessota: US Department of Agriculture Forest Service Technology & Develoipment Center. 2002; pp. 74-131.

[40] Polster DF. Soil Bioengineering for slope Stabilization and Site Restoration. In: Spiers GA, Beckett P, Conroy H, Eds. Proceedings of Subdury Mining and the Enviroment Conference Proceedings. 2003 May 25-28; 2003.

CHAPTER 6

Biofertilizers and Biopesticides: Sustainable Alternatives for Agriculture

Rafael Torres-Martínez[1], Yesica R. Cruz-Martinez[2], Ana K. Villagómez-Guzmán[3], Olivia Pérez-Valera[2], Héctor M. Arreaga-González[4] and Tzasna Hernández-Delgado[3,*]

[1] *Chemical Ecology and Agroecology Laboratory, Research Institute for Ecosystems and Sustainability, National Autonomous University of Mexico (UNAM), Morelia, Michoacan, Mexico*

[2] *Institute of Chemistry, National Autonomous University of Mexico (UNAM), Mexico City, Mexico*

[3] *Natural Products Bioactivity Laboratory, UBIPRO, Superior Studies Faculty (FES)-Iztacala, National Autonomous University of Mexico (UNAM), Tlalnepantla de Baz, Mexico State, Mexico*

[4] *Biotechnology Laboratory, Institute of Agroindustries, Technological University of Mixteca, Heroic City of Huajuapan de Leon, Oaxaca, Mexico*

Abstract: The escalating requirements for agricultural production systems to ensure global food security and mitigate environmental degradation necessitate a shift toward more sustainable approaches that reduce adverse effects and increase efficiency in crop productivity and profitability. Historically, the primary method used to achieve these goals has involved the application of chemical fertilizers and pesticides. Nevertheless, the persistent and excessive use of these substances has resulted in contamination, pest resistance, health issues, soil depletion, and diminished microbiota, consequently reducing crop yields. Therefore, the controlled use of pesticides and fertilizers has been recommended, advocating for reduced application amounts and site-specific, targeted administration. One promising solution lies in the use of advanced tools, such as biotechnology and nanotechnology, that have played an important role in agro-technological transformation. Microorganisms, along with biofertilizers and biopesticides, have the potential to enhance agricultural systems and safeguard food security. Nanoparticles are emerging as a cutting-edge technology poised for revolutionizing contemporary agrarian methodologies, balancing crop nutrients, and the supply of pesticides and fertilizers. Diverse nanoparticle-based formulations, including biofertilizers, biopesticides, and nanosized sensors, have been extensively researched for plant health management and soil quality improvement. A profound understanding of the interactions between plants and nanomaterials enhances agricultural techniques by monitoring water quality, improving disease resilience, crop output, pest control, and nutrient absorption. This examination underscores the pivotal

---

* **Corresponding author Tzasna Hernández-Delgado:** Natural Products Bioactivity Laboratory, UBIPRO, Superior Studies Faculty (FES)-Iztacala, National Autonomous University of Mexico (UNAM), Tlalnepantla de Baz, Mexico State, Mexico; Tel: +525527288013; E-mail: tzasna@unam.mx

Israel Valencia Quiroz (Ed.)

factors that must be considered in future studies on biofertilizers and biopesticides to enhance productivity and food security.

Keywords: Biofertilizer, Biopesticide, Biological control, Microorganisms, Formulations, Nano-biotechnology, Plant health, Sustainable agriculture.

INTRODUCTION

Environmental sustainability in agriculture refers to the prudent management of natural systems and resources essential for farm operations. Sustainable practices include crop rotation, promoting biodiversity, using cover crops, implementing zero- or reduced-tillage systems, adopting integrated pest management strategies, promoting synergy between livestock and crops, participating in agroforestry techniques, and using precision agricultural methods. The main objective of sustainable agricultural policies is to guarantee environmental sustainability and, at the same time, improve, or at least sustain, agricultural productivity because this represents one of the cornerstones of society and plays a fundamental role in supporting the food of millions of humans globally.

It is estimated that by 2050, the world population will be approximately 9 billion, increasing the demand for land dedicated to agriculture. Currently, approximately 5.5 billion hectares, or 38% of the world's land surface, are allocated to agriculture, of which approximately one-third is allocated to cropland. The global area of cropland per capita decreased continuously during the period from 1961 to 2016: from approximately 0.45 hectares per capita in 1961 to 0.21 hectares per capita in 2016 [1 - 3]. While, the use of fertilizers and pesticides shows an increasing trend, which has allowed, on the one hand, an increase in crop yields, having a positive impact on the producer economies. However, even though global food production has managed to keep pace with population growth, the 2024 report by Food Security Information Network (FSIN) and the Global Network Against Food Crises (UN) on Food Crises [4] reveals that 281.6 million people, equivalent to 21.5% of the population examined, are experiencing high levels of acute food insecurity in 59 countries/territories [5].

On the other hand, the extensive use of fertilizers and pesticides poses significant challenges, including soil contamination and degradation, reduced water oxygen levels, lake eutrophication, loss of species of native plants, animals, and microorganisms, imbalance of soil nutrients, and the accumulation of heavy metals that can be incorporated into the food chain, representing health risks and large amounts of pollution [3, 6, 7]. The above principles are essential for developing environmentally friendly biofertilizers and biopesticides while preserving the benefits of their chemical analogs [8]. Biotechnology plays a fundamental role in developing products based on microorganisms that produce secondary metabolites capable of eliminating pests or diseases in crops and compounds that favor the fixation of essential elements for the nutrition, development, and protection of plant species through symbiotic relationships. Despite the benefits of biotechnology, it is necessary to make responsible use of biofertilizers and biopesticides as improper management can lead to similar negative outcomes as those caused by chemical fertilizers and pesticides. It is essential, therefore, to generate legal frameworks regarding the selection of strains to be used, to increase quality in production, storage, and distribution, and to generate awareness among the population about their benefits, as well as regarding their application. These products are often specific to different soil types and crops, making informed and regulated use crucial for sustainable agricultural practices.

BIOFERTILIZERS

Classification and Mechanism of Action of Biofertilizers

Biofertilizers can be categorized based on various factors, including their mechanism of action and composition. Direct mechanisms include N2 fixers, P solubilizers, and phytohormone producers. In contrast, indirect mechanisms involve substances that facilitate siderophore production, induce systemic resistance, modulate plant stress, generate antibiotics, and stimulate the synthesis of glucanase and chitinase [9]. In terms of composition, biofertilizers are available in solid forms like granules, microgranules, hydratable powders, water-dispersible granules, and shots, as well as in liquid forms such as suspension concentrates, oil-miscible concentrated fluids, sprays, and oil dispersions [10].

The first commercial use of biofertilizers occurred with the introduction of "Nitragin" in 1895. Preparing biofertilizers involves the mass cultivation of microbes under controlled conditions, such as carefully regulating factors such as pH, temperature, and cell count. An appropriate carrier material is essential for ensuring optimal biofertilizer effectiveness. In the future, emphasis should be placed on developing mutant and genetically modified microbes as biofertilizers that offer superior benefits compared with wild-type microbes, thereby providing a sustainable solution to agricultural challenges [11].

Microbial biofertilizers enhance plant growth either by directly acquiring essential resources like growth hormones or indirectly by implementing regulatory mechanisms on various plant pathogens that act as biocontrol agents. The mechanism of action can be categorized into two types. Direct methods involve microorganisms aiding in nitrogen and phosphorus fixation, potassium solubilization, siderophore production, phytohormone modulation, and phytoprotection. Indirect methods involve microorganisms acting in the plant rhizosphere through biocontrol, exclusion mediation, systemic resistance, and antifungal compound production [3].

Advantages and Drawbacks of Biofertilizers Compared to Chemical Fertilizers

Biofertilizers are formulations consisting of beneficial microorganisms, including fungi, bacteria, algae, and their products. These are used in agriculture as inocula, either solid or liquid. Their primary functions revolve around preserving soil structure, colonizing rhizospheres and internal plant tissues, and fixing and solubilizing organic and inorganic compounds. This process involves the capture and conversion of nutrient-containing compounds [8, 12], facilitating nutrient distribution, and restoring nutrient balance in the soil. Additionally, they are involved in the formation of biopolymers and the production of secondary metabolites like growth hormones and antibiotic compounds that play a direct role in plant defense against pests and diseases [9, 13, 14]. Furthermore, they produce compounds that indirectly stimulate the secondary metabolism of plants, facilitating their adaptive response to environmental stresses, both chemical and biological, thereby enhancing crop fertility and yield. Finally, biofertilizer reduces the need for chemical fertilizers, consequently preventing the accumulation of chemical compounds and pollution (Fig. **1**) [15 - 17].

Despite the advantages of biofertilizers, it is imperative to note that their production necessitates a significant amount of water and chemical reagents for reproduction and development. Therefore, exploring alternative methods to mitigate these challenges is crucial. A practical approach involves using waste materials from the livestock industry, residues from primary food processing, and residual biomass from microorganisms that yield primary metabolites like lipids, proteins, and carbohydrates. These waste materials offer a rich reservoir of nutrients that can be effectively used for biofertilizer production [12]. The circular economy concept is also gaining prominence as a sustainable solution for reducing environmental impact and production costs. This approach involves the co-production of food, azomethane, and biofertilizers through the anaerobic digestion of agricultural raw materials, promoting resource efficiency and minimizing waste [18].

Another crucial aspect to consider is the specific production conditions required for biofertilizers, which involve well-trained personnel, appropriate storage conditions, and effective field application. Biofertilizers generally have a shorter shelf life than chemical fertilizers. Additionally, microorganisms must undergo a transitional phase to adapt to their new environment, which includes survival, acclimatization, competition with other microorganisms, and interaction with plants. Meanwhile, abiotic factors such as light, pH, and temperature also play crucial roles. These factors should be carefully considered when selecting biofertilizers [7].

Soil composition is shaped by geological processes that accumulate various materials in different geographical regions worldwide. This has contributed to the diverse microbiological environment of soils. This diversity encourages the dominance of well-adapted microorganisms and results in the formation of soils with varying richness levels. These factors directly influence soil productivity and nutrient availability [19].

Although the biofertilizer industry is experiencing growth, establishing regulatory frameworks and conducting comprehensive scientific research are essential to gaining a more profound understanding of microbiology across diverse geographical locations. Using biofertilizers containing non-native microorganisms may lead to competition with or displacement of native species, potentially resulting in the negative consequences of chemical fertilizers [17].

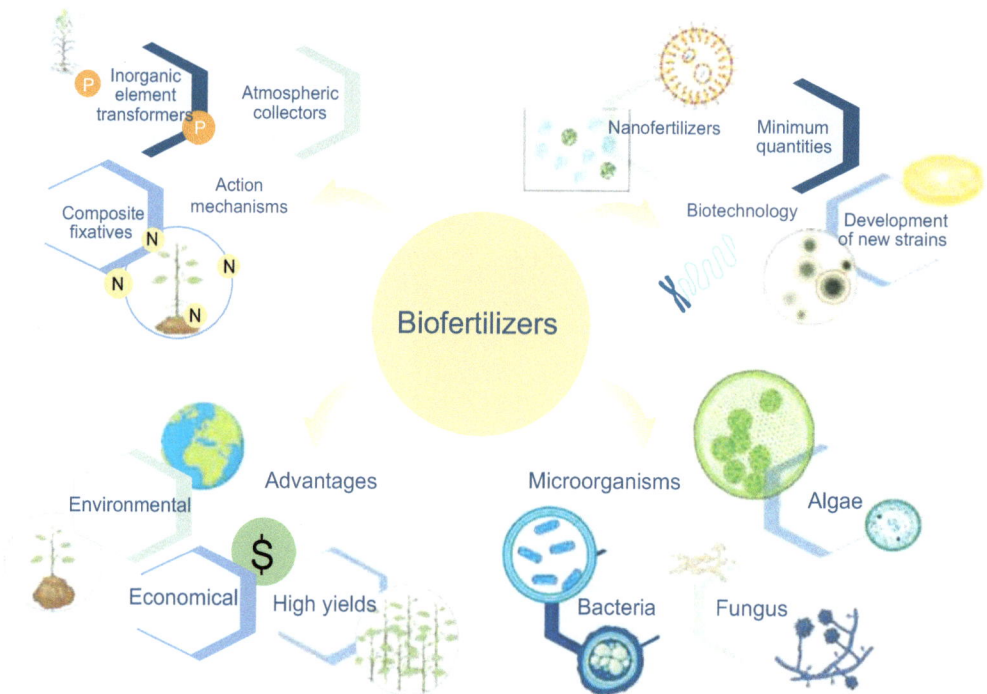

Fig. (1). Classification of Biofertilizers. Figure made with Biorender.

Biofertilizer Market, Development Costs, and Input Economic Impact

The biofertilizer market experienced substantial growth, reaching 1.06 billion USD in 2016, 2.1 billion USD in 2022, and 3.27 billion USD in 2024. It is projected to grow to 5.23–6.24 trillion dollars by 2029, reflecting an annual growth rate of 9.85% (Table 1) [8, 14, 18, 20]. Mycorrhizae, known as arbuscular mycorrhizal fungi, are widely used biofertilizers that enhance crop yields and plant surface areas and establish symbiotic relationships with plant roots, particularly in legumes. These biofertilizers are dominant in global markets, accounting for 74.6% of consumption in 2022, with a market value of 995.3 million dollars and a volume of 96.6 thousand tons.

Table 1. Estimation of the biofertilizer market.

Biofertilizer Market	
Study period	2017-2029
Market size (2024)	$3.27 billion
Market size (2029)	$5.23 billion
Greater participation	Mycorrhizas
Annual growth	(2024-2029) 9.85%
Highest proportion by region	North America

North America and Europe are the primary consumers of biofertilizers globally. In 2022, North America accounted for 35.9% of the global market, with the organic farming area in this region expanding by 13.5% from 1.4 million hectares in 2017 to 1.6 million hectares in 2021. Conversely, Europe accounted for 31.9% in 2022 and boasted the world's largest organic farming area, covering 6.5 million hectares in 2021. The Asia-Pacific region, particularly China and India, is a crucial producer of organic crops like rice, sugar cane, and fruits and vegetables. Although organic farming remains a small segment of agricultural land in these regions, the increase in organic farming from 2017 to 2021 is expected to boost the biofertilizer market. The growing global demand for organic products, particularly in Europe and North America, is likely to fuel the biofertilizer market during the forecast period [20].

The anticipated technological advancements in the economy are positioned to open new investment opportunities, tackling issues like resource depletion, food security, and climate change. The increasing worldwide need for sustainable agribusiness offers encouraging investment prospects in agro-industrial facilities that manufacture bioinputs like biofertilizers and biopesticides from microorganisms such as bacteria and fungi [21].

Biotechnology in the Development of Biofertilizers

Biotechnology has played an active role in the development of novel fertilization methodologies. Biofertilizers have emerged as a feasible option to address soil contamination and degradation issues in crop cultivation, aiming to reduce the environmental impact of agricultural practices and products. These biofertilizers contain live microorganisms that live in the rhizospheres or internal plant tissues. Upon application to seeds, plant surfaces, or soil, they promote plant growth by enhancing the availability of vital nutrients in it.

The typical microorganisms used as biofertilizers in agriculture include mycorrhizae, *Azospirillum spp.*, *Azotobacter spp.*, *Rhizobium spp.*, and phosphate-solubilizing bacteria. Diverse categories of microbes, such as rhizosphere, endophytic, and epiphytic, linked to plants exhibit characteristics that promote plant growth (PGP), establishing them as a valuable and promising resource for sustainable agriculture. These PGP microbes can enhance plant growth either directly or indirectly through various mechanisms, like the secretion of growth regulators, solubilization of crucial nutrients like phosphorus, potassium, and Zn, biological nitrogen fixation, and the production of siderophore, ammonia, HCN, and other secondary metabolites that can act antagonistically against pathogenic microorganisms [22].

Microorganisms that produce biofertilizers must adhere to specific selection criteria, which will be elucidated in the subsequent chapter, based on their distinct functions and requirements aligned with the crop type. Among the nitrogen-fixing microorganisms, several key bacteria are commonly used, including *Rhizobium meliloti*, *Bradyrhizobium sp.*, *Herbaspirillum spp.*, *Enterobacter cloacae*, *Rhizobium japonicum*, *Streptomyces griseoflavus*, *Kosakania sp.* KB117, *Lysinibacillus sphaericus*, *Pantoea agglomerans*, *Bacillus mojavensis*, *Stenotrophomonas maltophilia*, and *Azospirillum brasilense*, and cianobacterias: *Aulosira fertilissimus*, *Anabaena sphaerica*, *Nostoc hatei*, *Cylindrospermum majus*, and *Westiellopsis prolífica*, *Anabaena variabilis*, *Nostoc sp.*, *Nostoc entophytum*, *Oscillatoria angustissima*, and *Calothrix sp.*

In terms of phosphorus-solubilizing organisms, bacteria like *Rhizobium leguminosarum*, *Bacillus spp.*, *Pseudomonas spp.*, *Chryseobacterium spp.*, *Burkholderia spp.*, *Athrobacter spp.*, *Acinetobacter spp.*, *Enterobacter cloacae*, and *Erwinia,* play a significant role, along with cyanobacteria like *Anabaena doliolum*, *Cylindrospermum sphaerica*, and *Nostoc calcicola* and fungi like *Glomus versiforme*, *Glomus mosseae*, and *Glomus etunicatum*. Among the potassium-fixing organisms are bacteria such as *Acidothiobacillus ferrooxidans*, *Alcaligenes faecalis*, and *Pantoea agglomerans* [7, 8].

Corn, wheat, and rice are among the most cultivated cereals globally. Corn production has reached 1.24 million metric tons worldwide. The application of *Pseudomonas spp.*, *Saccharomyces spp.*, *Bacillus subtilis*, and *Lactobacillus spp.* resulted in increased grain yields from 4.45 to 8.60 tons per hectare [8].

Biofertilizers are developed from microorganisms and can be categorized based on the level of risk. Type 1 biofertilizers are considered low-risk, posing minimal to no threat to individuals or populations in causing diseases. Conversely, groups 2, 3, and 4 consisted of microorganisms with varying degrees of risk, ranging from moderate to high. The microorganisms in groups 3 and 4 can induce invasion and competitive reactions, leading to the production of antibiotic compounds that harm the natural microbiota and facilitate horizontal gene transfer. Hence, carefully selecting microbial strains intended for use as biofertilizers is crucial when developing new agrochemical products.

Formulation and Biofertilizer Application

Biofertilizers consist of living organisms, or dormant inocula, that are directly deposited at the intended application site. These formulations are specifically tailored to the soil type, application requirements, and specific plant species for which they are designed. Incorporating a carrier substance is crucial; it must possess inertness, non-toxicity, organic nature, affordability, and user-friendliness. It is imperative that the carrier effectively adheres to seeds, retains moisture, and facilitates the activity of the bacterial cells responsible for biofertilizer function. Common solid carriers include mineral charcoal, farmyard manure, vermiculite, and mud [3, 14].

The formulation of biofertilizers begins with the selection of the appropriate microbial strains. These microorganisms require specific handling conditions, and formulations are typically sensitive to environmental factors like temperature and storage. Therefore, novel formulations based on microorganism immobilization have been investigated.

Biofilms represent an innovative advancement based on biofilm formation. Different microorganisms can create these structures near plant tissues, serving as a shield against environmental factors and other competing microorganisms. In this way, they generate secondary metabolites that govern plant growth, serving as a valuable source of biofertilizers [14].

Microbial consortia constitute a significant portion of the biofertilizers currently available on the market, typically composed of isolated strains. Consequently, it is plausible to develop biofertilizers containing compatible strains that offer synergistic advantages. These strains may possess different functions but are ultimately beneficial [23].

Encapsulation is a prominent technique in which a wall material envelops microbial cells during the formulation process. Generally, natural polymeric materials are used to create microcapsules, with the capsule diameter determining the type of wall material employed. This wall material is susceptible to temperature and pH variations; thus, upon its introduction into the soil, the microcapsule layer ruptures, releasing its contents at the target site [14].

Nanofertilizers are essential in agriculture. Nanoparticles exhibit a size range of 1–100 nm, providing distinct advantages over alternative delivery methods. These particles enable the targeted and precise distribution of vital nutrients to plants within a specific area, ensuring slow and efficient dispersion [23].

The primary objective of bionanofertilizers is to trap these substances within a matrix of polymers, mainly derived from lipids and proteins. Depending on the type of wall material, different structures can be formed through mechanisms like self-assembly, micelles, liposomes, nanotubes, particles, nanospheres, polymeric nanoparticles, dendrimer, micelles, inverted fibers, microgels, and aerogels. These structures exhibit remarkable resistance to environmental conditions such as temperature, pH, humidity, and enzymatic activity, facilitating the controlled release of bioactive components at specific sites of action [7, 24 - 26].

Inorganic nanofertilizers, also known as nanobiofertilizers, have emerged as a financially feasible and eco-friendly agricultural solution that offers versatility and durability. The process involves the green synthesis of microbes, encapsulating inorganic nanoparticles like Si, Zn, Cu, Fe, Ni, Ti, and Ag, and organic materials such as chitosan, cellulose, and starch to develop nanobiofertilizers. This approach helps address pollution concerns associated with traditional fertilizers and ensures sustainable agricultural practices [7].

Biofertilizers can be applied through various methods. One common method is seed treatment, which involves immersion in a pre-dissolved biofertilizer solution followed by drying, emphasizing prompt sowing of seeds to prevent pathogenic microorganism contamination. Alternatively, the biofertilizer solution can be applied to a seedbed, where the seedling roots are submerged in the solution before planting. This technique is widely used for extensive vegetable, cereal, and flower crops. Both solid and liquid biofertilizers are commonly applied through diverse methods, including direct and scheduled application in crops [3, 6].

Microbial biofertilizers enhance plant growth by directly acquiring essential resources like growth hormones or indirectly by implementing regulatory mechanisms on various plant pathogens that act as biocontrol agents. The mechanism of action can be categorized into direct methods, which involve microorganisms aiding in nitrogen and phosphorus fixation, potassium solubilization, siderophore production, phytohormone modulation, and phytoprotection, and indirect methods, in which microorganisms act in the plant rhizosphere through biocontrol, exclusion mediation, systemic resistance, and antifungal compound production [3]. Table **2** shows the main biofertilizer formulations produced by companies worldwide whose active agents come from different origins.

Table 2. Companies and Trends in the Market of Biofertilizers and Bioplaguicides.

Market Trends	Active agent/Trade Name	Leading Companies
Biofertilizers	*Azospirillum brasilense* / AZOSPIRRILLUM-GREEN® *Azotobacter sp.* / Bio-Organik Azotobacter® Arbuscular Mycorrhizal Fungi (AMF) / Arpha-gold® Phosphate Solubilizing Bacteria / Green Growth® *Rhizobium spp.* / Katyayani Rhizobium®	Gujarat State Fertilizers & Chemicals Ltd Indian Farmers Fertilizer Cooperative Ltd Koppert Biological Systems Inc. Symborg Inc. T. Stanes and Company Ltd
Bioplaguicides	*Bacillus thuringiensis* / DiPel® *Bacillus subtilis* / FUNGISEI® Azadiractin (Neem oil) / Aza-Direct® *Beauveria bassiana* / BOTANIGARD® *Reynoutria sachalinensis* extract / REGALIA® *Cry1Ac* gene / 5345® *SfMNPV* gene / Spodovir® Silver-Copper / NANO PROTEX®	Seipasa Natural Technology Valent BioSciences Corporation Marron Bio Innovations Inc. Corteva Agriscience Certis Biologicals Monsanto Company Organic Crop Protectants Group Annadata Organic

Biopesticides

Although the use of chemical pesticides in agriculture is inevitable to meet the growing demand for food in the world, their excessive and inappropriate use has caused various problems. These include the presence of pesticide residues in food, contamination of soil and groundwater, and damage to the environment and human health. Therefore, the management and control of insect pests and diseases in crops is an important issue worldwide, causing farmers to seek effective, selective, biosecure, and environmentally friendly strategies and methodologies. Biopesticides have emerged as a viable alternative to address these challenges [27, 28].

Biopesticides are a type of pesticide derived from natural substances that control crop pests and diseases in an environmentally friendly manner. They operate through non-toxic mechanisms that minimally impact non-target organisms, exhibiting high specificity [29]. However, it is important to point out that biopesticides comprise a broad spectrum of toxicity. On one hand, their activity is directed to a single pest species at a specific moment in their life cycle. On the other hand, their effect is more extensive toward different species while maintaining the safety of beneficial species. This consideration reflects a more complex system of interactions between different individuals [30].

Biopesticides can be derived from plants, animals, bacteria, fungi, viruses, and certain minerals. These can be used individually or in combination with other crop protection measures, including cultural, mechanical, and chemical tools. This integrated approach is a key component of integrated pest management (IPM) [31].

Classification and Mechanisms of Biopesticide Action

Biopesticides are biological agents used for ecological pest management and crop protection, offering an alternative to chemical pesticides. They are generally less toxic, break down quickly, and are designed to target specific pests with greater precision. Different plant species have different modes of action, ranging from toxin production, predation, and parasitism to competition for nutrients, reproductive interference, and induction of plant resistance [32]. Biopesticides are classified according to their source of origin or extraction method of the molecules or compounds used in their preparation [29]. Biological pesticides can originate from compounds produced by plants and microorganisms such as bacteria, fungi, and viruses, as well as from biological nanoparticles [33]. The main categories of biopesticide are shown in Fig. **2** and are as follows:

- Microbiological pesticides (Origin from microorganisms).
- Biochemical pesticides (plant-derived, pheromones, enzymes, plant growth regulators, minerals).
- Plant-incorporated protectants (PIPs, known as genetically modified crops).
- Nanobiopesticides (Nanomaterials, such as gold, silver, copper nanoparticles.).

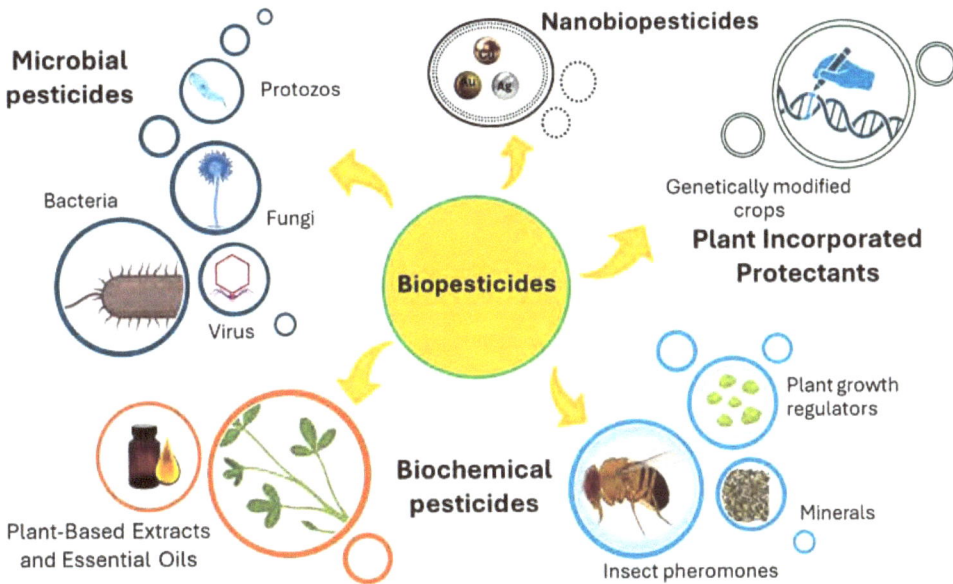

Fig. (2). Biopesticide classification. Figure made with Biorender.

Microbial Pesticides (Bacterium, Fungus, Virus or Protozoan)

Microbial biopesticides are promising pest and disease control strategies in modern agriculture, having the ability to selectively control pests while offering greater safety for human health and the environment. These agents can be derived from microorganisms such as bacteria, fungi, and viruses. Some examples of microbial biopesticide include *Bacillus thuringiensis* (Bt), a gram-positive bacteria capable of producing toxic proteins known as Cry toxins or δ-endotoxins. Cry-1 is the most common endotoxin, and its activity has been identified as specifically active for certain groups of insects, including lepidopterans (moths and butterflies), dipterans (flies and mosquitoes), and coleopterans (beetles). Bt is considered one of the most effective biological insecticides because of its rapid action and specificity toward the host [34]. Furthermore, it is important to highlight that Bt proteins are selective for insects and harmless to humans, vertebrates, and plants. Consequently, they represent approximately 90% of the production and application of biopesticides and are becoming the most widely used microbial pesticides globally [35]. Nanotechnology has facilitated significant advances in the research of biopesticides derived from Bt [36].

On the other hand, other microorganisms used as insecticides include some viruses of the family Baculoviridae. Approximately 90 species have been characterized within this family and classified into four genera. These viruses are characterized by the formation of occlusion bodies and are divided into two morphological groups: nucleopolyhedrovirus (NPVs) and granulovirus (GV). These occlusion bodies confer greater persistence in the environment, allowing the viruses to exert their action over extended periods. It is recognized that baculoviruses have a diversity of hosts and affect the larval stages of insects belonging to the orders Hymenoptera, Diptera, and Lepidoptera [37, 38].

Another biological control strategy involves using microorganisms such as yeasts, fungi, or bacteria to combat plant infections caused by phytopathogens. One of the most widely used microorganisms for these purposes is *Trichoderma spp.*; this genus effectively inhibits the growth of pathogenic fungi in many plants, such as *Rhizoctonia, Fusarium, Pythium, Sclerotium,* and *Phoma* genera [39, 40]. *Trichoderma spp.* can colonize and grow in plant roots to promote average growth; it can also compete for nutrients, suppressing the progression of infections. Moreover, it directly affects fungi by producing metabolites with antibacterial activity and lytic enzymes. Some species, like *T. atroviride, T. asperellum,* and *T. gamsii,* have been registered as commercial biopesticides [39]. Another challenge facing modern agriculture is the control of invasive plants or weeds. Until recently, burning or agrochemicals were the most common strategies for these purposes. However, concerns about health and the environment have promoted the use of bioherbicides for food production and increased crop yields. In this context, fungi are recognized as microorganisms that produce bioactive molecules with herbicidal effects [41]. For example, *Fusarium oxysporum* has been shown to exhibit activity against the growth of *Avena fatua*, a weed responsible for wheat loss [42]. Finally, microbial biopesticides contribute to biodiversity conservation by reducing the negative impacts of synthetic pesticides on natural ecosystems. Additionally, they are less likely to cause resistance in pest populations because of their specificity, and they provide prolonged control when used appropriately.

Biochemicals Pesticides

Biochemical pesticides are products, compounds of natural origin, or biochemical processes that are used in integrated pest management. They are compounds that use non-toxic mechanisms for pest control, excluding synthetic molecules that eliminate pests [33]. These biopesticides include various types of compounds, such as extracts or essential oils obtained from plants, pheromones (semiochemicals), and plant and insect growth regulators. Some biochemical pesticide constituents are described below:

Plant-Based Extracts and Essential Oils

In the world, approximately 300,000 plant species exist, which are an important source of specialized compounds synthesized in response to environmental stimuli or as protective agents. Plant pesticides can be applied as plant extracts (PEs) or essential oils (EOs). PEs are mixtures of compounds obtained from plants with organic solvents or water. They are used as insecticides, fungicides, or herbicides, such as the methanolic extract of *Azadirachta indica* (Meliaceae) or Neem tree, a recognized insecticide. The methanolic extract of *A. indica* has been tested for managing red flour beetles (*Tribolium castaneum*), which can affect stored grains [43]. Some plants belonging to the Fabaceae family are remarkable insecticides, such as *Retama raetam* and broom bush. The chloroform extract of this plant was tested for the control of *Aphis gossypii*, an important cotton pest. The major constituent of this extract is eugenol, which is responsible for insecticide activity [44]. One of the most devastating pests is phytopathogenic fungi, which are responsible for economic losses in sectors such as cereals, vegetables, and fruits. In its control, PE has been tested as an alternative; for example, the ethyl acetate extract of *Capsicum chinense* (Solanaceae) inhibits the mycelia growth of *C. gloeosporioides* (98.3%) [45]. The acetonic leaf extracts of *Ptaeroxylon obliquum* (Rutaceae) have been reported as good inhibitors of *Aspergillus niger*, *C. gloeosporioides*, and *Penicillium digitatum* with MICs lower than 80 µg/mL [46].

EOs are complex mixtures of volatile compounds responsible for the fragrance of flowers and plants. They are well known for their insecticidal activities, including functions as repellents and metabolic blockers. Some examples are the EOs of *Rosmarinus officinalis*, *Origanum majorana*, and *Cymbopogon winterianus*, with the EO of rosemary being the most toxic (LC_{50} 2526.08 mg/L) against *Spodoptera littoralis* larvae [47]. Interestingly, *C. winterianus* EO has also been reported as a fungicide against *Fusarium oxysporum*; citronella EO promoted the inhibition of this microorganism at IC_{50} 3.57 µL/mL and IC_{90} 2.58 µL/mL [48]. Furthermore, the mortality of *C. flexuosus* EO has been reported against *Sitophilus zeamais*, the experimentation conducted to LD_{50} 9.3 µL, and the bioactivity was attributed to β-citral and α-citral, the major constituents [49]. Allelopathic bioactivity is another biological property attributed to some EOs; for example, the antigerminative activity of *Allardia tridactylites* (Asteraceae) against the agricultural weed *Bidens pilosa* was recently reported, and the observed inhibition was 44.0% at a concentration of 0.1 mg/mL [50].

Insect Pheromones

Insect pheromones are chemicals that insects use to communicate with other insects of the same species. Insect sex pheromones have been used to control low-density pest populations with minimal impact on their natural enemies [51]. Unlike traditional insecticides, insect pheromones do not directly impact the target pest, as their common use is to interfere with mating or lure the insect into a trap where it will be in contact with a lethal pesticide [52]. Therefore, insect pheromones meet modern requirements for pest control because they are species-specific, environmentally friendly, and non-toxic to mammals [53].

Plant-Incorporated-Protectants (PIPs)

Plant-incorporated protectants (PIPs) are pesticidal substances produced by plants that involve the use of genetic engineering; hence, they are known as genetically modified crops (GMC) [54]. The ability of plants to produce substances with pesticidal activity is due to the insertion of genetic material from a pest control agent, such as a bacterium, fungus, or toxin-producing gene, into the genetic makeup of the plant, resulting in a transgenic crop [27]. Among the crops with incorporated protectants, the best are so-called Bt crops, which have been genetically modified to express a gene from the bacterium *Bacillus thuringiensis*. These crops produce a cryogenic protein that is toxic to some commercial crop pests, such as the codling moth (*Spodoptera frugiperda*) and corn borer (*Diatraea grandiosella*) [55]. The pesticide proteins produced by these modified crops are usually specific to a target pest; thus, using these crops with incorporated protectants will not affect beneficial plant organisms [56].

Nanobiopesticides

Nanobiopesticides are biological protection products formulated from nanomaterials to improve the efficacy of pesticides. Their use has become novel due to their size, structure, and nature as they are small biologically active particles of a size of 1-100 nm, which can prevent the development of pathogens or repel pests using smaller quantities than traditional pesticides [57, 54]. Nanobiopesticides are attached to substrates or nanotransporters to generate stimuli or enzymatic triggers. They are found in different formulations, such as nano-fibers, nano-encapsulation, nano-gels, nano-spheres, nano-containers, and nano-boxes, which are easily degradable and have high stability and efficacy in pest control [58]. Nanobiopesticides based on metallic nanoparticles of gold, silver, nickel, zinc, titanium, and copper are obtained from different organisms such as plants (*Olea europaea*), bacteria (*Photorhabdus luminescence*), fungi (*Agaricus arvensis*), and algae (*Sargassum muticum*), among others [33, 59].

The Market, Impact, and Advantages of Biopesticide over Synthetic Pesticides

Biopesticides are derived from natural sources, such as plants, microorganisms, and nanoparticles of biological origin [54], making them a sustainable and eco-friendly alternative to synthetic pesticides (chemical). Biopesticides are increasingly recognized worldwide for their safety for the environment and humans. Therefore, the US Environmental Protection Agency supports their use under the Federal Insecticide Act. As a result, the biopesticide industry has grown at a compound annual growth rate of 14.1% [60]. Globally, biopesticides comprise 10% of pesticides available on the market, which is expected to grow from approximately $5 billion today to $15 billion by 2029. The largest category of biological control agents is plant extracts, growth regulators, and pheromones. However, by 2029, microbial biopesticides are expected to reach similar levels [61].

The advantages of biopesticides over synthetic ones include their low toxicity, security, and specificity for pest management [33]. The problem with synthetic pesticides lies in the fact that they are one of the largest sources of pollution and promote resistance and accumulate in food crops [62]. The effectiveness of biopesticides compared with synthetics is related to their mode of action; they act by denaturalizing proteins, generating metabolic disorders, or causing paralysis in target pests [33].

However, each type of biopesticide has its advantages and limitations. Particularly, plant-based biopesticides are cost-effective and sustainable compared to synthetic pesticides, but their quality depends on the quality of the raw material used. On the other hand, microbiological agents are specific in their mode of action, and they are easy to make on a great scale. Still, one of their major disadvantages is the short lifespan of these agents, which is related to their low stability under different environmental conditions [63, 64].

Formulation and Biopesticide Application

Currently, biopesticide-based products are commercially available in different forms, with the primary objective of acquiring such products at an accessible price. Thus, they can be a basic tool in the control of pests and crop diseases [65]. Proper preparation of biopesticide formulations will help improve the efficacy of the biopesticide, providing a safe application method for consumers. Since many active components are derived from live microorganisms, it is crucial to ensure the viability of these species during processing, storage, transport, and application at target sites [66, 67]. Therefore, a successful formulation lies in its ability to maintain these properties, and its composition should be based on three main components: an active ingredient (or biopesticidal agent), inert carriers (excipient), and adjuvants (improve dispersion and protect the active ingredient) [68]. The physicochemical characteristics of the active ingredient determine the type of formulation in which the biopesticide product is presented. As such, various forms can be obtained, including liquid and solid inoculants encapsulated in biopolymers, effervescent tablets, and emulsions, among others that are less commonly used [68].

Liquid formulations include active ingredients in suspension, fluid concentrates that are miscible in oil, and dispersion in oil. The high effectiveness of suspension concentrates is due to the absence of powders and the ease of use compared with solid formulations. Solid formulations are low-cost and easy to produce; the main products are granules, microgranules, powders, wettable powders, and pellets [69]. Biopolymer encapsulation-based formulations engulf or trap microorganisms or nanomaterials, protecting them from adverse environmental conditions and enzymatic degradation, thus increasing the shelf life of the active ingredient until it reaches its target [70]. In the case of effervescent tablets, the active ingredient is compressed with excipients or in combination with a drug that helps the microbial inoculants achieve greater stability, easy dispersion, and less contamination during storage due to the low humidity of the product [68]. The main formulations of biopesticides currently used in agriculture are shown in Table **2**, where the companies that manufacture these products are listed [71].

Research into the development of new biopesticide formulations must continue to explore the various advantages and challenges associated with these formulations, in addition to agreements with companies developing them and their commercialization in the pest control product market [65].

CONCLUDING REMARKS

The future of biofertilizers and biopesticides appears promising for achieving more sustainable and environmentally friendly agriculture. Some challenges in determining the suitability of these products include the regulatory approval process, which delays their adoption; however, they should not be evaluated in the same way as synthetic products. Additionally, the acceptance and adoption of biofertilizers and biopesticides by farmers are crucial, requiring education and training on the benefits and proper management of these products. Investments in research and development are also necessary, and collaborations among academic institutions, companies, and governments are vital in advancing this field. The implementation of biofertilizers and biopesticides can increase agricultural productivity and food security and contribute to environmental preservation and soil health. Key aspects and future trends in this field involve bioprospecting diverse ecosystems to discover new microorganisms with beneficial properties for crops. Biotechnology and genomics can improve the efficiency and adaptability of these products to specific agroecological conditions. Finally, the incorporation of nanoparticles into formulations is expanding, allowing for the controlled release and increased efficacy of these treatments.

AUTHORS' CONTRIBUTION

The authors confirm their contribution to the chapter as follows:

Conceptualization, methodology, investigation, data curation, validation, writing-original draft preparation, writing-review, and editing: Rafael Torres-Martínez; Conceptualization, methodology, investigation, data curation, validation, writing-original draft preparation, writing-review, and editing: Yesica R. Cruz-Martinez; Conceptualization, methodology, investigation, data curation, validation, writing-original draft preparation, writing-review, and editing: Ana K. Villagómez-Guzmán; Conceptualization, Methodology, investigation, data curation, validation, writing-original draft preparation, writing-review, and editing: Olivia Pérez-Valera; Conceptualization, methodology, investigation, data curation, validation, writing-original draft preparation, writing-review, and editing: Héctor M. Arreaga-González; Conceptualization, supervision, validation, writing-review, and editing: Tzasna Hernández-Delgado.

All authors read and approved the final manuscript.

QUESTIONS RELATED TO THE TEXT

1.- What are your opinions on the optimal biofertilization choice for crops in the coming decades?
2.- What obstacles have been encountered in the widespread adoption of biofertilizers on a global scale?
3.- What benefits do biofertilizers offer compared to conventional fertilizers?
4.- According to this information, which biopesticide will predominate in the market by 2029?
5.- Indicate the classification of biopesticide and give three examples of each category according to the information provided throughout the chapter.
6.- Indicate the advantages of biopesticide *versus* synthetic (chemical) biopesticide.

ACKNOWLEDGMENT

A.K.V.G. is grateful to DGAPA-UNAM for the postdoctoral scholarship.

LIST OF ABBREVIATIONS

Ag   Silver

Bt   *Bacillus thuringiensis*

Cu   Copper

Cry-1   Toxin from *Bacillus thuringiensis*

EO	Essential oil
Fe	Iron
FSIN	Food Security Information Network
GMC	Genetically Modified Crops
GV	Granulovirus
HCN	Hydrogen Cyanide
IC$_{50}$	Inhibitory Concentration of 50%
IC$_{90}$	Inhibitory Concentration of 90%
IPM	Integrated Pest Management
LC$_{50}$	Lethal Concentration 50
LD$_{50}$	Lethal Dose 50
MIC	Minimum Inhibitory Concentration
N$_2$	Nitrogen
Ni	Nickel
NPV	Nucleopolyhedrovirus
P	Phosphorus
PE	Plant Extracts
PGP	Plant Growth Promoting
pH	Hydrogen Potential
PIPs	Plant-incorporated Protectants
Si	Silicon
Ti	Titanium
UN	United Nations
USD	United States Dollar
Zn	Zinc

REFERENCES

[1]   Simpson BK, Aryee ANA, Toldrá F. Bioproducts from Agriculture and Fisheries Adding Value for Food, Feed, Pharma and Fuels. 1st ed. Jhon Wiley & Sons Ltd. 2020; pp. 479-500.
[http://dx.doi.org/10.1002/9781119383956]

[2]   Food and Agriculture Organization of the United Nations (FAO), Land used in agriculture by the numbers, [updated 07th May 2020; cited 06 Jul 2024] available from: https://www.fao.org/sustainability/news/detail/en/c/1274219/#:~:text=Global%20trends,and%20pastures

[3]   Rakshit A, Meena VS, Parihar M, *et al.* Biofertilizers Advances in bio inoculants. 3-19.
[http://dx.doi.org/10.1016/C2019-0-03689-8]

[4]   Global Network Against Food Crises (GNAFC), 2024 Global report on food crises, join analysis for better decisions [updated 24th Apr 2024; cited 06 Jul 2024] available from: https://www.fightfoodcrises.net/events/grfc-2024

[5]   Wijerathna-Yapa A, Pathirana R. Sustainable Agro-Food Systems for Addressing Climate Change and Food Security. Agriculture 2022; 12(10): 1554.
[http://dx.doi.org/10.3390/agriculture12101554]

[6]   Rakshit A, Meena VS, Parihar M, *et al.* Biofertilizers Advances in bio inoculants. 183-90.
[http://dx.doi.org/10.1016/C2019-0-03689-8]

[7]   Rai PK, Rai A, Sharma NK, Singh T, Kumar Y. Limitations of biofertilizers and their revitalization through nanotechnology. J Clean Prod 2023; 418: 138194.
[http://dx.doi.org/10.1016/j.jclepro.2023.138194]

[8]   Soni R, Yadav AN, Suyal DC, *et al.* Trends of Applied Microbiology for Sustainable Economy A volume in Developments in Applied Microbiology and Biotechnology. 1st ed. Elsevier 2022; pp. 689-97.
[http://dx.doi.org/10.1016/C2020-0-02717-9]

[9]   Inamuddin, Ahamed MI, Biofertilizers. Study and impact 1st ed. Jhon Wiley & Sons Inc. 2020; pp 393-412. ISBN: 978-1-119-72498-8.

[10]   Allouzi MMA, Allouzi SMA, Keng ZX, Supramaniam CV, Singh A, Chong S. Liquid biofertilizers as a sustainable solution for agriculture. Heliyon 2022; 8(12): e12609.
[http://dx.doi.org/10.1016/j.heliyon.2022.e12609] [PMID: 36619398]

[11]   Inamuddin, Ahamed MI, Biofertilizers. Study and impact 1st ed. Jhon Wiley & Sons Inc. 2020; pp 575-586

[12]   Asraful A, Jing-Liang X, Zhongming W. Microalgae Biotechnology for Food, Health and High Value Products. 1st ed. Springer Nature Singapore Pte Ltd. 2020; pp. 397-415.
[http://dx.doi.org/10.1007/978-981-15-0169-2]

[13]   Hakeem KR, Dar GH, Mehmood MA, *et al.* Microbiota and Biofertilizers A Sustainable Continuum for Plant and Soil Health. 1st ed. Springer Nature Switzerland AG 2021; pp. 83-98.
[http://dx.doi.org/10.1007/978-3-030-48771-3]

[14]   Mącik M, Gryta A, Frąc M. Biofertilizers in agriculture: An overview on concepts, strategies and effects on soil microorganisms. Adv Agron 2020; 162: 31-87.
[http://dx.doi.org/10.1016/bs.agron.2020.02.001]

[15]   Chaillan F, Chaîneau CH, Point V, Saliot A, Oudot J. Factors inhibiting bioremediation of soil contaminated with weathered oils and drill cuttings. Environ Pollut 2006; 144(1): 255-65.
[http://dx.doi.org/10.1016/j.envpol.2005.12.016] [PMID: 16487636]

[16]   Malam Issa O, Défarge C, Le Bissonnais Y, *et al.* Effects of the inoculation of cyanobacteria on the microstructure and the structural stability of a tropical soil. Plant Soil 2007; 290(1-2): 209-19.
[http://dx.doi.org/10.1007/s11104-006-9153-9]

[17]   Yilmaz E, Sönmez M. The role of organic/bio–fertilizer amendment on aggregate stability and organic carbon content in different aggregate scales. Soil Tillage Res 2017; 168: 118-24.
[http://dx.doi.org/10.1016/j.still.2017.01.003]

[18]   Bose A, O'Shea R, Lin R, *et al.* The marginal abatement cost of co-producing biomethane, food and biofertiliser in a circular economy system. Renew Sustain Energy Rev 2022; 169: 112946.
[http://dx.doi.org/10.1016/j.rser.2022.112946]

[19]   Hartmann M, Six J. Soil structure and microbiome functions in agroecosystems. Nat Rev Earth Environ 2022; 4(1): 4-18.
[http://dx.doi.org/10.1038/s43017-022-00366-w]

[20]   Mordor intelligence Biofertilizer Market Size. Report on Industry Size & Market Share Analysis - Growth Trends & Forecasts Up To 2029. Cited 06 Jul 2024. available from: https://www.mordorintelligence.com/industry-reports/global-biofertilizers-market-industry

[21]   Bullor L, Braude H, Monzon J, *et al.* Bioinsumos: Oportunidades de inversión en América Latina -

Direcciones de inversión No. 9. Roma, FAO

[22] Yadav AN, Singh J, Rastegari AA, *et al.* Plant Microbiomes for Sustainable Agriculture: Current Research and Future Challenges. Springer 2020; pp. 475-82.
[http://dx.doi.org/10.1007/978-3-030-38453-1]

[23] O'Callaghan M, Ballard RA, Wright D. Soil microbial inoculants for sustainable agriculture: Limitations and opportunities. Soil Use Manage 2022; 38(3): 1340-69.
[http://dx.doi.org/10.1111/sum.12811]

[24] Sun C, Wei Z, Xue C, Yang L. Development, application and future trends of starch-based delivery systems for nutraceuticals: A review. Carbohydr Polym 2023; 308: 120675.
[http://dx.doi.org/10.1016/j.carbpol.2023.120675] [PMID: 36813348]

[25] Sharma B, Tiwari S, Kumawat KC, Cardinale M. Nano-biofertilizers as bio-emerging strategies for sustainable agriculture development: Potentiality and their limitations. Sci Total Environ 2023; 860: 160476.
[http://dx.doi.org/10.1016/j.scitotenv.2022.160476] [PMID: 36436627]

[26] El-Ghamry AM, Mosa AA, Alshaal TA, *et al.* Nanofertilizers *vs.* Biofertilizers: New Insights. Env Biodiv Soil Security 2018; 2: 51-72.
[http://dx.doi.org/10.21608/jenvbs.2018.3880.1029]

[27] Olson S. An Analysis of the Biopesticide Market Now and Where it is Going. Outlooks Pest Manag 2015; 26(5): 203-6.
[http://dx.doi.org/10.1564/v26_oct_04]

[28] Oguh CE, Okpaka CO, Ubani CS, *et al.* Natural Pesticides (Biopesticides) and Uses in Pest Management-A Critical Review. Asian J Biotechnol Genet Eng 2019; 2: 1-18.https://journalajbge.com/index.php/AJBGE/article/view/37

[29] Ruiu L. Microbial Biopesticides in Agroecosystems. Agronomy (Basel) 2018; 8(11): 235.
[http://dx.doi.org/10.3390/agronomy8110235]

[30] Dar SA, Wani SH, Mir SH, *et al.* Biopesticides: Mode of Action, Efficacy and Scope in Pest Management. J Adv Res Biochem Pharma 2021; 4(1): 1-8. https://www.medicaljournalshouse.com/index.php/ADR-Pharmacology_Biochemistry/article/view/536

[31] Hanif K, Zubair M, Hussain D, *et al.* Biopesticides and insect pest management. Int J Trop Insect Sci 2022; 42(6): 3631-7.
[http://dx.doi.org/10.1007/s42690-022-00898-0]

[32] Aioub AAA, Ghosh S, AL-Farga A, *et al.* Back to the origins: biopesticides as promising alternatives to conventional agrochemicals. Eur J Plant Pathol 2024; 169(4): 697-713.
[http://dx.doi.org/10.1007/s10658-024-02865-6]

[33] Ayilara MS, Adeleke BS, Akinola SA, *et al.* Biopesticides as a promising alternative to synthetic pesticides: A case for microbial pesticides, phytopesticides, and nanobiopesticides. Front Microbiol 2023; 14: 1040901.
[http://dx.doi.org/10.3389/fmicb.2023.1040901] [PMID: 36876068]

[34] Khan AR, Mustafa A, Hyder S, *et al. Bacillus* spp. as Bioagents: Uses and Application for Sustainable Agriculture. Biology (Basel) 2022; 11(12): 1763.
[http://dx.doi.org/10.3390/biology11121763] [PMID: 36552272]

[35] Li Y, Wang C, Ge L, *et al.* Environmental Behaviors of *Bacillus thuringiensis* (*Bt*) Insecticidal Proteins and Their Effects on Microbial Ecology. Plants 2022; 11(9): 1212.
[http://dx.doi.org/10.3390/plants11091212] [PMID: 35567212]

[36] Saxena AK, Kumar M, Chakdar H, Anuroopa N, Bagyaraj DJ. *Bacillus* species in soil as a natural resource for plant health and nutrition. J Appl Microbiol 2020; 128(6): 1583-94.
[http://dx.doi.org/10.1111/jam.14506] [PMID: 31705597]

[37]   Harrison RL, Herniou EA, Jehle JA, *et al.* ICTV Virus Taxonomy Profile: Baculoviridae. J Gen Virol 2018; 99(9): 1185-6.
[http://dx.doi.org/10.1099/jgv.0.001107] [PMID: 29947603]

[38]   Moore S, Jukes M. The History of Baculovirology in Africa. Viruses 2023; 15(7): 1519.
[http://dx.doi.org/10.3390/v15071519] [PMID: 37515205]

[39]   Ferreira FV, Musumeci MA. Trichoderma as biological control agent: scope and prospects to improve efficacy. World J Microbiol Biotechnol 2021; 37(5): 90.
[http://dx.doi.org/10.1007/s11274-021-03058-7] [PMID: 33899136]

[40]   Brizuela AM, Gálvez L, Arroyo JM, Sánchez S, Palmero D. Evaluation of *Trichoderma* spp. on *Fusarium oxysporum* f. sp. *asparagi* and *Fusarium* wilt Control in Asparagus Crop. Plants 2023; 12(15): 2846.
[http://dx.doi.org/10.3390/plants12152846] [PMID: 37571000]

[41]   Camargo AF, Bonatto C, Scapini T, *et al.* Fungus-based bioherbicides on circular economy. Bioprocess Biosyst Eng 2023; 46(12): 1729-54.
[http://dx.doi.org/10.1007/s00449-023-02926-w] [PMID: 37743409]

[42]   Dahiya A, Sharma R, Sindhu S, Sindhu SS. Resource partitioning in the rhizosphere by inoculated *Bacillus* spp. towards growth stimulation of wheat and suppression of wild oat (*Avena fatua* L.) weed. Physiol Mol Biol Plants 2019; 25(6): 1483-95.
[http://dx.doi.org/10.1007/s12298-019-00710-3] [PMID: 31736550]

[43]   Alhaithloul HAS, Alqahtani MM, Abdein MA, *et al.* Rosemary and neem methanolic extract: antioxidant, cytotoxic, and larvicidal activities supported by chemical composition and molecular docking simulations. Front Plant Sci 2023; 14: 1155698.
[http://dx.doi.org/10.3389/fpls.2023.1155698] [PMID: 37275255]

[44]   Kamel AI, El-Rokh AR, Dawidar AM, Abdel-Mogib M. Bioactive compounds from *Retama raetam* (Forssk.) Webb & Berthel. and their insecticidal activity against cotton pests *Aphis gosspyii* and *Amrasca biguttula*. Fitoterapia 2024; 172: 105749.
[http://dx.doi.org/10.1016/j.fitote.2023.105749] [PMID: 37972716]

[45]   Santos LS, Fernandes CC, Santos LS, Dias ALB, Souchie EL, Miranda MLD. Phenolic compounds and antifungal activity of ethyl acetate extract and methanolic extract from Capsicum chinense Jacq. ripe fruit. Braz J Biol 2024; 84: e258084.
[http://dx.doi.org/10.1590/1519-6984.258084] [PMID: 35195174]

[46]   Ramadwa TE, Makhubu FN, Eloff JN. The activity of leaf extracts, fractions, and isolated compounds from *Ptaeroxylon obliquum* against nine phytopathogenic fungi and the nematode *Meloidogyne incognita*. Heliyon 2024; 10(7): e28920.
[http://dx.doi.org/10.1016/j.heliyon.2024.e28920] [PMID: 38596024]

[47]   Awad M, Hassan NN, Alfuhaid NA, *et al.* Insecticidal and biochemical impacts with molecular docking analysis of three essential oils against Spodoptera littoralis (Lepidoptera: Noctuidae). Crop Prot 2024; 180: 106659.
[http://dx.doi.org/10.1016/j.cropro.2024.106659]

[48]   Peixoto PMC, Júlio AA, Jesus EG, *et al.* Fungicide potential of citronella and tea tree essential oils against tomato cultivation's phytopathogenic fungus *Fusarium oxysporum* f. sp. lycopersici and analysis of their chemical composition by GC/MS. Nat Prod Res 2024; 38(4): 667-72.
[http://dx.doi.org/10.1080/14786419.2023.2184358] [PMID: 36855252]

[49]   Mota Filho TMM, da Silva Camargo R, de Menezes CWG, *et al.* Fumigant toxicity of *Cymbopogon flexuosus* lemon grass (Poaceae) essential oil to *Sitophilus zeamais* maize weevil (*Coleoptera: Curculionidae*) and phytotoxicity to *Zea mays* (Poaceae). Cereal Res Commun 2023; 52: 215-20.
[http://dx.doi.org/10.1007/s42976-023-00389-z]

[50]   Zomba D, Sharma M, Jandrotia R, Singh HP, Batish DR. Chemical profiling, phytotoxicity,

cytotoxicity, and antibacterial activity of the essential oil of a high-altitude plant, *Allardia tridactylites*, from the Trans-Himalayan Region, Ladakh. Chem Zvesti 2024; 78(3): 1887-96.
[http://dx.doi.org/10.1007/s11696-023-03213-4]

[51]   Petkevicius K, Löfstedt C, Borodina I. Insect sex pheromone production in yeasts and plants. Curr Opin Biotechnol 2020; 65: 259-67.
[http://dx.doi.org/10.1016/j.copbio.2020.07.011] [PMID: 32866709]

[52]   Yew JY, Chung H. Insect pheromones: An overview of function, form, and discovery. Prog Lipid Res 2015; 59: 88-105.
[http://dx.doi.org/10.1016/j.plipres.2015.06.001] [PMID: 26080085]

[53]   Seybold SJ, Bentz BJ, Fettig CJ, Lundquist JE, Progar RA, Gillette NE. Management of western North American bark beetles with semiochemicals. Annu Rev Entomol 2018; 63(1): 407-32.
[http://dx.doi.org/10.1146/annurev-ento-020117-043339] [PMID: 29058977]

[54]   Kumar J, Ramlal A, Mallick D, Mishra V. An Overview of Some Biopesticides and Their Importance in Plant Protection for Commercial Acceptance. Plants 2021; 10(6): 1185.
[http://dx.doi.org/10.3390/plants10061185] [PMID: 34200860]

[55]   Razzaq A, Ali A, Zafar MM, *et al.* Pyramiding of *cry* toxins and methanol producing genes to increase insect resistance in cotton. GM Crops Food 2021; 12(1): 382-95.
[http://dx.doi.org/10.1080/21645698.2021.1944013] [PMID: 34193022]

[56]   Shahid M, Shaukat F, Shahid A, *et al.* Biopesticides: A potential solution for the management of insect pests. Agrobiological Records 2023; 13: 7-15.
[http://dx.doi.org/10.47278/journal.abr/2023.022]

[57]   Chaudhary P, Chaudhary A, Parveen H, *et al.* Impact of nanophos in agriculture to improve functional bacterial community and crop productivity. BMC Plant Biol 2021; 21(1): 519.
[http://dx.doi.org/10.1186/s12870-021-03298-7] [PMID: 34749648]

[58]   Pan X, Guo X, Zhai T, *et al.* Nanobiopesticides in sustainable agriculture: developments, challenges, and perspectives. Environ Sci Nano 2023; 10(1): 41-61.
[http://dx.doi.org/10.1039/d2en00605g]

[59]   Karunakaran G, Sudha KG, Ali S, Cho EB. Biosynthesis of Nanoparticles from Various Biological Sources and Its Biomedical Applications. Molecules 2023; 28(11): 4527.
[http://dx.doi.org/10.3390/molecules28114527] [PMID: 37299004]

[60]   Saddam B, Idrees MA, Kumar P, Mahamood M. Biopesticides: Uses and importance in insect pest control: A review. Int J Trop Insect Sci 2024; 44(3): 1013-20.
[http://dx.doi.org/10.1007/s42690-024-01212-w]

[61]   Marrone PG. Status of the biopesticide market and prospects for new bioherbicides. Pest Manag Sci 2024; 80(1): 81-6.
[http://dx.doi.org/10.1002/ps.7403] [PMID: 36765405]

[62]   Kalpana T, Anil T. A review of biopesticides and their plant phytochemicals information. Ann Rom Soc Cell Biol 2021; 25: 3576-88. http://annalsofrscb.ro/index.php/journal/article/view/468

[63]   Damalas CA, Koutroubas SD. Botanical pesticides for eco-friendly pest management: drawbacks and limitations In Pesticides in Crop Production. Physiological Biochemical Action 2020; 10: 181-93.
[http://dx.doi.org/10.1002/9781119432241.ch10]

[64]   Adeleke BS, Ayilara MS, Akinola SA, Babalola OO. Biocontrol mechanisms of endophytic fungi. Egypt J Biol Pest Control 2022; 32(1): 46.
[http://dx.doi.org/10.1186/s41938-022-00547-1]

[65]   Melanie M, Miranti M, Kasmara H, *et al.* Nanotechnology-Based Bioactive Antifeedant for Plant Protection. Nanomaterials (Basel) 2022; 12(4): 630.
[http://dx.doi.org/10.3390/nano12040630] [PMID: 35214959]

[66] Kumar S, Nehra M, Dilbaghi N, Marrazza G, Hassan AA, Kim KH. Nano-based smart pesticide formulations: Emerging opportunities for agriculture. J Control Release 2019; 294: 131-53.
[http://dx.doi.org/10.1016/j.jconrel.2018.12.012] [PMID: 30552953]

[67] Borges DF, Lopes EA, Fialho Moraes AR, *et al.* Formulation of botanicals for the control of plant-pathogens: A review. Crop Prot 2018; 110: 135-40.
[http://dx.doi.org/10.1016/j.cropro.2018.04.003]

[68] Hezakiel HE, Thampi M, Rebello S, *et al.* Biopesticides: A Green Approach Towards Agricultural Pests. Appl Biochem Biotechnol 2023; 196: 5533-62.
[http://dx.doi.org/10.1007/s12010-023-04765-7] [PMID: 37994977]

[69] López MD, Cantó-Tejero M, Pascual-Villalobos MJ. New Insights Into Biopesticides: Solid and Liquid Formulations of Essential Oils and Derivatives. Frontiers in Agronomy 2021; 3: 763530.
[http://dx.doi.org/10.3389/fagro.2021.763530]

[70] Taban A, Saharkhiz MJ, Khorram M. Formulation and assessment of nano-encapsulated bioherbicides based on biopolymers and essential oil. Ind Crops Prod 2020; 149: 112348.
[http://dx.doi.org/10.1016/j.indcrop.2020.112348]

[71] United States Enviromental Protection Agency. Index and part 180 tolerance information of Pesticide chemicals in food and feed commodities. Cited 27 September 2024. Available from: https://www.epa.gov/pesticide-tolerances/indexes-part-180-tolerance-inform-tion-pesticide-chemicals-food-and-feed

CHAPTER 7

Microalgae and Biotechnology: Water Purification and Biomass Production

Yuri Córdoba Campo[1,*] and **Israel Valencia Quiroz**[2]

[1] *Fundación Universitaria Navarra, Uninavarra, Neiva, Huila, Colombia*

[2] *Phytochemistry Laboratory, UBIPRO, Superior Studies Faculty (FES)-Iztacala, National Autonomous University of Mexico (UNAM), Tlalnepantla de Baz, Mexico State, Mexico*

Abstract: Water resources suitable for direct human consumption have become increasingly scarce due to the rapid expansion of the global economy and population. Water pollution arises from multiple sources, including municipal, agricultural, and industrial activities, which release a diverse array of toxic pollutants into water bodies daily. These pollutants include heavy metals, pesticides, dyes, pharmaceuticals, and other hazardous substances. Addressing water contamination is, therefore, a critical global challenge that requires sustainable, cost-effective, and efficient solutions. Microalgae offer a promising alternative for wastewater treatment due to their ability to thrive in various wastewater types, low energy consumption, and unique capacity to convert harmful pollutants into valuable compounds. Microalgae-based wastewater treatment systems are particularly attractive because of their low operational costs, minimal environmental impact, and ability to function effectively under a wide range of environmental conditions. Additionally, they can remove a broad spectrum of pollutants through mechanisms like biosorption, bioaccumulation, and biodegradation. The resulting biomass from these processes is of significant interest, as it can be used to produce high-value products such as biofuels, biofertilizers, pharmaceuticals, and food additives, contributing to an integrated and sustainable economic model. This chapter highlights the importance of microalgal wastewater treatment by exploring these three key remediation mechanisms in detail and discusses the potential for microalgae to produce bioenergy and other valuable bioproducts. Furthermore, it addresses the current challenges and future opportunities within the algae-based biotechnology sector, which has recently gained attention for its potential to significantly contribute to environmental sustainability and economic development worldwide.

Keywords: Biomass, Microalgae, Pollutants, Remediation, Wastewater, Water purification.

---

[*] **Corresponding author Yuri Córdoba Campo:** Fundación Universitaria Navarra, Uninavarra, Neiva, Huila, Colombia; Tel: +525556231136; E-mail: yuriscordoba@gmail.com

Israel Valencia Quiroz (Ed.)

INTRODUCTION

Various types of environmental pollution, such as land, air, and water pollution, are responsible for significant environmental imbalances and contribute to global warming. These forms of pollution negatively affect ecosystems, biodiversity, and human health, generating a range of impacts that require urgent attention to mitigate their adverse effects on the planet. Toxic substances include heavy metals, nuclear waste, chemical fertilizers, pesticides, hydrocarbons, and pharmaceutical by-products [1]. While water is abundant in some parts of the world, access to clean water remains a significant challenge in industrially based developing countries. Water pollution is primarily caused by industrial, municipal, and agricultural wastewater, which contains both organic and inorganic pollutants [2].

Various physicochemical and biological methods have been developed to remove or remediate pollutants, including filtration, ion exchange, electrochemical methods, and osmosis. However, most of these methods are neither commercially viable nor environmentally sustainable. Therefore, there is a need for integrated technologies to reduce energy consumption and costs [1].

Currently, biological treatments are preferred over chemical treatments, even though, the latter tend to be faster and can have negative effects on the environment and living organisms. Biological remediation is a low-cost, low-toxicity and a long-term alternative. This process uses microorganisms, including microalgae, fungi, and bacteria, to remove pollutants from the environment [1]. Microalgae, one of the oldest microorganisms, thrive in various hostile environments being particularly useful in wastewater treatment. Thus, microalgae ability to consume or absorb pollutants helps to reduce biological oxygen demand, total suspended solids, pathogenic organisms, nitrogen and phosphorus concentrations in wastewater [2, 3].

Microalgae are versatile organisms because they can continue to grow even under adverse conditions with low nutrient concentrations. In addition, they can adapt to a wide range of environmental applications, making them suitable for various biotechnological applications. This adaptability creates a dual system that enables both pollutant removal and biomass production for commercial purposes. In large-scale biomass production, no additional growth formulations are needed due to their high adaptability. Biomass production is particularly higher in warm regions of the world, where sunlight and favorable temperatures support their growth [4]. Microalgae biomass and its products are used in many industries, such as food, agriculture, fertilizers, pharmaceutical industries, biosurfactants, biofuels, and scientific research.

Importance of Water Purification

Global water consumption has multiplied sixfold in the past 100 years. It continues to increase by 1% annually due to population growth, economic development, and consumption. Climate change has worsened the water quantity situation by increasing the frequency and magnitude of extreme events such as heatwaves, storms, and tempest and unprecedented rainfall [5]. In addition, water is an essential resource for several industries, such as pharmaceuticals, electronics, petrochemicals, agrochemicals, and food, all of which have increasingly affected water quality and availability worldwide [6]. The disposal of water from these industries poses significant risks to both the environment and human health. Wastewater (WW) contains various organic and inorganic compounds that are released into the environment, raising chemical oxygen demand (COD) and biological oxygen demand (BOD). The eutrophication of aquatic environments caused by high concentrations of phosphorus and nitrogen induces the generation of solid waste and unpleasant emissions into the atmosphere [7]. Consequently, the phytoremediation process may be considered as a relevant method to cope with WW. (Fig. **1**).

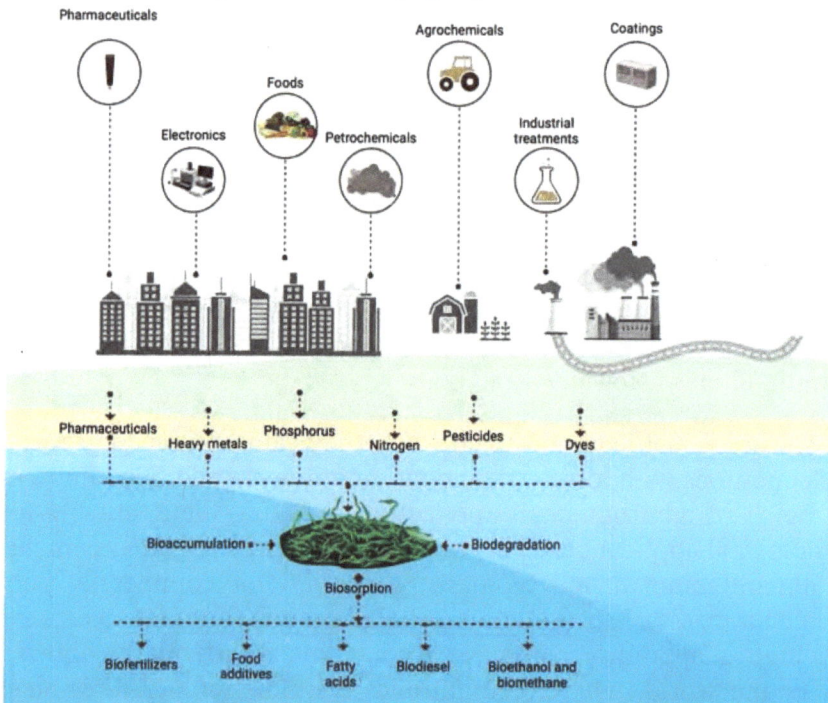

Fig. (1). Treatment of wastewater from different sources and the processes involved in remediation by microalgae. Pathways are associated with microalgae biomass conversion processes. Figure made with Biorender.

Heavy metals (mercury, lead, chromium, nickel, copper, zinc, cadmium, arsenic, manganese, iron, *etc.*) are among the most widespread and concentrated pollutants in wastewater. Their presence has increased due to rapid industrial and agricultural expansion, as these metals are non-biodegradable and tend to accumulate in living organisms [7]. Direct inhalation, ingestion, and contact with these substances are harmful to human health, causing damage to the skin, brain, liver, and kidneys, as well as leading to bone disease, weakened immune system, and an increased risk of cancer [7 - 9]. Guidelines and detection limits for heavy metals in water are presented in Table **1**.

Table 1. Guideline values and detection limits for heavy metals [9].

Heavy Metal	Guideline Level, mg/L	Detection Limit
Antimony, Sb	0.02	0.01 mg/L by EAAS; 0.1–1 mg/L by ICP/MS; 0.8 mg/L by graphite furnace atomic absorption spectrophotometry; 5 mg/L by hydride generation AAS
Arsenic, As	0.01	0.1 mg/L by ICP/MS; 2 mg/L by hydride generation AAS or FAAS
Barium, Ba	0.7	0.1 mg/L by ICP/MS; 2 mg/L by AAS; 3 mg/L by ICP–optical emission spectroscopy
Cadmium, Cd	0.003	0.01 mg/L by ICP/MS; 2 mg/L by FAAS
Copper, Cu	2	0.02–0.1 mg/L by ICP/MS; 0.3 mg/L by ICP–optical emission spectroscopy; 0.5 mg/L by FAAS
Chromium, Cr	0.05	0.05–0.2 mg/L for total chromium by AAS
Lead, Pb	0.01	1 mg/L by AAS
Manganese, Mn	0.4	0.01 mg/L by AAS; 0.05 mg/L by ICP/MS; 0.5 mg/L by ICP/optical emission spectroscopy; 1 mg/L by EAAS; 10 mg/L by FAAS
Mercury, Hg	0.006	0.05 mg/L by cold vapor AAS; 0.6 mg/L by ICP; 5 mg/L by FAAS
Molybdenum, Mo	0.07	0.25 mg/L by graphite furnace AAS; 2 mg/L by ICP–AES
Nickel, Ni	0.07	0.1 mg/L by ICP-MS; 0.5 mg/L by FAAS; 10 mg/L by ICP-AES
Selenium, Se	0.01	0.5 mg/L by AAS with hydride generation
Uranium, U	0.015	0.01 mg/L by ICP/MS; 0.1 mg/L by solid fluorimetry with either laser excitation or UV light; 0.2 mg/L by ICP *via* adsorption with chelating resin

ICP: inductively coupled plasma; MS: mass spectrometry; AAS: atomic absorption spectroscopy; FAAS: flame atomic absorption spectrometry; EAAS: electrothermal atomic absorption spectrometry.

When water becomes contaminated, the purification process becomes indispensable and mandatory, requiring the selection of the best treatment methods and meeting the desired purification objectives. Conventional methods include physical and chemical treatments. Persistent organic pollutants are chemicals that persist in the environment by resisting biodegradation; they bioaccumulate in the food web and can be transported over long distances. They can cause reduced reproductive growth, behavioral changes, birth defects, carcinogens, endocrine disruptors, and even death [10].

Some of these contaminants must be removed before water can be safely discharged into the environment or made suitable for human consumption. The nutrients in wastewater are essential for microalgae to perform their cellular functions and generate valuable biomass. In recent decades, several researchers have focused on developing promising methods for wastewater recycling [1].

Microalgae development depends on the availability of adequate nutrients in the medium, such as micronutrients, vitamins, trace minerals, and macronutrients like nitrogen and phosphorus, to ensure optimal microalgae growth and the formation of useful products. Most organic and inorganic compounds found in WW are beneficial to microalgae. Microalgae can resist contaminants at low concentrations due to their ability to adapt to the environment [11].

MICROALGAE REMEDIATION MECHANISMS

Microalgae are a heterogeneous group of unicellular photoautotrophic (chlorophyll a) microorganisms comprising over 30,000 species, which can be either prokaryotic or eukaryotic. They possess a wide range of metabolic and biochemical properties, including the production of fatty acids and lipids, hydrocarbons, and sterols, and bioactive compounds, including secondary metabolites [7, 12]. Microalgae biotechnology is a form of biomass production similar to conventional agriculture, with advantages such as photosynthetic efficiency, high growth rate, and greater CO_2 fixation compared to higher plants. It is feasible to grow them in variable climates and on non-arable land, even in marginal areas unsuitable for agricultural purposes; they can use less water than traditional crops and have the great advantage of using non-potable water, including wastewater, for growth [12].

Environmental remediation with microalgae uses different mechanisms (biosorption, bioaccumulation, and biodegradation):

Biosorption

This is a passive process that occurs when the sorbent is a biological material capable of binding and concentrating water pollutants. Biosorption includes physical, chemical, and metabolism-independent processes involving various mechanisms such as precipitation, complexation, ion exchange, sorption, adsorption, and electrostatic interactions. This mechanism requires a biosorbent (solid phase sorbent) and a target sorbate dissolved in water. The affinity between these two is driven by their overall ability to retain the molecules; this process continues until equilibrium is reached between these two. The degree of affinity of the biosorbent for a specific sorbate determines its distribution between the solid and liquid phases [7, 13].

The cell wall and surface charge of the microalgae are responsible for biosorption, and their chemical composition dictates the mechanism involved in the process. In the wall, there are different chemical groups, such as carboxyl, hydroxyl, and sulfate, which allow binding sites, ion exchangers, the formation of complexes with metal ions, and the adsorption of organic substances from contaminated water. A scheme of biosorption is shown in Fig. (**2**) [14].

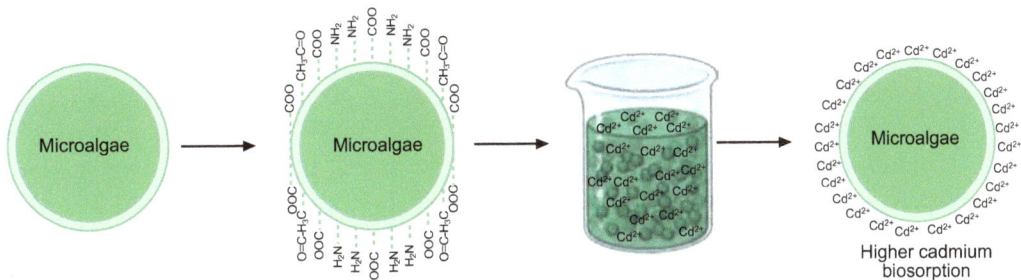

Fig. (2). The biosorption of toxic metals in microalgae depends on the presence of reactive groups in the cell wall that act as binding sites. Figure made with Biorender, adapted from reference [14].

The biosorption of metals can occur *via* various mechanisms, such as ion exchange, complexation, chelation adsorption, and microprecipitation. However, the biosorption process is quite complex, and a combination of several mechanisms can occur simultaneously. Among the factors that dominate this mechanism are (i) the type of functional group on the surface of the biosorbent, (ii) the metal ions present in the aqueous solution, and (iii) the characteristics of the solution (pH, concentration, presence of other inert or competing ions, *etc.*) [11, 14].

Lipids, proteins, and nucleic acids can be deposited on the cell surface, but these are predominantly in the cytoplasm and plasma membrane, where they bind to metal cations through various functional groups (carboxylic, amine, thiol, imidazole, thioesteric, nitrogen, and oxygen in peptide bonds) [11].

Bioaccumulation

This is an active process that requires energy and is slower than biosorption. In this process, substances are retained and then accumulated or metabolized. This process is an essential pathway to remove inorganic and organic pollutants (heavy metals, sulfates, nitrates, phosphates, pesticides, *etc.*) by transferring them into the cell. Bioaccumulation can be quantified by the bioconcentration factor, which is the ratio of the concentration of pollutants adsorbed on the medium. However, not all adsorbed molecules can undergo bioaccumulation. Although the processes of biosorption and bioaccumulation differ, quantifying pollutants in each process remains challenging because the two mechanisms are dynamically changing [7, 13].

Contaminants can also enter microalgae without requiring energy, moving along concentration gradients from areas of higher (external) to lower (internal) concentrations. Due to the hydrophobicity of the cell membrane, apolar and lipid-soluble low-molecular-weight contaminants can enter the cell membrane *via* passive diffusion. In contrast, polar, ionic, and high-molecular-weight contaminants will not be able to diffuse across the cell membrane. Such passive diffusion in microalgae has been reported with sulfamethoxazole, carbamazepine, nonylphenol, progesterone, norgestrel, trimethoprim, and florfenicol [7, 15].

External and internal physicochemical parameters influencing bioaccumulation include pH, contact time, temperature, and concentration of contaminants. In addition, some contaminants stimulate the production of reactive oxygen species (hydroxyl, alkoxy, superoxide, perhydroxyl, hydrogen peroxide, and singlet oxygen radicals), which can lead to oxidative damage to biomolecules, dysfunction, and cell death [7, 16].

Bioremediation can be enhanced by optimizing the physicochemical environment because this process depends on the speed and capacity of the microalgae. In addition, knowledge of microalgae species that are tolerant to high pollutant concentrations can generate higher capacities and bioaccumulation rates [7].

Biodegradation

This is an essential process for removing pollutants from WW, in which complex compounds are degraded into simple and safe compounds that can be used as nutrients for growth and development [7, 13]. This process can occur through the mineralization of pollutants into CO_2 and H_2O or *via* biotransformation involving a series of enzymatic reactions, thus generating different metabolic intermediates [7]. Furthermore, it has been observed that algae can degrade pollutants, transform them into intermediates, or increase the degradation potential of the microbiota. Matamoros, Uggetti, Garcia, and Bayona (2016) found that microalgae have an increased rate of caffeine degradation of more than 99% [17]. Ibrahim, Karam, EI-Shahat, and Adway (2014) demonstrated the ability of *Nostoc muscorum* and *Spirulina platensis* to biodegrade and use malathion as a phosphorus source [18].

Microalgae can biodegrade or biotransform organic pollutants by metabolic action with the accumulation of pollutants and then degrade them by transformation and mineralization; a microalga and blue-green alga can degrade naphthalene into trans-naphthalene dihydrodiol, cis-naphthalene dihydrodiol, 4-hydroxy-4-tetralone, and 1-naphthol, metabolites that are produced at a nontoxic concentration [13].

The two mechanisms of biodegradation are i) metabolic degradation and ii) co-metabolism. In the former, pollutants serve as electron donors/acceptors and carbon sources for microalgae. While in co-metabolism, pollutants serve as electron donors and carbon sources for non-living matter [7]. Biodegradation can occur extracellularly or intracellularly, or a combination of these processes, including extracellular breakdown followed by intracellular of the breakdown intermediates. Extracellular degradation relies on extracellular polymeric substances excreted by microalgae and is restricted to the outer cell walls. This process enables microalgae to mineralize contaminants in a dissolved state and establish an external digestive system. In addition, extracellular polymeric substances function as biosurfactants, increasing the bioavailability of pollutants in the environment and generating bioaccumulation in microalgae [7, 13].

The biodegradation mechanism comprises three enzymatic stages. In stage 1, cytochrome P450 enzymes such as hydrolases, carboxylases, decarboxylases, and monooxygenase are required to increase solubility by the addition or deprotection of a hydroxyl group *via* hydrolysis or oxidation-reduction reactions. In the next stage, the enzyme's glutathione S-transferase and glucosyltransferase can stimulate the conjugation of glutathione to different compounds with $CONH_2$, epoxide rings, and COOH to protect against oxidative damage in the cell. In the third stage, detoxification occurs with dehydrogenase, glutamyl-tRNA reductase, carboxylase, mono(di)oxygenase, transferase, pyrophosphatase, dehydratase, and hydrolases to transform the molecules into less toxic or nontoxic intermediates [16, 19, 20].

In the mechanism of co-metabolism, organic substrates function as electron donors in non-living matter and biomass production. The stimulation of microalgae-mediated biodegradation of recalcitrant pollutants has led to the addition of nutritional substances or organic substrates to generate a co-metabolic system. For example, the removal efficiency of ciprofloxacin by *Chlamydomonas mexicana* and sulfamethoxazole by *Chlorella pyrenoidosa* increased from 13% to 56% and from 12.6 to 34.6%, respectively, with the addition of sodium acetate [21 - 23]. Furthermore, the type and concentration of the nutrients have an effect on this mechanism. Studies using sugar-based carbon sources have shown that enzyme concentration and removal efficiency are determining factors for pollutant removal efficiency due to the presence of some organic substrates and the possible repression of catabolites [20, 24]. In general, these three mechanisms are coupled in the use of microalgae for wastewater treatment because compounds can be adsorbed on the surface, followed by their transfer to cells *via* active processes [20, 24].

In addition, this invention will increase biomass recovery and revenue for various sectors. In general, it has been reported that the three mechanisms can go hand in hand in WWT microalgae, as biodegraded particles can be adsorbed on their surface, followed by particle transfer to cells *via* active processes. Eventually, enzymatic reactions promote precipitation, biotransformation, and bioaccumulation within cells [7, 13, 20].

ENVIRONMENTAL APPLICATIONS OF MICROALGAE IN WATER PURIFICATION

Domestic Effluents

In domestic water, nitrates, phosphate, and organic compounds are present, primarily carbon (C), phosphorus (P), and nitrogen (N) are prevalent pollutants. These pollutants cause eutrophication and lead to oxygen depletion in aquatic environments. Detergents account for approximately 50% of the total phosphorus P in domestic water, whereas nitrogen is released by metabolic interconversion. They occur in the form of ammonium, nitrate, nitrite, and o-phosphate in water. Nitrogen derivatives serve as sources of nutrients for microalgae, which are assimilated into cells for growth, whereas phosphates are intermediates in the metabolic activity of carbohydrates. Microalgae use 1 ton of P and 5 tons of N to generate 100 tons of biomass [25 - 27].

Organic pollutants damage water quality by decreasing O_2 concentration, thereby damaging the aquatic environment. High concentrations of biodegradable organic pollutants are found in wastewater, resulting in increased pollution loads, such as biochemical oxygen demand (BOD), total suspended solids (TSS), and chemical oxygen demand (COD). Chemical and biological methods are used to oxidize the carbon atoms of organic pollutants into CO_2 and H_2O. Microalgae use CO_2 for photosynthesis, which generates oxygen, which is needed for the degradation of organic forms of carbon, nitrogen, and phosphorus, which are converted into CO_2

and inorganic nitrogen and phosphorus by heterotrophic bacteria [7]. Studies have reported that *Chlorella* can reduce 75% of organic substances in kitchen wastewater [13]. Table **2** presents the percentage removal efficiency of nitrogen and phosphorus by various microalgae species used to study the treatment of domestic wastewater.

Table 2. Efficiency of nitrogen and phosphorous removal efficiency by microalgae in domestic wastewater

Microalgae	Eliminated, P (%)	Eliminated, N (%)	Reference
Alginate-immobilized microalgae Chlorella vulgaris	100	83	[28]
Chlorella pyrenoidosa and *Scenedesmus abundans*	97	98,86	[29]
Chlorella	73	67	[30, 31]
Chlorella sorokiniana	86	99	[32]
Scenedesmus sp. ZTY1	97	90	[33]
Spirulina platensis	94,13	92,58	[34]
Oscillatoria tenuis	82,9	96,1	[35]
Micractinium reisseri and *Scenedesmus obliquus*	94	80	[36]

Industrial Effluents

Industrial waters are the main cause of pollution due to the direct or indirect release of toxic pollutants. Industrial wastewater originates from various sources, processes, cooling and heating operations, and sanitary wastes, among others. The composition of this wastewater is varied and includes pesticides, dioxins, phenolic compounds, dioxins, polyaromatic hydrocarbons (PAHs), microorganisms, petrochemicals, heavy metals, and polychlorinated biphenyls (PCBs). These processes cause an increase in COD, TSS, BOD, TDS (total dissolved solids), and toxic metals (Pb, Cr, Cd, and Ni) [7, 13].

Exfluents from the textile industry are the main source of water pollution due to their high pH levels, increased COD, elevated temperatures, and the presence of various dyes. The color and quality of fabrics are achieved using artificial dyes and chemicals. Synthetic dyes are widely used compared to natural dyes because they are light-stable, resistant to microbial attack and temperature changes, and easy to use and synthesize [13]. Microalgae are used for the treatment of wastewater because they can use nutrients and dyes for their growth in the bioconversion process. During the bioconversion process, the microalgae consume the dyes as a carbon source and are converted into metabolites. This process will depend on the species of the microalgae and the structure of the dye. During a 10-day incubation period, various species of microalgae showed different percentages of decolorization of synthetic dyes, *i.e.*, *Gloeocapsa sp.* (9-98%), *Oscillatoria sp.* (4-62%), *Phormidium sp.* (14-69%), *Aphanotheca sp.* (8-83%), *Synechocystis sp.* (19-58%), and *Synechococcus sp.* (8-53%) [37]. The removal of dyes in wastewater is associated with biosorption and bioconversion. *C. vulgaris* is an adsorbent for Orange-G dye removal in 10 min at 10°C and pH 5. Fourier Transform Infrared Spectroscopy analysis of the microalgae biomass indicated that the stress developed by the binding of dye molecules with microalgae changed the adsorption peaks and generated new bonds (C-Br, C-Cl, C=S, aromatic rings), which play a role in the multilayer model of the adsorption mechanism and may be responsible for dye binding [30].

Tannery Effluent

This wastewater contains a wide variety of chemical compounds, such as ammonium salts, sulfates, chlorides, surfactant polymers, resins, replenishing agents, dyes, pigments, organic nitrogen, fats, proteins, heavy metals, and so on. 90% of the water used in tanneries is released into effluents, which are highly toxic and must be treated with advanced methods before being discharged into water sources. Microalgae have been used to bioremediate these effluents with high performance for treating tannery effluents under dilute conditions. The highest removal rates for effluents 50RE50W, 50PE50W, and 50BE50W diluted in 50% water were 51.02%, 99.90%, 82.88%, and 91.75% for COD, N-NH3, TKN, and P-PO4, respectively. The most efficient results were obtained at 50% dilution, compared to 75% and 100%. This may be because at 75% dilution, there are few nutrients for microalgae growth, and at 100% dilution, the dark color of the effluent prevents the entry of light, resulting in a slow growth rate and low pollutant removal [38].

Immobilization of cells refers to the process by which all parts of an aqueous phase within a system are fixed in place by natural or artificial means. Microalgae are immobilized with various polymers and present alternatives, such as improved nutrient uptake, increased cell retention time within bioreactors, and increased metabolic activity. Jaysudha and Smapathkumar [39] found that phosphate removal with immobilized *Chlorella salina* resulted in a high accumulation of 99.39%, compared with 81.94% in free cells, and nitrogen removal with immobilized *C. salina* resulted in a higher accumulation of 98.71%, compared with 9.11% in free cells [39].

Effluents from the Pharmaceutical Industry

Wastewater, which is generated during the drug manufacturing process, presents a wide variety of compounds, *i.e.*, lipid regulators, antibiotics, antiepileptics, psychiatric drugs, anti-inflammatory drugs, analgesics, diuretics, and beta-blockers. These substances can be toxic to living organisms due to their ability to be absorbed by living organisms. Gokovic *et al.* reported that most of the lipophilic compounds biperiden, trihexyphenidyl, clomipramine, and amitriptyline accumulated in the biomass of most microalgae, with a correlation between their accumulation and their content in the total biomass. In contrast, the most persistent compounds in the growth were hydrophilic compounds, such as caffeine, fluconazole, trimethoprim, codeine, carbamazepine, oxazepam, and tramadol, with a removal of 60% in wastewater [40]. Escape *et al.* (2015) found that the removal kinetics of pharmaceuticals was 2.3 times higher for salicylic acid than for paracetamol, achieving a removal rate of 93% for salicylic acid in semi-continuous culture. Nutrient removal was almost complete at the end of the batch culture, with over 70% for nitrates and 89% for phosphates in the semi-continuous culture [41].

The concentration of pharmaceutical pollutants in urban wastewater can be significantly reduced during algae cultivation. Notable examples of pharmaceuticals that have shown reduced concentrations during algae cultivation include beta-blockers such as atenolol, bisoprolol, and metoprolol, as well as clarithromycin, bupropion, atracurium, diltiazem, and terbutaline [42]. Naproxen is one of the most commonly used nonsteroidal drugs that can affect mRNA expression and have gastrointestinal or renal effects in zebrafish. At high concentrations of naproxen, chlorophyll a, carotenoid content, and enzyme activities of *Cymbella sp.* and *Scenedesmus quadricauda* were affected, and growth inhibition was 100% at 100 mg/L. *Cymbella sp.* showed a more satisfactory effect on naproxen bioremediation than *Scenedesmus quadricauda*, with high naproxen removal efficiency and bioaccumulation of naproxen [43]. Climbazole is an imidazole fungicide that is widely used in personal care products such as gels, toothpaste, shampoo, conditioners, anti-dandruff, and antifungals. *S. obliquus* achieved 88% removal of climbazole after 12 days of incubation. Biotransformation was the predominant process to remove climbazole, whereas the contribution of bioaccumulation and bioabsorption was negligible [44].

BIOMASS PRODUCTION BY MICROALGAE

The biomass generated by microalgae in wastewater treatment can be extracted and converted into various value-added bioproducts, fuels, chemicals, and so on. Microalgae have higher oil yields than other biofuel feedstock crops, such as palm oil and soybeans. Microalgae cultivation is a modern biotechnological development, and it is estimated that by 2036, the global biomass market will have a production of 5000 tons of dry matter per year, which will be ca. $125 million [2, 4].

The microalgal biomass cultivation process generates a suspension with a low concentration of metabolites; therefore, it must be dewatered. In the first stage of dewatering, a thickening is generated, and in the second stage, the biomass is concentrated for further processing. Subsequent stages, such as drying and cell disruption, are crucial for selecting and using the harvested material [4].

Photosynthesis is the primary process through which renewable energy biomass is produced. During photosynthesis, photons from sunlight are absorbed by chlorophyll, and charge separation is generated that expelling electrons, which can dissociate water and generate protons and oxygen, eq (**1**). This process allows the reduction of CO_2 in organic materials, which can be converted into proteins, carbohydrates, and lipids [4].

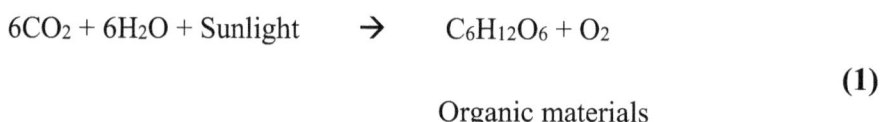

$$6CO_2 + 6H_2O + \text{Sunlight} \quad \rightarrow \quad C_6H_{12}O_6 + O_2$$

(1)

Organic materials

Organic Materials

This process can be separated into two phases: light-dependent (photochemical and redox) and light-independent (enzymatic) reactions. Light-dependent reactions last for a few milliseconds eq. (**2**), whereas light-independent reactions last for seconds to hours eq. (**3**). The notation [H] refers to the combination of reduced co-enzyme nicotinamide adenine dinucleotide phosphate (NADH) and an electron [4].

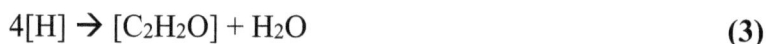

$$2H_2O \rightarrow 4[H] + O_2 + \text{Energy} \qquad \textbf{(2)}$$

$$4[H] \rightarrow [C_2H_2O] + H_2O \qquad \textbf{(3)}$$

Heterotrophic microalgae consume organic carbon and soluble carbonates dissociated from carbonic acids, either by direct ingestion or by conversion of CO_2 to carbonates with carbonic anhydrase eq. (**4**) [45].

$$CO_2 + H_2O \rightleftarrows H_2CO_3 \rightleftarrows HCO_3^- + H^+ \rightleftarrows CO_3^{2-} + 2H^+ \qquad \textbf{(4)}$$

Therefore, if heterotrophic microalgae use organic carbon ($[CH_2O]_n$) as their sole carbon source, they can omit the energy and replace the synthetic process of forming organic carbon *via* CO_2 assimilation during photosynthetic metabolism eq. (**5**). The algal biomass is represented by $C_{106}H_{263}O_{110}N_{16}P$.

$$16NH_4^+ + 92CO_2 + 92H_2O + 14HCO_3^- + HPO_4^{2-} \rightarrow C_{106}H_{263}O_{110}N_{16}P + 106O_2 \quad \textbf{(5)}$$

Microalgae have high CO_2 sequestration efficiency and have an additional environmental benefit in their ability to use nutrients to remediate wastewater. Microalgae are also known to have high oil yields compared to other biofuel feedstock crops such as palm oil and soybeans. Currently, there is increasing interest in high (*e.g.*, nutraceuticals) and low (*e.g.*, biofuels) value-added biofuel products, biofuels) added value from microalgae due to the scarcity of fossil fuels and climate change, which encourages the search for renewable, sustainable, and clean energy sources. Microalgae have several advantages, *i.e.*, high photosynthetic efficiency in biomass production, fast growth rate, and high lipid yield [4].

Nutritional Supplement

Microalgae biomass can be used as supplementary feed to improve product characteristics and disease resistance in ruminants, goats, fish, poultry, and other food sources. The total lipid content ranges from 20–50% of the cell's dry weight. Fatty acids with 14–20 carbon atoms can be used to make biodiesel, whereas PUFAs (polyunsaturated fatty acids) with more than 20 carbon atoms have been used in dietary supplements (docosahexaenoic acid and eicosapentaenoic acid) [16]. Fatty acids help in the detection and reduction of various diseases, *e.g.*, cardiovascular, hypertension, cholesterol, arteriosclerosis, rheumatoid arthritis, and cholesterol [16, 46]. Microalgae also contain carotenoids such as neoxanthin, β-carotene, astaxanthin, violaxanthin, canthaxanthin, and vitamins such as B1, B12, and B2, which can serve as antioxidants [16, 47].

Biodiesel Production

This is a sustainable fuel produced from renewable biomass that can replace petroleum diesel. It contains no sulfate and has lower particulate matter and gas emissions. Microalgae that produce biodiesel include *Chlorella sp.*, *N. oculata*, *Botryococcus sp.*, *Scenedesmus sp.*, *Picochlorum sp.*, and *Saccharomyces cerevisiae*, which contain oleic acid that helps improve the oxidative stability of biodiesel [1]. Biodiesel is obtained *via* the transesterification of lipids (especially triacylglycerols) in the presence of an alkali or acid, resulting in the formation of an ester of short-chain fatty acids and alcohol. Crude oil extracted from microalgae has a higher viscosity than diesel oil, so it cannot be used in engines; therefore, to reduce viscosity and increase fluidity, a transesterification process is required. In Malaysia, biodiesel is obtained from *Chlorella sp.,* which has a high fatty acid and biodiesel content (34.53-230.38 mg/L/day) [1, 48].

Bioethanol and Biomethane Production

Microalgae contain enormous amounts of lipids and carbohydrates, such as starch, sulfated glycans, and cellulose, which can be reduced by chemical or enzymatic means to simple sugars, which are then converted to bioethanol under anaerobic conditions [1]. Methane is produced by microalgae through anaerobic digestion. The anaerobic biodegradation of microalgae is carried out by three groups of bacteria: (a) hydrolytic and fermentative bacteria that hydrolyze polymers to generate carboxylic acids and alcohols; (b) acetogenic bacteria that convert these acids and alcohols to acetates, hydrogen, and CO_2; and (c) methanogenic bacteria that convert the end products of b into methane and CO_2. *Macrocystis pyrifera* and *D. antarctica* are rich sources of renewable energy that can generate biogas with high methane concentrations in a two-stage anaerobic digestion system, with very similar compositions of biogas per gram due to the very similar elemental compositions (C, H, N) [49]. Microalgae are more effective for bioethanol production than corn, soybean, or sugarcane crops; it is almost twice that of sugarcane and five times that of corn [1]. Among the species found were *Chlorococcum spp.*, *Spirogyra spp.*, *C. sorokiniana*, *Gelidiumamansii*, *Sargassum spp.*, *Gracilaria spp.*, *Laminaria spp.*, and *Prymnesium parvum*.

CONCLUDING REMARKS

The use of microalgae in wastewater treatment offers numerous benefits, as microalgae can serve as a valuable source of food and biomass. Microalgae can remove, by different mechanisms, various pollutants generated from different water sources, *e.g.*, industrial, domestic, pharmaceutical, and tannery processes. Bioremediation of wastewater with microalgae does not generate secondary contamination because the biomass produced can be used to produce medicines, biofuels, and additives, among others. This technology offers many advantages over conventional technologies; however, its application still presents many challenges that require further research.

AUTHORS' CONTRIBUTION

Yuri Córdoba Campo contributed to the design and implementation of the chapter, the writing of the manuscript, and the edition of the figure.

Israel Valencia Quiroz performed the English language grammar revision and verification of the contents.

QUESTIONS RELATED TO THE TEXT

1. What are the primary sources of water pollution mentioned in the chapter, and

what types of pollutants do they typically release into water bodies?

2. How do microalgae contribute to wastewater treatment, and what are the key mechanisms by which they remove pollutants from contaminated water?

3. What advantages do microalgae-based systems have over conventional water treatment methods in terms of cost, energy consumption, and environmental impact?

4. What are the potential applications of biomass produced through microalgae-based wastewater treatment, and how do these applications contribute to a sustainable economic model?

5. What challenges does the chapter identify in the implementation of microalgae-based water treatment technologies, and what opportunities are highlighted for future development in this field?

6. How do different environmental factors, such as nutrient availability and temperature, affect the growth and pollutant removal efficiency of microalgae in wastewater treatment processes?

LIST OF ABBREVIATIONS

AAS	Atomic absorption spectroscopy
BOD	Biological oxygen demand
COD	Chemical oxygen demand
EAAS	Electrothermal atomic absorption spectrometry
FAAS	Flame atomic absorption spectrometry
ICP	Inductively coupled plasma
MS	Mass spectrometry
NADH	Adenine dinucleotide phosphate
PAHs	Polyaromatic hydrocarbons
PCBs	Polychlorinated biphenyls
PUFAs	Polyunsaturated fatty acid
TDS	Total dissolved solids
TSS	Total suspended solids
WW	Wastewater
WWT	Wastewater treatment

REFERENCES

[1]   Ahmad A, Banat F, Alsafar H, Hasan SW. Algae biotechnology for industrial wastewater treatment, bioenergy production, and high-value bioproducts. Sci Total Environ 2022; 806(Pt 2): 150585. [http://dx.doi.org/10.1016/j.scitotenv.2021.150585] [PMID: 34597562]

[2]   Peter AP, Khoo KS, Chew KW, *et al.* Microalgae for biofuels, wastewater treatment and environmental monitoring. Environ Chem Lett 2021; 19(4): 2891-904.

[http://dx.doi.org/10.1007/s10311-021-01219-6]

[3]   Khoo KS, Chia WY, Chew KW, Show PL. Microalgal-bacterial consortia as future prospect in wastewater bioremediation, environmental management and bioenergy production. Indian J Microbiol 2021; 61(3): 262-9.
[http://dx.doi.org/10.1007/s12088-021-00924-8] [PMID: 34294991]

[4]   Musa M, Ayoko GA, Ward A, Rösch C, Brown RJ, Rainey TJ. Factors affecting microalgae production for biofuels and the potentials of chemometric methods in assessing and optimizing productivity. Cells 2019; 8(8): 851.
[http://dx.doi.org/10.3390/cells8080851] [PMID: 31394865]

[5]   Unesco, World Water Assessment Programme (United Nations), UN-Water. Water and climate change. n.d.

[6]   Morseletto P, Mooren CE, Munaretto S. Circular economy of water: definition, strategies and challenges. Circ Econ Sustain 2022; 2(4): 1463-77.
[http://dx.doi.org/10.1007/s43615-022-00165-x]

[7]   Abdelfattah A, Ali SS, Ramadan H, *et al.* Microalgae-based wastewater treatment: Mechanisms, challenges, recent advances, and future prospects. Environ Sci Ecotechnol 2023; 13: 100205.
[http://dx.doi.org/10.1016/j.ese.2022.100205] [PMID: 36247722]

[8]   Abidli A, Huang Y, Ben Rejeb Z, Zaoui A, Park CB. Sustainable and efficient technologies for removal and recovery of toxic and valuable metals from wastewater: Recent progress, challenges, and future perspectives. Chemosphere 2022; 292: 133102.
[http://dx.doi.org/10.1016/j.chemosphere.2021.133102] [PMID: 34914948]

[9]   Usepa, Region, Foia. Guidelines for Drinking-water Quality Third edition incorporating the first and second addenda Recommendations Geneva 2008 WHO Library Cataloguing-in-Publication Data. 2008; 1.

[10]   Muir DCG, Howard PH. Are there other persistent organic pollutants? A challenge for environmental chemists. Environ Sci Technol 2006; 40(23): 7157-66.
[http://dx.doi.org/10.1021/es061677a] [PMID: 17180962]

[11]   Pennesi C, Rindi F, Totti C, Beolchini F. Marine macrophytes: Biosorbents. Springer Handbook of Marine Biotechnology, Springer Berlin Heidelberg 2015; pp. 597-610.
[http://dx.doi.org/10.1007/978-3-642-53971-8_24]

[12]   Johansen MN. Microalgae: Biotechnology, Microbiology and energy. Illustrated 2012.

[13]   Mustafa S, Bhatti HN, Maqbool M, Iqbal M. Microalgae biosorption, bioaccumulation and biodegradation efficiency for the remediation of wastewater and carbon dioxide mitigation: Prospects, challenges and opportunities. J Water Process Eng 2021; 41: 102009.
[http://dx.doi.org/10.1016/j.jwpe.2021.102009]

[14]   Soto-Ramírez R, Lobos MG, Córdova O, Poirrier P, Chamy R. Effect of growth conditions on cell wall composition and cadmium adsorption in Chlorella vulgaris: A new approach to biosorption research. J Hazard Mater 2021; 411: 125059.
[http://dx.doi.org/10.1016/j.jhazmat.2021.125059] [PMID: 33450636]

[15]   Song C, Wei Y, Qiu Y, Qi Y, Li Y, Kitamura Y. Biodegradability and mechanism of florfenicol *via* Chlorella sp. UTEX1602 and L38: Experimental study. Bioresour Technol 2019; 272: 529-34.
[http://dx.doi.org/10.1016/j.biortech.2018.10.080] [PMID: 30391846]

[16]   Dayana Priyadharshini S, Suresh Babu P, Manikandan S, Subbaiya R, Govarthanan M, Karmegam N. Phycoremediation of wastewater for pollutant removal: A green approach to environmental protection and long-term remediation. Environ Pollut 2021; 290: 117989.
[http://dx.doi.org/10.1016/j.envpol.2021.117989] [PMID: 34433126]

[17]   Matamoros V, Uggetti E, García J, Bayona JM. Assessment of the mechanisms involved in the removal of emerging contaminants by microalgae from wastewater: a laboratory scale study. J Hazard

Mater 2016; 301: 197-205.
[http://dx.doi.org/10.1016/j.jhazmat.2015.08.050] [PMID: 26364268]

[18]   Ibrahim WM, Karam MA, El-Shahat RM, Adway AA. Biodegradation and utilization of organophosphorus pesticide malathion by Cyanobacteria. BioMed Res Int 2014; 2014: 1-6.
[http://dx.doi.org/10.1155/2014/392682] [PMID: 24864237]

[19]   Ding T, Yang M, Zhang J, *et al.* Toxicity, degradation and metabolic fate of ibuprofen on freshwater diatom Navicula sp. J Hazard Mater 2017; 330: 127-34.
[http://dx.doi.org/10.1016/j.jhazmat.2017.02.004] [PMID: 28214648]

[20]   Xiong Q, Hu LX, Liu YS, Zhao JL, He LY, Ying GG. Microalgae-based technology for antibiotics removal: From mechanisms to application of innovational hybrid systems. Environ Int 2021; 155: 106594.
[http://dx.doi.org/10.1016/j.envint.2021.106594] [PMID: 33940395]

[21]   Xiong JQ, Kurade MB, Kim JR, Roh HS, Jeon BH. Ciprofloxacin toxicity and its co-metabolic removal by a freshwater microalga Chlamydomonas mexicana. J Hazard Mater 2017; 323(Pt A): 212-9.
[http://dx.doi.org/10.1016/j.jhazmat.2016.04.073] [PMID: 27180206]

[22]   Xiong Q, Liu YS, Hu LX, *et al.* Co-metabolism of sulfamethoxazole by a freshwater microalga Chlorella pyrenoidosa. Water Res 2020; 175: 115656.
[http://dx.doi.org/10.1016/j.watres.2020.115656] [PMID: 32145399]

[23]   Xiong JQ, Kim SJ, Kurade MB, *et al.* Combined effects of sulfamethazine and sulfamethoxazole on a freshwater microalga, Scenedesmus obliquus: toxicity, biodegradation, and metabolic fate. J Hazard Mater 2019; 370: 138-46.
[http://dx.doi.org/10.1016/j.jhazmat.2018.07.049] [PMID: 30049519]

[24]   Vo HNP, Ngo HH, Guo W, *et al.* Micropollutants cometabolism of microalgae for wastewater remediation: Effect of carbon sources to cometabolism and degradation products. Water Res 2020; 183: 115974.
[http://dx.doi.org/10.1016/j.watres.2020.115974] [PMID: 32652348]

[25]   Cai T, Park SY, Li Y. Nutrient recovery from wastewater streams by microalgae: Status and prospects. Renew Sustain Energy Rev 2013; 19: 360-9.
[http://dx.doi.org/10.1016/j.rser.2012.11.030]

[26]   Jalilian N, Najafpour GD, Khajouei M. Macro and micro algae in pollution control and biofuel production – A Review. ChemBioEng Rev 2020; 7(1): 18-33.
[http://dx.doi.org/10.1002/cben.201900014]

[27]   Sepúlveda C, Acién FG, Gómez C, Jiménez-Ruíz N, Riquelme C, Molina-Grima E. Utilization of centrate for the production of the marine microalgae Nannochloropsis gaditana. Algal Res 2015; 9: 107-16.
[http://dx.doi.org/10.1016/j.algal.2015.03.004]

[28]   Mujtaba G, Lee K. Treatment of real wastewater using co-culture of immobilized Chlorella vulgaris and suspended activated sludge. Water Res 2017; 120: 174-84.
[http://dx.doi.org/10.1016/j.watres.2017.04.078] [PMID: 28486168]

[29]   Lekshmi B, Joseph RS, Jose A, Abinandan S, Shanthakumar S. Studies on reduction of inorganic pollutants from wastewater by Chlorella pyrenoidosa and Scenedesmus abundans. Alex Eng J 2015; 54(4): 1291-6.
[http://dx.doi.org/10.1016/j.aej.2015.09.013]

[30]   Kumar S, Ahluwalia AS, Charaya MU. Adsorption of Orange-G dye by the dried powdered biomass of Chlorella vulgaris Beijerinck. Curr Sci 2019; 116(4): 604-11.
[http://dx.doi.org/10.18520/cs/v116/i4/604-611]

[31]   Kumar PK, Krishna SV, Naidu SS, Verma K, Bhagawan D, Himabindu V. Biomass production from

microalgae Chlorella grown in sewage, kitchen wastewater using industrial CO2 emissions: Comparative study. Carbon Resources Conversion 2019; 2(2): 126-33.
[http://dx.doi.org/10.1016/j.crcon.2019.06.002]

[32] Rani S, Chowdhury R, Tao W, Srinivasan A. Tertiary treatment of municipal wastewater using isolated algal strains: treatment efficiency and value-added products recovery. Chem Ecol 2020; 36(1): 48-65.
[http://dx.doi.org/10.1080/02757540.2019.1688307]

[33] Zhang TY, Wu YH, Hu HY. Domestic wastewater treatment and biofuel production by using microalga Scenedesmus sp. ZTY1. Water Sci Technol 2014; 69(12): 2492-6.
[http://dx.doi.org/10.2166/wst.2014.160] [PMID: 24960012]

[34] Zhai J, Li X, Li W, *et al.* Optimization of biomass production and nutrients removal by Spirulina platensis from municipal wastewater. Ecol Eng 2017; 108: 83-92.
[http://dx.doi.org/10.1016/j.ecoleng.2017.07.023]

[35] Cheng Q, Deng F, Li H, Qin Z, Wang M, Li J. Nutrients removal from the secondary effluents of municipal domestic wastewater by *Oscillatoria tenuis* and subsequent co-digestion with pig manure. Environ Technol 2018; 39(24): 3127-34.
[http://dx.doi.org/10.1080/09593330.2017.1375020] [PMID: 28859537]

[36] Abou-Shanab RAI, El-Dalatony MM, EL-Sheekh MM, *et al.* Cultivation of a new microalga, Micractinium reisseri, in municipal wastewater for nutrient removal, biomass, lipid, and fatty acid production. Biotechnol Bioprocess Eng; BBE 2014; 19(3): 510-8.
[http://dx.doi.org/10.1007/s12257-013-0485-z]

[37] Sasikala C, Sudha SS. Phycoremediation of textile dying effluents with algal species from aquatic origin. Scrutiny International Research Journal of Microbiology and Bio Technology 2015; 2: 7-18.

[38] Pena ACC, Bertoldi CF, Fontoura JT, Trierweiler LF, Gutterres M. Consortium of microalgae for tannery effluent treatment. Braz Arch Biol Technol 2019; 62: e19170518.
[http://dx.doi.org/10.1590/1678-4324-2019170518]

[39] Jaysudha S, Sampathkumar P. Nutrient removal from tannery effluent by free and immobilized cells of marine microalgae Chlorella salina. International Journal of Environmental Biology 2014; 4: 21-6.

[40] Gojkovic Z, Lindberg RH, Tysklind M, Funk C. Northern green algae have the capacity to remove active pharmaceutical ingredients. Ecotoxicol Environ Saf 2019; 170: 644-56.
[http://dx.doi.org/10.1016/j.ecoenv.2018.12.032] [PMID: 30579165]

[41] Escapa C, Coimbra RN, Paniagua S, García AI, Otero M. Nutrients and pharmaceuticals removal from wastewater by culture and harvesting of Chlorella sorokiniana. Bioresour Technol 2015; 185: 276-84.
[http://dx.doi.org/10.1016/j.biortech.2015.03.004] [PMID: 25780903]

[42] Gentili FG, Fick J. Algal cultivation in urban wastewater: an efficient way to reduce pharmaceutical pollutants. J Appl Phycol 2017; 29(1): 255-62.
[http://dx.doi.org/10.1007/s10811-016-0950-0] [PMID: 28344390]

[43] Ding T, Lin K, Yang B, *et al.* Biodegradation of naproxen by freshwater algae Cymbella sp. and Scenedesmus quadricauda and the comparative toxicity. Bioresour Technol 2017; 238: 164-73.
[http://dx.doi.org/10.1016/j.biortech.2017.04.018] [PMID: 28433904]

[44] Pan CG, Peng FJ, Ying GG. Removal, biotransformation and toxicity variations of climbazole by freshwater algae Scenedesmus obliquus. Environ Pollut 2018; 240: 534-40.
[http://dx.doi.org/10.1016/j.envpol.2018.05.020] [PMID: 29758527]

[45] Chai WS, Tan WG, Halimatul Munawaroh HS, Gupta VK, Ho SH, Show PL. Multifaceted roles of microalgae in the application of wastewater biotreatment: A review. Environ Pollut 2021; 269: 116236.
[http://dx.doi.org/10.1016/j.envpol.2020.116236] [PMID: 33333449]

[46] Masojidek J, Torzillo G. Mass cultivation of freshwater microalgae. Encyclopedia of Ecology 2008;

2226-35.
[http://dx.doi.org/10.1016/B978-008045405-4.00830-2]

[47]   Ding Y, Wang Y, Liu X, Song X. Improving nutrient and organic matter removal by novel integration of a high-rate algal pond and submerged macrophyte pond. Pol J Environ Stud 2019; 29(1): 997-1001.
[http://dx.doi.org/10.15244/pjoes/99824]

[48]   Vello V, Phang SM, Chu WL, Abdul Majid N, Lim PE, Loh SK. Lipid productivity and fatty acid composition-guided selection of Chlorella strains isolated from Malaysia for biodiesel production. J Appl Phycol 2014; 26(3): 1399-413.
[http://dx.doi.org/10.1007/s10811-013-0160-y]

[49]   Vergarafernández A, Vargas G, Alarcón N, Velasco A. Evaluation of marine algae as a source of biogas in a two-stage anaerobic reactor system. Biomass Bioenergy 2008; 32(4): 338-44.
[http://dx.doi.org/10.1016/j.biombioe.2007.10.005]

Biological Desalination: Biotechnological Alternatives for Freshwater Extraction

Ana K. Villagómez-Guzmán[1]**, Héctor M. Arreaga-González**[2] **and Tzasna Hernández-Delgado**[1,*]

[1] *Natural Products Bioactivity Laboratory, UBIPRO, Superior Studies Faculty (FES)-Iztacala, National Autonomous University of México (UNAM), Tlalnepantla de Baz, Mexico State, Mexico*

[2] *Biotechnology Laboratory, Institute of Agroindustries, Technological University of Mixteca, Heroic City of Huajuapan de León, Oaxaca, Mexico*

Abstract: Water is essential for life, and accessibility has been a global problem for many years. The estimations provided by the United Nations suggest that the worldwide population will increase to 10.2 billion individuals by the year 2050. As a result, a notable rise in water consumption is projected due to changes in lifestyle and demographic expansion. Numerous innovations have arisen within desalination technology over the past few decades. Thus, the desalination of seawater plays a crucial role in enhancing the accessibility of potable water to populations in need. However, this process requires considerable energy to produce freshwater. Various desalination methods exist for seawater, including thermal and membrane processes. Nevertheless, this practice poses risks to the environment and entails substantial costs. With the swift advancement of industrialization and urbanization on a global scale, there is a rising need for an enhanced clean water supply. Sustainability is becoming a focal point in desalination and wastewater management to meet the increasing global demand for clean water. Given the limited availability of freshwater resources, the sector is progressively considering using recycled water as a crucial approach to guaranteeing sustainable business operations. Biological desalination involves a novel approach that employs diverse salt-tolerant organisms such as bacteria, microalgae, halophyte and halotolerant plants, microbial electrochemical systems, biological membranes, and biopolymers. Compared with conventional desalination techniques, biological tools require less energy and have fewer environmental impacts. Consequently, they are recognized as a more environmentally friendly and sustainable desalination method.

Keywords: Biological desalination, Desalination, Phytodesalination, Seawater desalination, Water desalination.

---

* **Corresponding author Tzasna Hernández-Delgado:** Natural Products Bioactivity Laboratory, UBIPRO, Superior Studies Faculty (FES)-Iztacala, National Autonomous University of Mexico (UNAM), Tlalnepantla de Baz, Mexico State, Mexico; Tel: +525527288013; E-mail: tzasna@unam.mx

Israel Valencia Quiroz (Ed.)

INTRODUCTION

Water scarcity represents an urgent challenge in our time, impacting around four billion people who face severe water deficits for at least a month a year [1]. Numerous public health emergencies are presently being impacted by water sources, such as floods, droughts, and waterborne illnesses. This problem is intensified by climate change, amplifying floods and droughts, altering precipitation patterns, changing water sources, hastening the melting of glaciers, and rising sea levels. Access to clean water is crucial for human well-being, industry, agriculture, and energy generation, making water a significant humanitarian concern [2 - 5]. Despite the vast water resources on Earth's surface, more than 99% are currently inaccessible for human consumption. Over 97% of Earth's water exists as saltwater in oceans, bays, seas, and saline aquifers, acting as significant reservoirs. Various assessments showed that the usable freshwater supply was approximately 0.7% [6 - 8].

Projections from the United Nations indicate that the global population will increase from 9.4–10.2 billion by 2050 [9, 10]. Consequently, a significant increase in water consumption is anticipated due to lifestyle shifts and population growth. In addition, projections from the United Nations indicated that by 2025, around 2.7 billion people could face water scarcity challenges. The study also forecasts that between 2.7 and 3.2 billion individuals will live under severe water scarcity conditions by 2050 [11]. Although there are adequate freshwater reserves globally, the primary obstacle is the need for additional infrastructure to produce and distribute potable water in specific regions [12]. Among the many difficulties in the quest to produce high-quality drinking water, seawater desalination has been recognized as a crucial step in advancing the development of sustainable freshwater sources [1, 7].

Desalination is a technological procedure for eliminating salts and dissolved impurities from brackish or saltwater [13], with seawater commonly serving as the primary source. Desalination is a widely used method to address water scarcity in some areas of the world where brackish or saline water is present [14]. The estimated value of the global water desalination market is USD 17.47 billion in 2024, with a projected increase of USD 31.32 billion by 2031. Concerning the technological aspects, reverse osmosis (RO) is projected to have a market share of 36.7% by 2024 [15].

Significant advances are currently being made in desalination technologies to mitigate water scarcity. Various membrane distillation methodologies are gaining traction owing to their ability to desalinate seawater with significantly lower energy requirements than conventional thermal distillation approaches. For instance, combining solar thermal desalination facilities with membrane distillation has proven highly efficient in dry coastal areas. These facilities have been established in countries such as Spain, Australia, and Mexico to supply clean drinking water to local populations [15].

Ensuring global access to safe water is crucial, and climate change requires energy efficiency, minimal greenhouse gas emissions, sustainable water consumption, and pollution reduction. Biological desalination is a novel approach that exploits the absorption and adsorption of salts by diverse salt-tolerant organisms [16]. A wide array of different species of organisms, including plants, microalgae, and bacteria, have undergone adaptations to thrive in environments characterized by fluctuating levels of salinity, evolving intricate mechanisms aimed at expelling excess intracellular NaCl. This phenomenon has sparked the interest of the scientific community, prompting a thorough investigation into the feasibility of harnessing biological processes for seawater desalination [17].

Biodesalination is emerging as a cutting-edge technology designed for the targeted extraction of Na^+ and Cl^- from saline water bodies, leveraging the specialized capabilities of organisms with remarkably minimal energy expenditure. Within this realm, the scope includes microbial desalination cells, which facilitate the simultaneous desalination of water resources while concurrently executing wastewater treatment procedures, thereby combining two essential processes [17, 18]. Bacteria serve as bioelectricity producers that supply the necessary energy for desalination in microbial desalination cells. Additionally, numerous species of algal cells can thrive in high salt concentrations, absorbing and storing them internally, thus enabling their direct application in seawater desalination without prior processing. Biopolymers play a role in seawater treatment by enhancing water evaporation in solar steam generators [18].

The direct use of living organisms in the desalination of water resources presents a promising approach for further exploration. Nevertheless, the advancement of these innovative technologies and their eventual practical implementation hinge on the precise selection of living organisms that are most suitable for desalinating seawater [17]. In contrast to traditional desalination methods, biological desalination systems exhibit reduced energy requirements, a limited environmental footprint, and fewer engineering intricacies. As a result, it has been acknowledged as a more sustainable and environmentally friendly approach [16]. This chapter presents a concise overview of existing and emerging advancements in water desalination biotechnology, particularly focusing on the attainment of biological states in the desalination process.

Current Distribution State of Water Worldwide

The earth contains a significant amount of water, but how much of it is actually available to us? Our planet's surface is approximately covered by 71% of water, with a total volume of water available estimated to be around 1,386,000,000 km³, with the majority being held in the oceans with salt water (96.5%) and other saline water 0.9%. The 2.5% of freshwater, about 10,633,450 km³, can be found in various forms, such as water glaciers and icecaps (68.7%), groundwater (30.1%), and freshwater surface (1.2%), about 93,113 km³. The surface water includes ground ice and permafrost (69%), lakes (20.9%), soil moisture (3.8%), swamps, marshes (2.6%), rivers (0.49%), atmosphere (3%), and freshwater for living things (0.26%) [6, 19, 20].

Although the amount of freshwater may seem substantial, several factors complicate its availability. These include population growth, increasing water demands for agriculture and industry, potential shifts in climate patterns, deteriorating freshwater quality due to global contamination from industrial, municipal, and agricultural sources, and rising energy requirements. These factors cause a rapid increase in water withdrawal rates relative to total water resources, leading to a significant part of the world's population currently facing severe water shortages. Water scarcity has become an essential global concern due to population expansion, industrial development, and pollution, all of which have intensified the problem [7, 20].

New findings from the Aqueduct Water Risk Atlas developed by the World Resources Institute (WRI) indicate that 25 nations, home to a quarter of the Earth's inhabitants, are currently exposed to extreme water stress and experience significant levels of water scarcity annually. The nations confronting "extreme water stress" denote that it is utilizing a minimum of 80% of their accessible supply, while "high water stress" signifies a withdrawal of 40% of their supply [21].

The risk of water scarcity is heightened in these regions, particularly during short-term droughts, leading to potential water shortages and sometimes requiring government intervention to restrict water access. Water stress has been observed at multiple locations worldwide, including England, India, Iran, Mexico, and South Africa. On the other hand, the countries experiencing the most severe water stress are Bahrain, Cyprus, Kuwait, Lebanon, Oman, and Qatar. The primary factors contributing to water stress in these nations are inadequate water supply levels and high demand from the residential, agricultural, and industrial sectors [21].

The statistics presented by the World Health Organization (WHO) and UNICEF in 2023 reveal that an estimated 2 billion individuals globally face challenges in accessing safe drinking water services that are managed safely. The World Bank's findings in 2023 underscore that roughly 2 billion individuals worldwide encounter difficulties in securing safe access to safe drinking water services. Among them 771 million lack access to essential drinking water services. Data published by WHO and UNICEF in 2019 show that over half of the global population (4.2 billion people) lacks access to safe sanitation services. As the Stockholm International Water Institute (SIWI) reported, roughly two-thirds of international transboundary rivers lack a cooperative management structure. The Food and Agriculture Organization (FAO) highlights that agriculture accounts for 70% of all water extraction in the world [22].

Without improvements in water infrastructure and management, water stress is poised to exacerbate, especially in regions characterized by rapidly growing populations and expanding economies. By the year 2050, an additional 1 billion individuals will be residing in areas experiencing exceedingly high levels of water stress despite global efforts to restrict the escalation of average temperatures to a range between 1.3 °C and 2.4 °C (equivalent to 2.3 °F to 4.3 °F) by the year 2100. The most significant change in water demand from this period until 2050 is anticipated to occur in sub-Saharan Africa. Although most nations in Sub-Saharan Africa are not facing severe water scarcity challenges, the surge in demand is more rapid in this region than elsewhere. Projections suggest that by 2050, water demand in Sub-Saharan Africa will experience a remarkable surge of 163% — a growth rate four times higher than that of Latin America, the second-highest region, where a 43% increase in water demand is expected [21].

Attainment of universally managed services will necessitate a sixfold escalation. The utilization of wastewater for water reclamation is emerging as a significant tactic. While wastewater and sludge use is widespread globally, a considerable portion is utilized informally or without adequate treatment and controls to safeguard human and environmental well-being [23].

Salinity and Conventional Desalination Process

Salinity is defined as the presence of at least 35 g.L^{-1} of dissolved salts in water. The primary contributors to salinity are sodium and chloride ions [24, 25]. Salinity has been categorized into various water classifications based on the concentration of chloride ions. For instance, the concentration of chloride ions ranges from 30 to 150 mg.L^{-1} in freshwater, 300 to 1000 mg.L^{-1} in brackish water and 30,000 to 35,000 mg.L^{-1} in seawater. The World Health Organization established the chloride ion concentrations in drinking water at a maximum of 250 mg.L^{-1}. Achieving a removal efficiency of around 99.28% is necessary to provide consumers with drinking water, considering seawater's prevailing chloride ion concentration [24].

Several advancements have been made in the field of desalination technology over the past decades. Desalination denotes the procedure of extracting salt particles from water, a method commonly utilized for the purification of seawater and brackish water sources. The salts eliminated in the desalination process encompass $NaSO_4$, $NaCl$, $MgCl_2$, $LiCl$, and $MgSO_4$ [26].

In 2015, nearly 18,426 desalination plants had been implemented globally in 130 nations, with a combined production capacity of 22,870 MGD (million gallons per day) or 86.55 million cubic meters per day [26]. In 2020, there were approximately 16,876 operational desalination plants worldwide, with 270 plants under construction and 3,825 plants offline. Reverse osmosis (RO) represented 85% of total desalination plants (14,360). Moreover, out of the 270 plants under construction, 247 contracts were awarded for membrane-based desalination technologies, indicating that RO represents around 91% of total desalination capacity. Desalination capacities for producing freshwater reached 97.2 million m^3/day for installed capacity and 114.9 million m^3/day for cumulative capacity across 20,971 projects. The largest operational desalination plants will be found across the continents of the Middle East. This region is responsible for 39% of the total global desalination capacity, followed by Asia at 21%, the Americas at 18%, Europe at 11%, Africa at 8%, and Oceania at 3% [27].

The Middle East region has been the predominant force in the global water desalination market for an extended period; this is a result of severe water scarcity challenges experienced by numerous countries in the area. It is projected that the region will hold a market share of 37.6% by the year 2024. Nations such as Saudi Arabia, the UAE, and Qatar experience arid climate conditions, leading to heightened reliance on desalination facilities to meet water demands for both municipal and industrial purposes. On the other hand, the Asia-Pacific region emerges as a promising area for future desalination market growth. Numerous coastal cities and island nations within this region, such as China, India, Indonesia, and Australia, are progressively encountering challenges related to saltwater intrusion in aquifers and excessive groundwater extraction [15].

There are various considerations for desalination water classification. Traditional standards have been used up to this point to establish desalination methods that have been employed for an extended period, such as thermal, electrical, and pressure [26]. Another proposed classification divides desalination techniques into physical and chemical methods [1]. The technologies predominantly utilized for desalination include reverse osmosis (RO), multistage flash distillation (MSF), multi-effect distillation (MED), and thermal distillation [28, 29].

Reverse Osmosis (RO)

This process uses semipermeable membranes to separate salts from water. Permeate water is obtained when the feed water is directed into the feed channel; here, minuscule water molecules pass through the membrane pores, yielding pure water while eliminating impurities.

Multi-stage Flash Distillation (MSF)

A procedure wherein seawater entering is elevated in pressure and heated close to the boiling point. By progressing through several stages, the pressure of seawater is reduced to produce vapor, which is then condensed by the incoming seawater.

Multiple-effect Distillation (MED)

The multiple-effect distillation process employs steam as a heat source and a sequence of evaporators operating at progressively lower pressures to yield water. The feedwater is heated in the tubes using steam for each stage, usually achieved by spraying salt water over them. Some of the water is vaporized, allowing the steam to enter the tubes of the next stage. As a result, the water heats up and causes more evaporation. Each stage primarily recycles energy from the previous phase, with successively decreasing pressures and temperatures.

Vapor-compression Distillation

Within this framework, the necessary heat for water evaporation on one side of the heat transfer interface is provided by the condensation of vapor-producing distillate on the opposing side of the interface [1, 14, 20, 26, 30 - 37].

Conventional methodologies are characterized by high energy consumption and the necessity of substantial chemical supplements, which contribute to a rise in the desalination facility's operational expenses [28]. Moreover, apart from the costs associated with energy consumption, significant concerns arise with the desalination above technologies, such as the emission of greenhouse gases, the substantial heat discharge, and the accumulation of high levels of brine in the surrounding ecosystem [16]. A typical RO desalination requires 1.2–10 kWh/m^3 of electricity for seawater desalination, contributing to the escalating costs of potable water treatment due to the extensive energy-intensive processes involved in water [28].

Biodesalination Concept

Biodesalination is an innovative concept that refers to the process of extracting salt from seawater or brackish water by using salt-tolerant microorganisms, predominantly algae, cyanobacteria, halophyte plants, and biological materials, either directly or indirectly [28, 38]. The fundamental concept underlying the process entails the requirement for adequate growth of microorganisms to adsorb or engulf ions present in high-salinity water, thereby ensuring efficient removal of salt. In addition, the biomass is nurtured sustainably, with minimal nutrient supplementation and the utilization of ambient solar radiation [39].

Biosorption and bioaccumulation are two mechanisms to remove salt from a saline biodesalination process. Bioadsorption refers to the rapid attachment of ions or molecules to a moist or dry algal surface. This phenomenon is a metabolic-independent and reversible process that occurs on the surface, driven by the surface properties of the substances involved, without energy consumption. In contrast, bioaccumulation is an active energy-dependent biological mechanism occurring within living cells, with swift passive salt adsorption on the external cell surface and transitioning into the slow movement of salt molecules inside the cell [39]. Bioadsorption has been identified as the predominant mechanism contributing significantly to the biodesalination process. This method is both straightforward and intriguing, as it provides an economical means of obtaining fresh water from saltwater, minimizing environmental impact [38]. These processes require minimal energy consumption, rendering the method highly energy efficient. In comparison to alternative seawater desalination methods, desalination using algae and cyanobacteria entails a reduced need for chemicals and operational components. Various methodologies are available for cultivating these microorganisms in seawater and brackish water [28].

Biodesalination through Salt-tolerant Organisms

Halophytes Plants

Halophytes are a vital plant group that exhibits a natural capacity to thrive in saline environments with salt levels of approximately 200 mM of NaCl or higher. These plants possess the ability to decrease the ion concentration of the solution within the xylem and exhibit the exclusion of Na^+ and Cl^- under conditions of high salinity. Halophytes have been categorized into succulent, non-succulent, and accumulating halophytes [40]. These species employ distinct physiological processes to manage salt stress and facilitate the equilibrium and expulsion of salt excess. The mechanisms include bioaccumulation, osmoregulation ability to draw Na^+ and Cl^- from the soil and store Na^+ in the vacuole through vacuolar Na^+/H^+ antiporter activity; exclusion, ultrafiltration mechanism to regulate the uptake of

Na$^+$, Cl$^-$ and heavy metals to prevent the escalation of harmful ions to toxic levels within sensitive plant tissues; and extrusion, salt excretion by developing different morphological adaptations (Fig. **1**) [40 - 43].

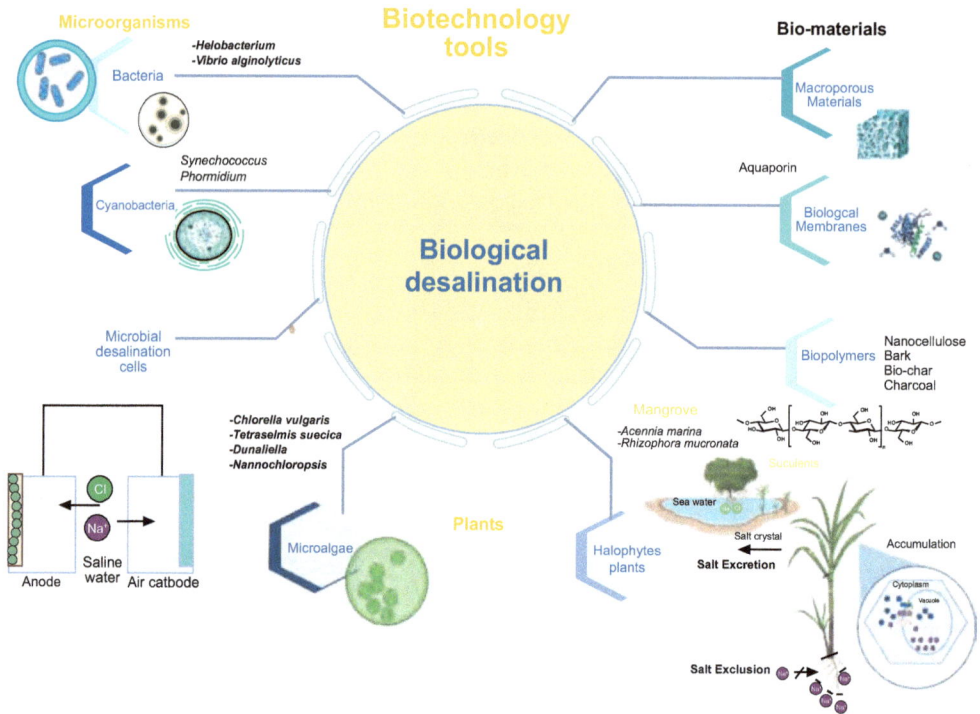

Fig. (1). Biological desalination techniques employed in water desalination include organisms and biomaterials. Figure made with Biorender.

Most halophytes possess various specialized structures that excrete surplus salts. These specialized structures include salt glands, salt bladders, and potentially other adaptations to facilitate the expulsion of excess salt. Salt glands play a crucial role in salt excretion by leaving through tiny holes. These structures are intricately integrated into the leaf surface and are comparable in size to stomata, with densities reaching up to 1000 per cm^2 on the leaf surface. These salt glands exhibit variations in the number of cells they comprise, serving as notable examples in various genera, such as *Avicennia, Frankenia, Limonium*, and *Tamarix*. In contrast, salt bladders are predominant on *Atriplex* leaf surfaces. Noteworthy are fleshy leaves in species like *Limonium* and *Atriplex*, which aid in the extrusion of surplus salts [43]. Conversely, accumulating halophytes lack a mechanism to expel salt from their plant body, leading to the buildup of excess salt in vacuoles and cell walls, eventually causing the shedding of leaves, like *Suaeda fruticosa, Juncus gerardii*, and others [40].

Halophytes are crucial for environmental management, wastewater treatment, and soil desalination. In Qatar, for example, halophytes like *Halopeplis perfoliata*, *Salicornia europaea*, *Salsola soda*, and *Tetraena qatarensis* have been observed to play crucial roles in phytoremediation, particularly in the remediation of polluted soils and waters in the region [43].

Mangroves are a member of the halophyte family of plants and are believed to help reduce salinity in coastal regions and areas exposed to saline conditions through their adaptive capabilities. Specifically, the root system of these plants is capable of uptaking nutrients to nourish the leaves, with the goal of reducing salt levels during the desalination process [44]. This activity is achieved through various mechanisms such as osmolyte accumulation, salt exclusion at the root level, airborne attraction of plants, and sequestration in older leaves or epidermis. Moreover, these plants are capable of ultrafiltration systems, which can eliminate approximately 90% of Na^+ ions in the adjacent seawater *via* the roots [45].

Chimayati and Titah 2019 [44] studied the impact of salinity reduction on the mangrove species *Rhizophora mucronata* and *Avicennia marina* through the introduction of *Vibrio alginolyticus* bacteria in a biodesalination procedure implemented in a reed bed system. Their findings revealed a reduction in salinity levels during the experiment, reaching 25% to 64.68% until the final day. In addition, the work developed by Puspaningrum and Titah [45] showed that the sample (20.09%) at the entry point decreased to 1.99% at the exit point after the application of *Rhizophora mucronata* treatment.

A bio-thermic desalination system was implemented using halophytes for seawater and saltwater desalination, incorporating a combination of *Bruguiera cylindrica* and *Rhizophora* mangle. In this system, the halophytes function similarly to selective membranes that generate freshwater through transpiration and condensation. Plants use transpiring water to acquire adequate nutrients and regulate the temperature of their leaf surfaces. The findings underscored the capacity to generate fresh water using mangroves as halophytes within a bio-thermic seawater desalination setup [46].

The potential of brackish water biodesalination has also been explored for halophyte succulents. Euhalophyte *Sesuvium portulacastrum* removes salt from brackish water through transpiration. Additionally, this halophyte can be utilized for extended periods by replenishing the transpired water with new brackish water [47].

Another documented strategy is using halophyte succulents, such as *Salicornia* plants, in environmental management and economic ventures. These plants can help convert brine byproducts produced during inland industrial desalination processes. The utilization of brine presents both an environmental concern and a financial challenge and serves as a critical barrier impeding the broader implementation of sophisticated desalination treatment technologies [48]. A novel approach proposes the reuse of the brine in multiple tasks that include the development of a hybrid system for desalination that integrates reverse osmosis, forward osmosis, and the cultivation of halophytes as a vehicle for enhanced environmental management in groundwater treatment [49]. This system can also facilitate the production of nutritious protein sources, such as fish, as well as a supplementary food source like duckweed plants [48].

Halophytes, salt-tolerant plants, can be cultivated next to inland desalination plants for potential growth and to repurpose the concentrated brine for alternative applications. *Salicornia, Ocimum, Atriplex,* and related species are commonly favored salt-tolerant plants for utilizing the remaining brine from inland desalination (Table **1**) [48].

Table 1. Plant salt-tolerant vegetation to use surplus brine produced by inland desalination [48].

Genus	Specie
Salicornia	S. alpini
	S. ambigua
	S. andina
	S. bigelovii
Ocimum	O. basilicum L.
Atriplex	A. nummularia Lindl
	A. lentiformis

Halophytes play a crucial role in salt removal from soil and water, simultaneously offering opportunities for food production. Their effectiveness in nutrient uptake and environmental remediation in natural settings and constructed wetlands has yielded promising outcomes. Nevertheless, the success of constructed wetlands largely relies on carefully selected plant species. Despite this, there remains a lack of understanding regarding the morphological, physiological, and biochemical attributes that dictate plant efficiency [35].

Algae

A diverse array of plants and algae presents numerous possibilities for utilization in engineered biotechnological systems designed for the treatment of saline water. Halophile algae exhibit salt tolerance and the ability to concentrate salt levels multiple times higher than the surrounding water they inhabit [50]. Algae represent a significant category of organisms present in a diverse array of environments, such as oceans, rivers, freshwater bodies, and various other settings. Algae encompass both prokaryotic and eukaryotic divisions, a wide array of pigment systems, a three-phase life cycle, and a lengthy evolutionary background. These minuscule life forms not only contribute to oxygen production, food sources, and medicinal products, and it is anticipated that power plants will utilize algae for carbon dioxide capture [51].

In salinity adaptation, algae can be broadly categorized as halophiles and halotolerants. Halophyte algae necessitate salt for optimal growth, while halotolerant algae possess response mechanisms that enable their survival in saline environments. Alterations in salt levels impact algae through three primary mechanisms: First, osmotic stress directly alters the cell's water potential. Secondly, ionic stress is caused by the absorption or release of ions, leading to acclimatization. Lastly, there are variations in cellular ion ratios due to selective membrane permeability. Algae can adapt to biochemical mechanisms such as producing osmolytes, accumulating them, or utilizing the Na^+/K^+ pump system to cope with salinity changes [52].

Algae thrive in aquatic environments characterized by significantly elevated salinity levels, reaching 70–80 parts per thousand (ppt). Examples of such algae include *Stephanoptera*, *Chlamydomonas ehrenbergii*, *Oscillatoria*, and *Ulothrix* [51]. *Chlorella vulgaris*, *Scenedesmus obliquus*, and *Chlorella sorokiniana* are among the most prevalent halotolerant freshwater algae, known to withstand salinity levels reaching up to 11 g.L^{-1} [24]. In aquatic ecosystems, algae, or macrophytes, play a crucial role in sequestering salts within their bodies and effectively removing them from the solution [53]. The halotolerant algae demonstrated a rising pattern in removing chloride ions, exhibiting an elimination capacity of 7.5 g.m^3.h^{-1}. In addition, the rate of nitrate absorption in halophytic algae surpasses that of phosphate by a factor of 10, independent of the salinity level. It can be inferred that microalgae may offer advantages in ion and nutrient uptake processes when dealing with highly saline water [24].

Microalgae

Microalgae are photosynthesis-active microorganisms capable of thriving in diverse aquatic habitats and generating oxygen and biomass. Within this category, microalgae offer numerous benefits, including superior photosynthetic efficacy, rapid growth rates, the capacity to acclimate to varying salinity levels, the potential for biomass generation, and desalination of water.

Microalgae can desalinate water *via* osmosis, ion exchange, or biosorption processes. Additionally, they can be utilized for purposes including wastewater treatment, biofuel production, and aquaculture [54]. Two approaches can be employed to elucidate the mechanisms by which microalgae eliminate salt from seawater or extract metal from industrial wastewater, metabolic and non-metabolic pathways. The bioaccumulation process entails the assimilation of salts into the metabolic activities of algal cells by utilizing algal biological systems with biomass byproducts that are capable of being repurposed in various manners [16].

Several research studies have focused on cultivating algae and cyanobacteria through the suspension method, which involves direct exposure of cells to seawater or brackish water for the biodesalination of water [28]. In the study by Sahle-Demessie *et al.* [50], the investigation examined salt tolerance and absorption of salt (NaCl) in algae species. Specifically, *Scenedesmus* species and *Chlorella vulgaris* were analyzed to assess their potential for biodesalination of brackish water utilizing batch photobioreactors. The initial salt uptake was observed to reach a maximum of 30% within a single cycle, prompting further exploration through pilot-scale experiments aimed at desalinating brackish waters. The species *Scenedesmus sp.* has the capability to remove 25% of the total salt present to thrive in conditions of high salinity [55].

Multiple systems for biodesalination have been developed as optimizers based on their ability to lower the salt levels in seawater, thus minimizing the need for additional RO treatment. This approach helps decrease the energy needed for RO membrane processes to remove salt from seawater [56]. Biodesalination provides a low-energy substitute for traditional desalination techniques. Microalgae, which depend on carbon dioxide for photosynthesis with the help of carbonic anhydrase (pCA) enzyme, are essential in this process, resulting in a minimal energy impact and diminishing greenhouse gases like CO_2 through biodesalination [56]. An additional instance involves utilizing the *Phormidium keutzingianum* strain with zeolites as the support material for attached formation in a natural outdoor setting, characterized by a constant airflow and growth biofilm-packed bed reactor, for real-time seawater treatment. In the first reactor, the most substantial removal of chloride ions was approximately 21% on day 32. Conversely, reactor 2 demonstrated a notable removal of chloride ions of 29% on day 11 [28].

Utilizing algae as a preliminary process in biodesalination can yield notable environmental advantages and diminish energy consumption in traditional desalination facilities. Furthermore, this approach brings about a substantial decrease in salt concentration, representing a significant progression in the biological desalination of seawater (Table **2**) [56].

Table 2. Investigations of microalgae species as potential desalination water.

Specie	Highlights	Reference
Scenedismus arcuatusa, Chlorella vulgaris, Spirulina maxima	-*Scenedismus arcuatusa*, chloride removal from 32.42 to 48.93%.	[16]
Scenedesmus obliquus	-The rate of growth declined with the increase in salinity. -The rate of desalination and the removal of NaCl showed a significant increase with the rise in salinity.	[57]
Scenedesmus obliquus	-Effectively eliminated K^+, Na^+, Ca^{2+}, and Mg^{2+} ions from saline-alkaline water. - Under alkaline stress, there was a significant decrease in the concentrations of Na^+ and K^+, while the ratios of K^+/Na^+, Ca^{2+}/Na^+, and Mg^{2+}/Na^+ were notably higher in all treatments compared to the control.	[58]
Scendesmus algae	-Continuous flow algae pond high removal for Total Dissolved Solids (TDS) from seawater. -The growth of species was successfully obtained in saline water as it absorbs salts and utilizes them in its metabolic processes.	[59]
Dunaliella salina, Chlorella vulgaris, Nannochloropsis oculata, Scenedesmus quadricauda	-The alga *D. salina* decreases chlorine, sodium, and bicarbonate concentrations within identical circumstances.	[60]
Dunaliella salina	-Salt absorption at 130 mS/cm concentration. -Substantial decrease in chlorine, sodium, and bicarbonate concentrations.	[61]
Desmodesmus subspicatus LC172266, *Desmodesmus armatus* LC172263, *Dictyosphaerium spp* LC172264	The levels of salinity in the seawater underwent a reduction from 37.5 g/L to values of 26.25 ± 1.33 mS/cm, 27.19 ± 1.33 g/L, and 30.0 ± 0.00 g/L.	[62]

Techniques for cultivating algae are financially advantageous and have potential applications in desalination technology if designed appropriately; nevertheless, the expense of biomass collection is equal to the costs associated with current desalination facilities [28]. Furthermore, a critical challenge in the manufacturing procedure lies in effectively eliminating the salt-laden cells from desalinated water while preserving the cell membrane integrity. Hence, a comprehensive exploration of the cell surface properties and the potential adoption of biopolymers as coagulants is deemed necessary [63].

Bacterial

Bacteria are frequently used in purifying municipal and industrial wastewater [64] and have also been applied in the process of water desalination. Halophilic bacteria utilize Na^+ to generate electrochemical gradients and enhance membrane stability through physiological processes. The activities of Na^+/H^+ antiporters exhibit strong pH dependence, playing a crucial role in internal pH regulation. Additionally, these bacteria leverage the H^+ gradient to extrude Na^+ ions from the cell [25, 65].

Two isolates of purple nonsulfur bacteria NW16 and KMS24, which exhibited resistance to metals, demonstrated the ability to reduce heavy metal concentrations such as Pb, Cu, Cd, Zn, and 31% for Na in water containing elevated heavy metal levels and 3% NaCl when subjected to either microaerobic-light or aerobic-dark conditions [66]. On the other hand, in the study of the integration of plants and bacteria, *Vibrio alginolyticus* was found in artificial saline desalination procedures. The findings revealed a reduction in salinity levels, reaching its lowest point by the conclusion of the experimental period. The elements present in botanical specimens are assimilated as either cations or anions, with the incorporation of bacteria being implemented to facilitate the uptake of salts by the plants. Helobacterium, *V. alginolyticus,* are microorganisms that inhibit in salt crystals, displaying a preference for environments characterized by high salinity levels and demonstrating resistance to radiation when introduced into the reactor reed bed system; they exhibit the ability to increase and endure within the submerged root zone of aquatic [44, 45].

Cyanobacterial

Cyanobacteria are photosynthetic bacteria that produce oxygen and thrive in extensive blooms in freshwater and marine environments. Furthermore, these organisms can thrive with minimal nutrient demands and rely on natural sunlight for energy. The capability of cyanobacterial populations to thrive at elevated cell concentrations with limited nutritional needs, including sunlight, carbon dioxide, and minerals, presents numerous prospects for sustainable water treatment initiatives [63].

Employing solar energy to power desalination using photosynthetic organisms presents a promising avenue for utilizing biological mechanisms for this objective. Cyanobacteria can produce a significant amount of biomass in brackish and marine environments, creating a reservoir with a low salt content within the saline solution. This reservoir could be utilized as an ion exchanger by controlling the activity of transport proteins in the cellular membrane [64]. Cyanobacteria have been documented to accumulate intracellular or extracellular osmolytes in response to ionic stress. Furthermore, it has been noted that cyanobacteria facilitate the transportation of Na^+ *via* diffusion in the presence of ionic stress. *Synechococcus* sp. PCC 7002 and *Synechocystis* sp. PCC 6803 have been applied in cyanobacteria desalination water treatment [39, 63].

According to Amezaga *et al.* [64], some steps are necessary to advance a novel biodesalination technique that employs biological membranes derived from cyanobacteria. The principal characteristic involves the development of a low-salinity biological compartment in seawater, which can function as an ion exchanger. Its development process can be disintegrated into various supplementary stages.

1. The initial phase encompasses the identification of a cyanobacterial variety capable of thriving at high cell concentrations in seawater with minimal need for energy inputs besides those naturally occurring. In this step, cultivation can be controlled to amplify the innate extrusion of sodium ions (Na^+) by cyanobacteria.
2. In the second phase, manipulating the ion transport mechanisms of cyanobacteria is imperative to cultivate cells where the process of exporting sodium is substituted with the accumulation of sodium within the cell.
3. Moving on to the third phase, the focus shifts towards developing an effective method to separate cyanobacterial cells from the desalinated water.
4. The fourth phase aims to amalgamate the initial three phases into a feasible and expandable engineering procedure.
5. Finally, the fifth and ultimate phase involves evaluating the potential hazards and societal acceptance concerns associated with this novel technology.

Microbial Desalination Cells

The microbial desalination cell (MDC) represents a form of microbial electrochemical system designed to leverage electro-active microorganisms to facilitate the energy required for the desalination process of water [18]. The desalination mechanism of microbial desalination cells (MDCs) fundamentally relies on generating electrical energy by microorganisms that break down the organic components in wastewater within the anode compartment, facilitating the transfer of electrons to the anode [17]. This system comprises three distinct compartments: the anodic chamber, the desalination chamber, and the cathodic chamber. Within a conventional MDC, the anodic and cathodic chambers are separated by ion-exchange membranes, with the desalination chamber in between. Water production occurs as oxygen engages with the electrons transferred from the anode to the cathode. Cl^- anions move toward the anodic compartment through an anion exchange membrane in this process. At the same time, Na^+ cations migrate to the cathodic compartment *via* a cation exchange membrane, leading to the elimination of salt ions from the seawater (Figure **1**) [67]. Consequently, these processes concurrently address wastewater treatment, electricity generation, and the elimination of total dissolved solids (TDS) from saline water (Fig. **1**) [17]. Different configurations of MDC reactors have been documented in Table **3**.

Table 3. Different microbial desalination cells, microorganisms, and highlights.

System	Highlights	Reference
Air cathode MDC	Exhibit constrained performance attributable to elevated activation energy associated with the oxygen reduction reaction (ORR), which adversely impacts the kinetics of the cathode and diminishes the overall efficiency of the reactor; hence, necessitating the implementation of effective electrocatalysis. Platinum-based catalysts represent the most appropriate materials, as they facilitate enhanced ORR activity in air-cathodes.	[68]
Stack structure microbial desalination cell MDC	In the cathodic section, an electrode, electrolyte, and oxidant collaborate to accept the electrons originating from the anodic chamber, thus establishing a potential gradient along the electrical circuit. The intermediary chamber is composed of stacks of ion exchange membranes to enhance the efficiency of desalination.	[67]
Recirculation of microbial desalination cell (rMDC)	The system was designed to circulate the solutions within the electrode chambers through narrow tubing and an external pump, thus preventing drastic alterations in pH levels.	[69]
Microbial electrolysis desalination and chemical-production cell (MEDCC)	It is designed to generate alkalis within the cathode chamber alongside various other functionalities, such as desalination and acid production.	[70]
Capacitive microbial desalination cells (cMDCs)	Integrating the concept of capacitive deionization into the framework while considering the elimination of salt.	[67]
Up-flow microbial desalination cell	Characterized by the presence of nested cylinders. Within this setup, the inner cylinder serves as the desalination chamber, while the outer cylinder houses the cathode and anode electrodes.	[71]
Osmotic microbial desalination cell (OsMFC)	The methodology was devised through the substitution of PEM, AEM, or CEM with FO. Collaborative integration of MFC and FO serves to enhance each other's capabilities and leverage the strengths of both techniques.	[72]
Bipolar membrane microbial fuel cell	Elevated perm-selectivity is observed. Minimal resistance and voltage decline are evident. An enhancement in desalination efficiency is noted, and mitigation of the pH imbalance is achieved.	[17]
Decoupled microbial desalination cell	The electrode compartments are consolidated within a unified brine solvent. Flexibility in modifying and altering the liquid volume ratios as required.	[67]

MDC has gained increased recognition in the field of desalination for both seawater and brackish water treatment, which is attributed to its minimal energy consumption. Investigations conducted at the laboratory scale have demonstrated that MDC could attain 98% elimination of salt by utilizing 35 gL^{-1} NaCl in artificial seawater [73]. MDCs demonstrate remarkable proficiency in extracting salt from saline solutions and producing electricity, thus offering the potential for application in extensive industrial operations. However, further exploration and modifications are indispensable to enhance the efficiency of MDCs and facilitate the production of valuable outputs, thereby bolstering their sustainability and effectiveness [67].

Biological Membranes and Biomaterials for Desalination

Aquaporin proteins have garnered significant attention recently due to their exceptionally high water permeability and solute rejection attributes. This material has positioned them as promising components for biomimetic membrane development, particularly in water desalination and reuse [74]. Aquaporins exhibit two advantageous characteristics that make them suitable for application in water desalination: first, a high flux rate, and second, effective rejection of solutes. Their ability to transport water surpasses that of artificial reverse osmosis membranes. Aquaporins facilitate water transfer *via* hydrophobic pores, differing from the jump-diffusion mechanism observed in current synthetic membranes [18].

Biomaterials also hold great potential as effective materials for water desalination. The advantages of utilizing biomaterials include their extensive availability, non-toxic nature, and exceptional capability in water desalination. Applying bio-based materials or biomaterials in water desalination presents a relatively recent and cost-efficient substitute for conventional techniques used in wastewater treatment. Various substances, including bark, biochar, and activated charcoal sourced from plants, can serve as constituents in water refinement procedures, acting as proficient filtration agents. One of the most promising materials for water purification is nanocellulose, a substance derived from cellulose, which represents the predominant polymeric structure on Earth. Nanocellulose emerges as an exceptionally effective absorbent that is easily accessible and poses no environmental risks [75].

CONCLUDING REMARKS

Biological desalination, characterized by the use of absorption and adsorption mechanisms by salt-tolerant organisms, is emerging as a novel approach to desalination. Halophytic species offer potential solutions for areas contaminated with salt by effectively reducing salt levels in soil and water to facilitate food production in challenging environments. However, despite their potential, limited research has been devoted to exploring their application in desalination, resulting in a knowledge gap that hinders a comprehensive evaluation of their economic and ecological impact compared to conventional desalination systems. In contrast to traditional methods, biodesalination systems are more energy-efficient, environmentally friendly, and less complex in design, making them a sustainable choice for long-term desalination endeavors. Nevertheless, challenges arise during the implementation of biodesalination due to its complex nature, the need for optimization, and the difficulties associated with scaling up the process. Moreover, the technology's economic viability needs to be established before it can be widely adopted. Currently, research on biodesalination revolves around the selection of suitable strains, the requirements for grown strains (nutrients and minerals), microbial desalination system development, and the scaling process.

AUTHORS' CONTRIBUTION

The authors confirm their contribution to the chapter as follows:

Writing-original draft, conceptualization, investigation, data curation, and editing: Ana K. Villagómez-Guzmán; Conceptualization, writing-original draft, investigation, data curation, and editing: Héctor M. Arreaga-González; Conceptualization, supervision, validation, review, and editing: Tzasna Hernández-Delgado.

All authors have reviewed and granted their approval for the final version of the manuscript.

QUESTIONS RELATED TO THE TEXT

1. What is the status of water globally?
2. Provide the categorization of conventional desalination methods and offer three instances.
3. What is the explanation of biodesalination and the instruments utilized?
4. What advantages does biodesalination present in contrast to the traditional desalination procedure?
5. Elucidate the primary feature of salt-tolerant organisms in biodesalination.
6. What are the current limitations and challenges in the implementation and scalability of biodesalination technologies?

ACKNOWLEDGMENTS

This work was supported by the UNAM Postdoctoral Program (POSDOC).

LIST OF ABBREVIATIONS

°C	Degrees Celsius
°F	Degrees Fahrenheit
AEM	Anion Exchange Membrane
Cd	Cadmium
CEM	Cation Exchange Membrane
cm^2	Square Centimeter
cMDCs	Capacitive Microbial Desalination Cell
CO_2	Carbon Dioxide
Cu	Copper
FAO	Food and Agriculture Organization
FO	Forward Osmosis
$g.L^{-1}$	Gram/Liter
$g.m^{-3}.h^{-1}$	Gram/Cubic Meter/Hour
km^3	Cubic Kilometers
kWh/m^3	Kilowatt Hours/Cubic Meter
MDC	Microbial Desalination Cell
MED	Multiple-Effect Distillation
MEDCC	Microbial Electrolysis Desalination and Chemical-Production Cell
$mg.L^{-1}$	Milligram/Liter
$MgCl_2$	Magnesium Chloride
MGD	Million Gallons per Day
$MgSO_4$	Magnesium Sulfate
mM	Millimolar
mS/cm	Millisiemens/Centimeter
MSF	Multi-Stage Flash Distillation
NaCl	Sodium Chloride
$NaSO_4$	Sodium Sulfate
ORR	Oxygen Reduction Reaction
Pb	Lead
pCA	Carbonic Anhydrase (Pca) Enzyme
PEM	Proton Exchange Membrane
ppt	Parts per Thousand
rMDC	Recirculation of Microbial Desalination Cell

RO	Reverse Osmosis
SIWI	Stockholm International Water Institute
TDS	Total Dissolved Solids
UAE	United Arab Emirates
UNICEF	United Nations Children's Fund
WHO	World Health Organization
WRI	World Resources Institute
Zn	Zinc

REFERENCES

[1]   Bartzis V, Sarris IE. Advanced technology for desalination and water purification. Water 2024; 16(8): 1094.
[http://dx.doi.org/10.3390/w16081094]

[2]   World Resources Institute: Aqueduct. Using cutting-edge data to identify and evaluate water risks around the world 2024. Available from: https://www.wri.org/aqueduct

[3]   European Environment Agency. EEA: Responding to climate change impacts on human health in Europe: focus on floods, droughts and water quality. EEA Report 2024. Available from: file:///C:/Users/Ana/Downloads/TH-AL-24--05-ENN_Responding_to_climate_change_0805%20(1).pdf

[4]   Jung Y-J, Khant NA, Kim H, Namkoong S. Impact of climate change on waterborne diseases: Directions towards sustainability. Water 2023; 15(7): 1298.
[http://dx.doi.org/10.3390/w15071298]

[5]   United Nations: Climate action. Water – at the center of the climate crisis Water. Available from: https://www.un.org/en/climatechange/science/climate-issues/water

[6]   USGS. How Much Water is There on Earth? Available from: https://www.usgs.gov/special-topics/water-science-school/science/how-much-water-there-earth#overview

[7]   Prakash S, Bellman K, Shannon MA. Recent advances in water desalination through biotechnology and nanotechnology. In: Reisner DE, Ed Bionanotechnology II: Global Prospects, 1st Edition: CRC Press. 2011; pp. 365-82.

[8]   Bank of America. Global Water scarcity: H2O no! Available from: https://institute.bankofamerica.com/content/dam/bank-of-america institute/sustainability/global-water-scarcity.pdf

[9]   United Nations: World Population Prospects 2017. Available from: https://www.un.org/development/desa/pd/sites/www.un.org.development.desa.pd/files/files/documents/2020/Jan/un_2017_world_population_prospects-2017_revision_databooklet.pdf

[10]   Boretti A, Rosa L. Reassessing the projections of the World Water Development Report. NPJ Clean Water 2019; 2(1): 15.
[http://dx.doi.org/10.1038/s41545-019-0039-9]

[11]   Musie W, Gonfa G. Fresh water resource, scarcity, water salinity challenges and possible remedies: A review. Heliyon 2023; 9(8): e18685.
[http://dx.doi.org/10.1016/j.heliyon.2023.e18685] [PMID: 37554830]

[12]   Fowler SJ, Smets BF. Microbial biotechnologies for potable water production. Microb Biotechnol 2017; 10(5): 1094-7.
[http://dx.doi.org/10.1111/1751-7915.12837] [PMID: 28905496]

[13] Belessiotis V, Kalogirou S, Delyannis E. Desalination Methods and Technologies—Water and Energy. In: Belessiotis V, Kalogirou S, Delyannis E, Eds. Thermal Solar Desalination Methods and Systems. UK: Elsevier Ltd 2016; pp. 1-19.
[http://dx.doi.org/10.1016/B978-0-12-809656-7.00001-5]

[14] Curto D, Franzitta V, Guercio A. A review of the water desalination technologies. Appl Sci (Basel) 2021; 11(2): 670.
[http://dx.doi.org/10.3390/app11020670]

[15] Coherent Market Insights. Global Water Desalination Market Analysis. Available from: https://www.coherentmarketinsights.com/industry-reports/global-water-desalination-market

[16] Ghobashy MOI, Bahattab O, Alatawi A, Aljohani MM, Helal MMI. A novel approach for the biological desalination of major anions in seawater using three microalgal species: A kinetic study. Sustainability 2022; 14(12): 7018.
[http://dx.doi.org/10.3390/su14127018]

[17] Martínez EO. Biological Seawater Desalination. In: Elnashar MMM, Karakuş S, Eds. Water Purification Present and Future Rijeka: Intech Open. 2024; pp. 1-16.
[http://dx.doi.org/10.5772/intechopen.113984]

[18] Danaeifar M, Ocheje OM, Mazlomi MA. Exploitation of renewable energy sources for water desalination using biological tools. Environ Sci Pollut Res Int 2023; 30(12): 32193-213.
[http://dx.doi.org/10.1007/s11356-023-25642-0] [PMID: 36725802]

[19] USGS. The distribution of water on, in, and above the Earth. Available from: https://www.usgs.gov/media/images/distribution-water-and-above-earth

[20] Wiener J, Khan MZ, Shah K. Performance enhancement of the solar still using textiles and polyurethane rollers. Sci Rep 2024; 14(1): 5202.
[http://dx.doi.org/10.1038/s41598-024-55948-z] [PMID: 38433241]

[21] World Resources Institute (WRI). 25 Countries, Housing One-quarter of the Population, Face Extremely High Water Stress. Available from: https://www.wri.org/insights/highest-water-stress-d-countries

[22] Naciones Unidas. Agua. Available from: https://www.un.org/es/global-issues/water

[23] World Health Organization. Drinking-water. Available from: https://www.who.int/news-room/fact-sheets/detail/drinking-water

[24] Zafar AM, Javed MA, Aly Hassan A, Mehmood K, Sahle-Demessie E. Recent updates on ions and nutrients uptake by halotolerant freshwater and marine microalgae in conditions of high salinity. J Water Proc. Engineering 2021; 44(102382): 102382. [http://dx.doi.org/10.1016/j.jwpe.2021.102382]

[25] Taheri R, Razmjou A, Szekely G, Hou J, Ghezelbash GR. *Biodesalination* —On harnessing the potential of nature's desalination processes. Bioinspir Biomim 2016; 11(4): 041001.
[http://dx.doi.org/10.1088/1748-3190/11/4/041001] [PMID: 27387607]

[26] Bhoj Y, Pandey G, Bhoj A, Tharmavaram M, Rawtani D. Recent advancements in practices related to desalination by means of nanotechnology. Chemical Physics Impact 2021; 2(100025): 100025.
[http://dx.doi.org/10.1016/j.chphi.2021.100025]

[27] Eke J, Yusuf A, Giwa A, Sodiq A. The global status of desalination: An assessment of current desalination technologies, plants and capacity. Desalination 2020; 495(114633): 114633.
[http://dx.doi.org/10.1016/j.desal.2020.114633]

[28] Zafar AM, Aly Hassan A. Seawater biodesalination treatment using Phormidium keutzingianum in attached growth-packed bed continuous flow stirred tank reactor. Environ Res 2023; 236(Pt 2): 116784.
[http://dx.doi.org/10.1016/j.envres.2023.116784] [PMID: 37517498]

[29] Almasoudi S, Jamoussi B. Desalination technologies and their environmental impacts: A review.

Sustainable Chemistry One World 2024; 1(100002): 100002.
[http://dx.doi.org/10.1016/j.scowo.2024.100002]

[30] Mandev E, Muratçobanoğlu B, Manay E, Şahin B. Desalination performance evaluation of a solar still enhanced by thermoelectric modules. Sol Energy 2024; 268(112325): 112325.
[http://dx.doi.org/10.1016/j.solener.2024.112325]

[31] Rashid A, Ayhan T, Abbas A. Natural vacuum distillation for seawater desalination – A review. Desalin Water Treat 2016; 57(56): 26943-53.
[http://dx.doi.org/10.1080/19443994.2016.1172264]

[32] Aybar HS. Analysis of a mechanical vapor compression desalination system. Desalination 2002; 142(2): 181-6.
[http://dx.doi.org/10.1016/S0011-9164(01)00437-4]

[33] Folley M, Whittaker T. The cost of water from an autonomous wave-powered desalination plant. Renew Energy 2009; 34(1): 75-81.
[http://dx.doi.org/10.1016/j.renene.2008.03.009]

[34] Shaffer DL, Werber JR, Jaramillo H, Lin S, Elimelech M. Forward osmosis: Where are we now? Desalination 2015; 356: 271-84.
[http://dx.doi.org/10.1016/j.desal.2014.10.031]

[35] Amidpour M, Khoshgoftar Manesh MH. Desalinated water production in cogeneration and polygeneration systems. In: Amidpour M, Khoshgoftar Manesh MH, Eds. Cogeneration and Polygeneration Systems. UK: Elsevier Academic Press 2021; pp. 115-35.
[http://dx.doi.org/10.1016/B978-0-12-817249-0.00008-2]

[36] Aladwani SH, Al-Obaidi MA, Mujtaba IM. Performance of reverse osmosis based desalination process using spiral wound membrane: Sensitivity study of operating parameters under variable seawater conditions. Clean Eng Technol 2021; 5(100284): 100284.
[http://dx.doi.org/10.1016/j.clet.2021.100284]

[37] Bragg-Sitton S. Woodhead Publishing Series in Energy. In Carelli MD, Ingersoll DT Eds Woodhead Publishing Handbook of Small Modular Nuclear Reactors. 2015; pp. 319-50.
[http://dx.doi.org/10.1533/9780857098535.3.319]

[38] Kumar Patel A, Tseng YS, Rani Singhania R, Chen CW, Chang JS, Di Dong C. Novel application of microalgae platform for biodesalination process: A review. Bioresour Technol 2021; 337: 125343.
[http://dx.doi.org/10.1016/j.biortech.2021.125343] [PMID: 34120057]

[39] Ahmed ME, Zafar AM, Hamouda MA, Aly Hassan A, Arimbrathodi S. Biodesalination research trends: A bibliometric analysis and recent developments. Sustainability (Basel) 2022; 15(1): 16.
[http://dx.doi.org/10.3390/su15010016]

[40] Saddhe AA, Manuka R, Nikalje GC, Penna S. Halophytes as a potential resource for phytodesalination. In: Grigore, ed Handbook of Halophytes Switzerland: Springer Cham. 2020; pp. 1-21.
[http://dx.doi.org/10.1007/978-3-030-17854-3_92-1]

[41] Turcios AE, Miglio R, Vela R, *et al.* From natural habitats to successful application - Role of halophytes in the treatment of saline wastewater in constructed wetlands with a focus on Latin America. Environ Exp Bot 2021; 190: 104583.
[http://dx.doi.org/10.1016/j.envexpbot.2021.104583]

[42] Yensen NP. Halophyte uses for the twenty-first century. In: Khan MA, Weber DJ, Eds. Ecophysiology of High Salinity Tolerant Plants. Netherlands: Springer Dordrecht 2006; pp. 367-96.
[http://dx.doi.org/10.1007/1-4020-4018-0_23]

[43] Yasseen BT, Al-Thani RF. Endophytes and halophytes to remediate industrial wastewater and saline soils: Perspectives from Qatar. Plants 2022; 11(11): 1497.
[http://dx.doi.org/10.3390/plants11111497] [PMID: 35684269]

[44]   Chimayati R, Titah HS. Removal of salinity using Interaction mangrove plants and bacteria in batch reed bed system reactor. J Ecol Eng 2019; 20(4): 84-93.
        [http://dx.doi.org/10.12911/22998993/102792]

[45]   Titah H, Puspaningrum T. The removal of salinity in a reed bed system using mangroves and bacteria in a continuous flow series reactor. J Ecol Eng 2020; 21(6): 212-23.
        [http://dx.doi.org/10.12911/22998993/124075]

[46]   Finck C. A bio-thermic seawater desalination system using halophytes. Water Sci Technol Water Supply 2014; 14(4): 657-63.
        [http://dx.doi.org/10.2166/ws.2014.020]

[47]   Alharbi A, Rabhi M, Alzoheiry A. Brackish Water Phytodesalination by the Euhalophyte *Sesuvium portulacastrum*. Water 2024; 16(13): 1798.
        [http://dx.doi.org/10.3390/w16131798]

[48]   Oron G, Appelbaum S, Guy O. Reuse of brine from inland desalination plants with duckweed, fish and halophytes toward increased food production and improved environmental control. Desalination 2023; 549(116317): 116317.
        [http://dx.doi.org/10.1016/j.desal.2022.116317]

[49]   Park K, Mudgal A, Mudgal V, Sagi M, Standing D, Davies PA. Desalination, water re-use, and halophyte cultivation in salinized regions: A highly productive groundwater treatment system. Environ Sci Technol 2023; 57(32): 11863-75.
        [http://dx.doi.org/10.1021/acs.est.3c02881] [PMID: 37540002]

[50]   Sahle-Demessie E, Aly Hassan A, El Badawy A. Bio-desalination of brackish and seawater using halophytic algae. Desalination 2019; 465: 104-13.
        [http://dx.doi.org/10.1016/j.desal.2019.05.002] [PMID: 32704185]

[51]   Sahoo D, Baweja P. General Characteristics of Algae. In: Sahoo D, Seckbach J, Eds. The Algae World Cellular Origin, Life in Extreme Habitats and Astrobiology. Dordrecht: Springer 2015; 26: pp. 1-8.
        [http://dx.doi.org/10.1007/978-94-017-7321-8_1]

[52]   Gautam S, Kapoor D. Application of halophilic algae for water desalination. In: El-Sheekh M, El-Fatah A, Eds. Handbook of algal biofuels Aspects of Cultivation, Conversion, and Biorefinery. Elsevier 2022; pp. 167-79.
        [http://dx.doi.org/10.1016/B978-0-12-823764-9.00002-9]

[53]   Omar M, Balla D. Drainage water purification in saline detention ponds with duckweeds. In: Drainage I-CTICoIa, Ed 23rd European Regional Conference, "Progress in Managing Water for Food and Rural Development" Lviv Ukraine.

[54]   Balasubramaniyan M, Kasiraman D, Amirtham S. *Chlorella vulgaris* in biodesalination: a sustainable future from seawater to freshwater. Marine Development 2024; 2(1): 7.
        [http://dx.doi.org/10.1007/s44312-024-00019-0]

[55]   El-Sayed A, El-Fouly M, Abou El-Nour E. Immobilized microalga Scenedesmus sp. for biological desalination of Red Sea water: I. Effect on growth. Nat Sci 2010; 8: 69-76.

[56]   Zafar AM, Mohamed BA, Wang Q, Aly Hassan A. Life cycle analysis of seawater biodesalination using algae. Desalination 2024; 578: 117433.
        [http://dx.doi.org/10.1016/j.desal.2024.117433]

[57]   Gan X, Shen G, Xin B, Li M. Simultaneous biological desalination and lipid production by Scenedesmus obliquus cultured with brackish water. Desalination 2016; 400: 1-6.
        [http://dx.doi.org/10.1016/j.desal.2016.09.012]

[58]   Yao Z, Ying C, Lu J, *et al.* Removal of K^+, Na^+, Ca^{2+}, and Mg^{2+} from saline-alkaline water using the microalga Scenedesmus obliquus. Chin J Oceanology Limnol 2013; 31(6): 1248-56.
        [http://dx.doi.org/10.1007/s00343-013-2116-0]

[59]   Nagy AM. Nadi El, Hussein HM. Determination of the best water depth in desalination algae ponds. El Azhar Univ., Faculty of Eng., CERM of Civil Eng 2019; 38(4): 1.

[60]   Moayedi A, Yargholi B, Pazira E, Babazadeh H. Investigation of bio-desalination potential algae and their effect on water quality. Desalination Water Treat 2021; 212: 78-86.
[http://dx.doi.org/10.5004/dwt.2021.26638]

[61]   Moayedi A, Yargholi B, Pazira E, Babazadeh H. Investigated of Desalination of Saline Waters by Using *Dunaliella Salina* Algae and Its Effect on Water Ions. Civil Engineering Journal 2019; 5(11): 2450-60.
[http://dx.doi.org/10.28991/cej-2019-03091423]

[62]   Ahamefule CS, Ugwuodo CJ, Idike PO, Ogbonna JC. Application of photosynthetic microalgae in the direct desalination pretreatment of seawater. Water Environ J 2021; 35(2): 657-69.
[http://dx.doi.org/10.1111/wej.12659]

[63]   Minas K, Karunakaran E, Bond T, *et al.* Biodesalination: an emerging technology for targeted removal of Na$^+$ and Cl$^-$ from seawater by cyanobacteria. Desalination Water Treat 2015; 55(10): 2647-68.
[http://dx.doi.org/10.1080/19443994.2014.940647]

[64]   Amezaga JM, Amtmann A, Biggs CA, *et al.* Biodesalination: a case study for applications of photosynthetic bacteria in water treatment. Plant Physiol 2014; 164(4): 1661-76.
[http://dx.doi.org/10.1104/pp.113.233973] [PMID: 24610748]

[65]   Hänelt I, Müller V. Molecular mechanisms of adaptation of the moderately halophilic bacterium halobacillis halophilus to its environment. Life 2013; 3(1): 234–43.
[http://dx.doi.org/10.3390/life3010234]

[66]   Panwichian S, Kantachote D, Wittayaweerasak B, Mallavarapu M. Isolation of purple nonsulfur bacteria for the removal of heavy metals and sodium from contaminated shrimp ponds. Electron J Biotechnol 2010; 13(4): 3-4.
[http://dx.doi.org/10.2225/vol13-issue4-fulltext-8]

[67]   Odunlami OA, Vershima DA, Tagbo CV, Ogunlade S, Nkongho S. Microbial desalination cell technique - A review. S Afr J Chem Eng 2023; 46: 312-29.
[http://dx.doi.org/10.1016/j.sajce.2023.07.011]

[68]   Borràs E, Aliaguilla M, Bossa N, *et al.* Nanomaterials-based air-cathodes use in microbial desalination cells for drinking water production: Synthesis, performance and release assessment. J Environ Chem Eng 2021; 9(4): 105779.
[http://dx.doi.org/10.1016/j.jece.2021.105779]

[69]   Qu Y, Feng Y, Wang X, *et al.* Simultaneous water desalination and electricity generation in a microbial desalination cell with electrolyte recirculation for pH control. Bioresour Technol 2012; 106: 89-94.
[http://dx.doi.org/10.1016/j.biortech.2011.11.045] [PMID: 22200556]

[70]   Li H, Liu H, Lin W, *et al.* Improved microbial electrolysis desalination and chemical-production cell for a high-value product: Hydrogen peroxide. J Environ Chem Eng 2022; 10(3): 107683.
[http://dx.doi.org/10.1016/j.jece.2022.107683]

[71]   Aber S, Shi Z, Xing K, *et al.* Microbial desalination cell for sustainable water treatment: A critical review. Glob Chall 2023; 7(10): 2300138.
[http://dx.doi.org/10.1002/gch2.202300138] [PMID: 37829683]

[72]   Sibi R, Sheelam A, Gunaseelan K, Jadhav DA, Gangadharan P. Osmotic microbial fuel cell for sustainable wastewater treatment along with desalination, bio-energy and resource recovery: A critical review. Bioresour Technol Rep 2023; 23(101540): 101540.
[http://dx.doi.org/10.1016/j.biteb.2023.101540]

[73]   Carmalin Sophia A, Bhalambaal VM, Lima EC, Thirunavoukkarasu M. Microbial desalination cell technology: Contribution to sustainable waste water treatment process, current status and future

applications. J Environ Chem Eng 2016; 4(3): 3468-78.
[http://dx.doi.org/10.1016/j.jece.2016.07.024]

[74]   Tang CY, Zhao Y, Wang R, Hélix-Nielsen C, Fane AG. Desalination by biomimetic aquaporin membranes: Review of status and prospects. Desalination 2013; 308: 34-40.
[http://dx.doi.org/10.1016/j.desal.2012.07.007]

[75]   Raza S, Ghasali E, Orooji Y, *et al.* Two dimensional (2D) materials and biomaterials for water desalination; structure, properties, and recent advances. Environ Res 2023; 219(114998): 114998.
[http://dx.doi.org/10.1016/j.envres.2022.114998] [PMID: 36481367]

CHAPTER 9

Biological Treatment of Wastewater: Use of Microorganisms for Purification

Román Adrián González Cruz[1,*], Alonso Ezeta Miranda[2] and Israel Valencia Quiroz[2]

[1] *Natural Products Bioactivity Laboratory, UBIPRO, Superior Studies Faculty (FES)-Iztacala, National Autonomous University of México (UNAM), Tlalnepantla de Baz, México State, México*

[2] *Phytochemistry Laboratory, UBIPRO, Superior Studies Faculty (FES)-Iztacala, National Autonomous University of Mexico (UNAM), Tlalnepantla de Baz, Mexico State, Mexico*

Abstract: Biological treatment of wastewater leverages the power of microorganisms to purify domestic and industrial effluents. Key processes include the conversion of carbon, organic, and nitrogen compounds, sulfate reduction, denitrification, and the fermentation of methane and volatile fatty acids. These microbial activities clean the water but also produce valuable biosolids, biogases, and protein compounds that can be recovered and reused. Microorganisms such as bacteria, fungi, algae, and archaea are central to these processes. They thrive in diverse nutrient and physicochemical conditions, enabling the degradation and remodeling of pollutants. The advantages of biological treatment span technical, economic, and environmental aspects, offering a sustainable solution to manage a wide range of organic contaminants. Wastewater treatment involves a sequence of physical, chemical, and biological steps to effectively separate and remove impurities. The importance of this treatment is universally recognized, particularly for water destined for human consumption. Rapid industrial growth and population increases have intensified the demand for clean water, yet a significant portion of available resources is saline or contaminated. Pollution from organic and inorganic substances further complicates conventional treatment methods, making biological processes crucial for efficient contaminant removal. Innovative biotechnological approaches inspired by natural microbial processes have expanded the potential for environmental decontamination. These methods offer significant benefits for human health, agriculture, energy, and overall environmental protection. Understanding and harnessing the biological capacities of microorganisms is key to developing more effective and sustainable wastewater treatment strategies.

Keywords: Biosolids, Biological purification, Biodegradation, Denitrification, Environmental decontamination, Industrial effluents, Microorganisms, Methane fermentation, Sustainable water management, Wastewater treatment.

* **Corresponding author Román Adrián González Cruz:** Natural Products Bioactivity Laboratory, UBIPRO, Superior Studies Faculty (FES)-Iztacala, National Autonomous University of México (UNAM), Tlalnepantla de Baz, México State, México; Tel: +525556231137; E-mail: romanagc23@gmail.com

INTRODUCTION

The present chapter aims to discuss the methodological aspects associated with the treatment of domestic and industrial wastewater using biological methods. It highlights the microbial group and metabolic pathways employed in the process while also discussing the possibility of recovering by-products of biotechnology interest, in addition to the monitoring and control of these treatment systems for the mitigation of environmental and public health [1, 2].

In biological processes, the microorganisms utilized for the purification of water carry out the processes of conversion of carbon, organic, and nitrogen compounds, sulfate reduction, denitrification, and methane and volatile fatty acid fermentations. They also capture some heavy metals and precipitate phosphates. These modifications lead to the formation of new biosolids, biogases, and cellular and extracellular protein compounds that can be recovered for specialized synthesis, while the purification products received are removed from the environment and can be recycled [3, 4].

In addition to the methods previously mentioned, biological treatment is another existing method used for the purification of water, presenting various advantages from technical, economic, and environmental points of view in the attenuation of a wide range of organic-contaminated compounds present in water. The use of microorganisms to degrade and/or remodel pollutants is a fundamental component for the maintenance of soil and water quality in the natural environment and is a significant parameter in domestic, urban, and industrial wastewater treatment systems. Differences in the diversity of nutrients (concentrations and types of carbon sources, nitrogen, phosphorus, vitamins, mineral ions, and other essential nutrients), cell concentrations, and physicochemical conditions (oxygen, pH, temperature, chemical toxicity, and other related factors) have led to the identification of various microbial groups and the contrasting structures used for strict or facultative responses to changes in physicochemical conditions [5, 6].

Treatment and purification of waters contaminated by some substances (industrial discharges and urban effluents) are usually carried out using a treatment sequence consisting of a series of steps, each one of which employs physical, chemical, or biological processes to accomplish the separation and complete or partial removal of inorganic and organic impurities. For example, physical processes include sedimentation and flotation, while chemical processes encompass precipitation, coagulation, maceration, adsorption, ion exchange, and oxidation/reduction [7, 8].

The importance of treating and purifying water, especially water intended for human consumption, is universally recognized. In the absence of appropriate measures, the drainage of domestic and industrial effluents readily leads to a significant environmental problem. Rapid growth in the world's population and in industry over the last few decades has contributed to the increasing demand for water. However, a significant percentage of the resources available to mankind consists of saline or contaminated water, whose physical and chemical characteristics render them unsuitable for various intended uses. Furthermore, pollution of natural water sources with organic and inorganic substances makes conventional methods of water treatment, such as coagulation and maceration, and disinfection with chlorine or ozone addition, inefficient for minimizing the risks of human contamination during the water cycle since these compounds have carcinogenic effects. The efficient removal of contaminants from water is important not only to protect human health but also for use in agriculture, energy, industry, and general environmental protection. The development of water and wastewater treatment methods, particularly those inspired by life's attitudes toward protein functions in the enzyme protein system of living beings, has provided numerous possibilities for use in environmental decontamination [9, 10].

Types of Wastewater

Municipal wastewater comprises both domestic sewage and industrial wastewater.

Industrial wastewater originates from producing, commercial, or storage sites. The composition of water from industrial sources depends on the manufacturing processes involved and may contain a toxic component and unusually high concentrations for some substances.

Domestic wastewater generated from private houses, schools, businesses, and public places is known as sewage.

Water serves various purposes, including for drinking, washing, food preparation, recreation, or the manufacturing of goods. Wastewater is water that has been used and is contaminated with organic and inorganic materials because of washing, food preparation, recreation, supplying toilet water, or other uses. Sources of wastewater include agricultural, industrial, commercial, and human waste. Wastewater has different degrees of contamination according to its source [11, 12].

Biological Treatment Methods

In the activated sludge treatment method, treatment is conducted in a treatment machine like that depicted in Fig. (**1**). Wastewater introduced here is mixed with activated sludge in a tank, where purification is performed. Air is then blown into this tank to activate the sand. The dissolving oxygen is consumed in the treatment process, and so it must be supplemented by blowing air into the tank to maintain oxygen levels at a level that allows proper treatment. The use of an air compressor for supplying air is depicted in this drawing as well. Once treated, the water is led to sorption or pressurization tanks, where residual sludge is separated from the water. The activated sludge returning from these tanks is mixed with incoming wastewater in the sedimentation tank, where, again, the activated sludge is separated from the liquid. The treated wastewater is sent to a river, oceanic area, or the next purification process [13, 14].

There are two major methods that use microorganisms for purifying wastewater: the activated sludge method and the microorganism film method. The former is employed in the treatment of domestic wastewater and is well-known due to its successful results. This method will be discussed here. The latter is used for the treatment of industrial wastewater containing a high concentration of organic compounds and is highly efficient in terms of space-saving since a biological film is laid on packing materials such as pieces of concrete or plastic foam. The microorganism film method is gaining popularity due to its convenience and high efficiency in treating complex wastewater streams [15].

Aerobic Treatment Processes

Identification of Possible Microorganisms

The following bacterial groups could be engaged in the mineralization processes of aromatic compounds: *Pseudomonas*, *Bacillus*, and *Clostridium*. *Pseudomonas sp.*, *Mycobacterium sp.*, *Flavobacterium sp.*, and *Aerobacter sp.* are usually found among the bacteria able to nitrify NH_4 to NO_3^- [16]. In environments with elevated concentrations of nitrogenous compounds, such as seawater and marine sediments, *Pseudomonas aeruginosa* can detach excessive cellular material by changing its cellular wall surface properties.

Sulfate-reducing bacteria such as *Desulfovibrio*, *Desulfotomaculum*, *Desulfonema*, *Desulfopila*, *Desulfobacter*, *Desulfomonas*, *Desulfococcus*, *Desulfobacterium*, *Desulfoluna*, and *Desulfoarculus* play a crucial role in sulfur metabolism [17, 18]. Additionally, purple sulfur bacteria (PSB) affiliated with the Alpha and Beta-Proteobacteria are involved in this process [19]. Other bacteria, including *Chromatium, Beggiatoa, Thiopedia*, and *Thiomicrospira*, are also capable of assimilating sulfur compounds.

Molecular techniques have been employed to track the dynamics of a denitrifying microbial community developing within a compost biofilter. It was observed that the composition of the denitrifying community was dependent on the rate of the nitrogen loading, leading to significant changes in diversity throughout the operation. Small changes in microbial community structure resulted in substantial functional changes. The study concluded that the presence of a diverse denitrifying community, especially the occurrence of *Ralstonia sp.* and other less well-known denitrifiers, is required to ensure a well-functioning bioreactor [20, 21].

Anaerobic Treatment Processes

Anaerobic treatment technologies are older than aerobic ones, having been used up until the middle of the 20th century. At that time, the engineers did not know about microorganisms that live and develop under conditions of absence of oxygen. Then, the great boom of aerobic treatment was triggered by advances in knowledge in the fields of biochemistry and bacteriology during the second half of the 20th century. At first, the microbiological mechanisms were considered empirically, leading to uncontrolled adjustments complementing the oxidation of the organic matter. Small advancements made about substrate balance were not consolidated due to the predominant focus on oxygen's role in treatment processes. Observations made in anaerobic conditions were initially linked to a possible deterioration of the conditions of cell respiration, in which inhibition of organic matter decomposition took place by oxygen action or by the alteration of substrate balance. These phenomena frequently resulted in odorous issues and other anaerobic conditions arising from the activity of facultative bacteria [22].

Aerobic treatment systems eliminate various organic compounds and NH_4^+, usually transforming them into CO_2, NO_3^-, H_2O, and H_2S. The initial organic load (COD) in domestic wastewater is about 1.2 kg/m^3, and BOD_5 ranges from 400 to 600 mg/L. Effluent BOD_5 values, as described by various organizations, are usually in the range of 25 to 50 mg/L according to different standards. The concentration of NH_3 in the effluent varies from 8 to 10 mg/L [23 - 25]. Methods for wastewater treatment are summarized in Table **1**.

Role of Microorganisms in Wastewater Treatment

Biodegradation is a process that consists of a central metabolic transformation of the polluting substance. Biological treatment not only prevents the introduction of additional pollutants into the environment but also facilitates the recovery of organic materials present in wastewater. Typically, around 99% of these substances can be eliminated by bacteria. For example, materials like diapers are largely organic; therefore, the sludge produced in the biological treatment in the activated sludge system (which is practically a piece of water with the same hydrological characteristics as domestic sewage) can be easily used for agricultural purposes. Biological treatment has been utilized for over a century and continues to evolve as our understanding of wastewater microorganisms and pollution processes advances [26].

Table 1. Efficiency of current methods applied in wastewater treatment.

Method	Description	Efficiency	Advantages	Limitations
Activated Sludge	Uses activated sludge in aeration tanks for wastewater purification.	High efficiency in organic matter removal.	High removal of organic matter.	Requires high energy input for aeration.
Biological Film	Uses biofilms of microorganisms attached to solid surfaces to degrade high-concentration organic pollutants.	High efficiency in treating industrial wastewater with organic loads.	Space-efficient, low maintenance.	Sensitive to clogging, lower efficiency for highly variable flows.
Aerobic Treatment	Involves oxygen supply to microorganisms for the breakdown of organic pollutants.	High removal of organic compounds.	Fast biodegradation, reliable for treating municipal sewage.	High-energy consumption for aeration.
Anaerobic Treatment	Operates in the absence of oxygen for anaerobic microorganism's organic matter degradation.	High efficiency in oxygen-free environments.	Low-energy consumption, biogas production.	Slower degradation rates.
Physical-Chemical	Includes methods such as precipitation, coagulation, maceration, adsorption, ion exchanges, and chemical reactions.	Efficient in separating inorganic contaminants and suspended solids, it is also efficient in breaking down recalcitrant contaminants.	Can remove difficult-to-treat pollutants.	High operational expenses and the potential formation of toxic by-products.
Bacterial Bioremediation	Uses bacteria to degrade hydrocarbons and other organic compounds.	High efficiency in removing complex organic contaminants.	Suitable for removing complex pollutants, supports natural cycles.	Limited by environmental conditions.
Microalgae Bioremediation	Uses microalgae to absorb heavy metals and degrade organic pollutants.	Efficient in removing heavy metals.	Produces useful biomass, supports sustainable biofuel production.	Requires large surface areas and is sensitive to light and environmental changes.

The processes used to improve the quality of wastewater involve the polluted water coming into contact with air or some microorganisms in the presence of air. The traditional chemical treatment consists of treating the water with some chemical substance, usually chloride, to induce a chemical reaction. This system, known as the physicochemical method, tends to consume a significant amount of chemicals and generates a high concentration of toxic or challenging residues. This approach may be suitable for small industrial activity, but it proves inefficient for addressing the predominant pollution issues faced by small communities and their domestic sewage. The biological treatment of wastewater is one that uses a high concentration of some groups of microorganisms, especially bacteria, that degrade the water pollutants, remedying the problems caused by them [7, 27].

Types of Microorganisms Used

Co-flocculation occurs when biodegradable or non-degradable organic materials adhere to form a nutritious layer for floc-forming bacteria and protozoa. The bacteria present in this process are heterotrophic (dominantly water floc-forming bacteria and floc-associated nitrifying bacteria). They attach themselves to the floccules by a bridge of very long mucilaginous extracellular polysaccharide polymers, which serve as a food source and as binding sites for soluble organics. The next most important type of bacteria is nitrifying bacteria. These bacteria are involved in the breakdown of organic materials and simultaneously convert the nitrogenous waste into less toxic nitrogen (N_2) or nitrate. Some aerobic protozoa, like ciliates and flagellates, also consume organic solids and are involved in floc formation. These organisms are protozoa, which graze on the bacteria on the floccule surface. At the next higher trophic level of bacteria and protozoa in the food chain is the filter feeder zooplankton. However, because most of the wastewater parameters and removal processes are biological, such as the suspension of clay and organic resins, nutrients (nitrogen & phosphorus) uptake by algae, bacteria, and fungi, biological or adsorption coagulation, and flocculation in the floc, much of the hydrophobicity of the organic molecules are offset, allowing the food chain may extend up to zooplankton. Algae can carry out photosynthesis, which leads to the stabilization of pH on the surface of the floc. Encapsulation is the process where the carbonaceous materials appreciably adsorb to the floc, leading to an improvement in efficiency due to a pH drop [28].

Some of the microorganisms that are used for the biological treatment of wastewater are bacteria, fungi, algae, and protozoa. A group of bacteria belonging to the genera *Pseudomonas*, *Bacillus*, *Achromobacter*, *Acinetobacter*, *Moraxella*, and *Aerobacter* can attack and digest organic compounds. Within these genera, certain subsets of these bacterial varieties have developed the ability to completely mineralize many organic compounds at the expense of molecular hydrogen. Another group of bacteria can couple the oxidation of reduced inorganic elements like iron, sulfur, or nitrogen with the reduction of oxygen. They use the energy produced in organic nutrient assimilation or oxidation to drive this reaction in the reverse direction. This group of bacteria is mixed in the presence of a nitrate ion, which produces anaerobic bacteria (Nitrifying bacteria). These bacteria are capable of oxidizing ammonia and producing nitrite. The nitrite can be oxidized to nitrate by some nitrifying bacteria [29, 30].

Microorganisms used in Water Bioremediation

Bacteria

Role and Examples: Aerobic and anaerobic bacteria are crucial in water bioremediation due to their capacity to remove toxic pollutants and transform them into harmless substances [31]. For instance, bacteria species such as *Pseudomonas* and *Mycobacterium* are capable of degrading organic pollutants such as petroleum hydrocarbons into water, carbon dioxide, and other compounds [32]. Additionally, some bacteria species, such as *Streptomyces* and *Arthrobacter,* can remove inorganic pollutants such as ammonium or phosphate [33].

Mechanisms of Action: Bacteria degrade pollutants through aerobic or anaerobic degradation. In aerobic degradation, bacteria use organic pollutants as carbon sources and produce water, carbon dioxide, and minerals; simultaneously, anaerobic degradation uses organic pollutants as carbon sources, but the products are carbon dioxide, methane, and minerals [34]. Furthermore, anaerobic bacteria can use inorganic pollutants as electron donors or acceptors [35].

Fungi

Role and Examples: Several fungi species, mostly from the phyla Ascomycota and Basidiomycota, are effective organic pollutant degraders. *Phanerochaete chysosporium* is the best-known pollutant degradation species due to its capability to degrade benzene, toluene, N-heterocyclic explosives, organochlorines, and synthetic polymers [36].

Mechanisms of Action: The capacity of pollutant degradation is due to the low specificity of several fungal extracellular enzymes, mainly oxidoreductases, which allows them to co-metabolize several contaminant chemical substances [36].

Algae

Role and Examples: Microalgae are used in bioremediation due to their capability to remove organic and inorganic pollutants [37, 38]. In water bioremediation, microalgae are used to remove or neutralize pharmaceutical pollutants and heavy metals [39, 40]. Microalgae *Parachlorella kessleri* is used in bioremediation to degrade several toxic compounds, and the biomass produced during the process can also be used in biofuel production [41].

Mechanisms of Action: Microalgae exploit two mechanisms that algae exploit to remove organic and inorganic pollutants from water: biosorption and bioaccumulation. The pollutants can be absorbed onto the cell surface or accumulate on micropearls inside the cell [40, 42].

Archaea

Role and Examples: Archaea play a critical role in water bioremediation of extreme environments such as halophilic and acidophilic. Archaea are employed to degrade halophilic and acidophilic hydrocarbons, and they are also used to dehalogenate heavy metals [43].

Mechanisms of Action: Archaea, mainly methanogens, degrade hydrocarbons in the absence of oxygen, using them as a carbon source. Also, archaea can use inorganic pollutants like heavy metals as electron acceptors, reducing them into a less toxic form [43].

Understanding the biological capacities of each microorganism to be used in bioremediation enables the scientific community to design more efficient, sustainable, and cost-effective bioremediation methods. An overview of microorganisms used in water bioremediation is base for future research to keep understanding microbial ecology and metabolic capacities.

Overview of Microbial Species that Play a Pivotal Role in Bioremediation.

Bacteria

Key Species and their Roles

Pseudomonas spp.: These bacteria are essential in bioremediation due to their adaptation to chemical pollutants and heavy metals. *Pseudomonas* present aerobic degradation of organic pollutants such as hydrocarbons, aromatic compounds, and pesticides; they can also degrade or remove inorganic pollutants such as nitrates, sulfates, and heavy metals [4, 44 - 47].

Deinococcus radiodurans: This bacterium is the most radiation-tolerant organism known. Its resistance to radioactive substances is due to its highly efficient DNA reparation. This organism limits DNA degradation and increases DNA repair protein activity to compensate for the damage caused by radiation [48, 49]. *D. radiodurans* is used in bioremediation to degrade radioactive pollutants such as iodine, ionic mercury, and uranium [49 - 51].

Shewanella oneidensis: This bacterium is used in bioremediation due to its multicomponent, branched electron transport systems, which confer on it the remarkable ability to reduce heavy metals such as cobalt, uranium, and chromium, transforming them into less toxic forms [52, 53].

Fungi

Key Species and their Roles

Phanerochaete chrysosporium: It is a white-rot fungus capable of degrading organic pollutants, including aromatic hydrocarbons such as benzene, toluene, ethylbenzene, and xylene. It can also degrade nitroaromatic and N-heterocyclic explosives, organochlorines, pesticides, and synthetic polymers [36]. The pollutant degradation is due to oxidative and reductive mechanisms carried out by extracellular enzymes [54].

Aspergillus spp.: Several species from the genera *Aspergillus* present biosorption and bioaccumulation of heavy metals, and it can also transform heavy metals into a less toxic form through reduction reactions [55]. *Aspergillus* is capable of degrading aliphatic hydrocarbons, chlorophenols, polycyclic aromatic hydrocarbons, pesticides, and synthetic dyes [56].

Algae

Key Species and their Roles

Chlorella spp.: Through biosorption and bioaccumulation processes, *Chlorella* species are capable of removing inorganic pollutants such as ammonium, phosphate, and heavy metals such as Cu, Zn, As, Cd, and Fe from wastewater. Furthermore, these microalgae are used in textile wastewater treatment to degrade textile dye [57, 58].

Spirulina spp.: These microalgae degrade and mineralize organic and inorganic pollutants such as textile effluent [58]. Also, *Spirulina* can remove inorganic pollutants such as ammonium and phosphate from eutrophic water bodies [58].

Archaea

Key Species and their Roles

Methanogens: These anaerobic organisms play a key role in the carbon cycle, reducing carbon dioxide to methane. Methanogens are useful organisms in bioremediation, employed in wastewater treatment to degrade organic pollutants such as polycyclic aromatic hydrocarbons, and their capacity for methane production is exploited as a biofuel source [59, 60].

Halophiles: Some archaea have developed the mechanism of high salt-in, which consists of accumulating inorganic ions inside the cell to compensate for environmental osmotic pressure. Because halophile archaea can grow in extremely saline environments, they are used to degrading organic pollutants such as alkane, benzene, naphthalene, toluene, and crude oil from hypersaline environments [61].

Role of Microbial Consortia

Microbial consortia consist of communities of different microbial species working synergistically through coupled metabolic processes. On several occasions, microbial consortia are more effective in bioremediation than a monoculture treatment because they are more stable and adaptable to different and fluctuating polluted environments. They also enhance biomass production and provide better conditions for catalytic enzymes; therefore, microbial consortia can degrade a wide range of complex pollutant compounds such as pesticides, petroleum, antibiotics, colorants, and plastic [41, 62, 63].

Technological and Research Advances

Genetic engineering is a fundamental tool to enhance microorganisms' pollutant degradation through editing metabolic pathways with the goal of improving microorganism survival and degradation. Also, engineering microbial consortia is another approach that apport information of microbial interactions and allow designing microbial consortia to degrade a specific and complex pollutant such as petroleum, considering environmental characteristics [61, 64].

Case Study: the use of Phytoremediation in Radionuclide-contaminated Sites

Radionuclide contamination presents a significant threat to human health and the environment due to its use in nuclear energy generation and industrial, agricultural, and medical activities [65]. The main radioactive elements released into the environment, such as cesium (Cs), strontium (Sr), uranium (U), plutonium (Pu), radium (Ra), technetium (Tc), and yttrium (Y), have direct negative effects on human health and the environment [65] (Fig. **1**).

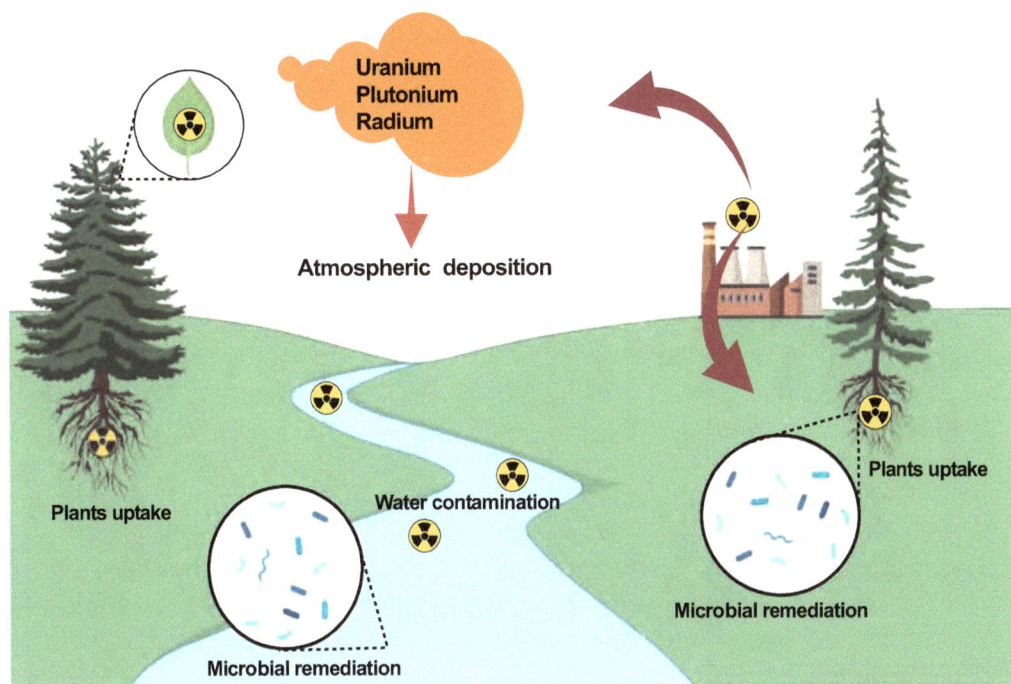

Fig. (1). Impact of radionuclide contamination on human health and the environment, highlighting the main radioactive elements released into the environment—such as uranium (U), plutonium (Pu), and radium (Ra)—and their associated risks. These contaminants, originating from nuclear energy production and industrial, agricultural, and medical activities, pose serious threats due to their long-lasting and harmful effects. The figure also illustrates the potential of microbial bioremediation as a strategy to mitigate these contaminants, reducing their harmful effects on ecosystems and human health. Figure made with Biorender.

A notable case study on the use of biotechnology for soil remediation is the application of phytoremediation techniques in areas affected by radionuclides. Phytoremediation is a biotechnological process that uses various types of plants to remove, transfer, stabilize, and destroy contaminants in soil and groundwater. This approach has shown promise in reducing radionuclide content in the topsoil [66].

Application of Phytoremediation

Following nuclear incidents, such as the Chernobyl disaster in 1986, scientists have been studying the effectiveness of using various hyperaccumulator plants to absorb radioactive isotopes from the soil. These plants, such as sunflowers (*Helianthus*) and *Brassica* species, are grown in contaminated soil and absorb radioactive elements like cesium-137 and strontium-90 through their roots. After a period of growth, the plants are harvested and disposed of as radioactive waste [67, 68].

Procedure: Implementing Microbial Bioremediation

To implement microbial bioremediation for contaminated soils, the following steps are generally involved:

1. Site Assessment: Conduct a thorough analysis of the contaminated site to determine the extent and type of contamination.
2. Microorganism Selection: Identify and select appropriate microorganisms capable of degrading the specific contaminants present in the soil.
3. Laboratory Testing: Perform laboratory tests to evaluate the efficacy of the selected microorganisms in breaking down the contaminants under controlled conditions.
4. Site Preparation: Prepare the contaminated site for treatment, which may involve adjusting soil pH, moisture, and nutrient levels to optimize conditions for microbial activity.
5. Bioaugmentation and/or Biostimulation: Introduce the selected microorganisms into the contaminated soil (bioaugmentation) and/or apply nutrients and other substances to stimulate the activity of indigenous microorganisms (biostimulation).
6. Monitoring: Regularly monitor the site to assess the progress of the remediation process, including changes in contaminant levels, microbial population dynamics, and soil conditions.
7. Post-Treatment Assessment: Upon completion of the treatment, conduct a final assessment to determine the effectiveness of the remediation and ensure that contaminant levels have been reduced to acceptable standards.
8. Site Restoration: Implement measures to restore the treated site to its intended use, which may involve landscaping, replanting, and other ecological restoration activities.

Challenges and Future Directions in Biological Treatment

In contrast, we explored the problems and challenges of biological wastewater treatment and possible future directions. Among these challenges, they point out that variations in different climates are somewhat common and can adversely affect treatment efficiency. The other challenges are related to the increasing restrictions in relation to the concentrations of contaminants that consume the oxygen present in the water and the generation of excessive amounts of sludge. Within this framework, the scientific community is increasingly seeking ways to direct research to solve these problems that remain unresolved, and one of the ways that is presenting good results is the use of new members and consortia of the scientific community [69, 70].

Biological treatment remains the most widely used technique for domestic and industrial wastewater. Studies reported some of the latest trends and advances in technology, including the ones that are environmentally friendly, to obtain better results and generate less sludge [71, 72].

Emerging Technologies

Each of the aforementioned technologies presents different advantages and disadvantages, and their selection is generally based on individual or combined processes. In a large-scale treatment plant, the most common combination of processes is primary treatment followed by secondary treatment (activated sludge) that has been modified to Nutrient Removal (German EBPR, Enhanced Biological Phosphorus Removal). The effluent from these processes contains a significant number of nutrients in the form of dissolved P and N compounds. These compounds, even at very low concentrations (0.6 mg/L of P), will promote the eutrophication and contamination of the water bodies in which they are discharged. This effluent needs to be subjected to a tertiary treatment to obtain an effluent that complies with the legal limits of these nutrient compounds [73, 74].

Bio-stimulated aerobic reactors (B.A.S.): This process is based on the combined action of the aerobic and anaerobic microorganisms added systematically at specific flow points of the WW. Ozone: It consists of the generation of ozonized water that produces the simultaneous precipitation of the suspended matter in the wastewater, favoring a higher action of the biodegradable organic matter on the microorganisms in the activated sludge bacteria field. This technique is based on the oxidation of the organic matter in the liquid phase. Other organic matter oxidation methods are advanced oxidation processes (AOPs) and biological membrane processes [75, 76].

CONCLUDING REMARKS

The biological treatment of wastewater offers a sustainable and efficient method for purifying domestic and industrial effluents. By leveraging the metabolic capabilities of microorganisms, various organic and inorganic contaminants can be effectively degraded or transformed. This process purifies water and generates valuable by-products such as biosolids, biogases, and protein compounds that can be recovered and reused.

The diverse microbial communities involved in these processes— including bacteria, fungi, algae, and archaea—operate under varying nutrient and physicochemical conditions, making biological treatment versatile and adaptable. The integration of biological methods with physical and chemical treatment steps enhances the overall efficiency of wastewater purification, addressing a broad spectrum of pollutants.

As the demand for clean water continues to rise due to population growth and industrial expansion, the importance of effective wastewater treatment becomes increasingly critical. Biological treatment methods provide a viable solution, especially when traditional methods fall short due to the complexity and toxicity of certain contaminants. These methods are crucial not only for protecting human health but also for maintaining environmental quality and supporting agricultural and industrial activities.

Future research and technological advancements in this field are essential to further optimize and expand the use of biological treatment methods. Continued exploration of microbial capacities, along with the development of innovative biotechnological approaches, will enhance the sustainability and effectiveness of wastewater treatment systems. By harnessing the natural processes of microorganisms, we can achieve significant progress in environmental protection and resource recovery, paving the way for a cleaner and healthier future.

AUTHORS' CONTRIBUTION

The authors confirmed their contribution to the chapter as follows:

Conceptualization, methodology, investigation, data curation, writing-original draft preparation, writing review and editing, supervision: Román Adrián González Cruz; Methodology, investigation, formal analysis, data curation, writing-original draft preparation, writing review, and editing: Alonso Ezeta Miranda; Conceptualization, supervision, validation, writing review, and editing: Israel Valencia Quiroz.

QUESTIONS RELATED TO THE TEXT

1. What are the key microbial processes involved in the biological treatment of wastewater, and how do they contribute to water purification?
2. How do different types of microorganisms, such as bacteria, fungi, algae, and archaea, contribute to the biodegradation and transformation of contaminants in wastewater?
3. What are the advantages of using biological treatment methods over traditional chemical and physical methods for wastewater purification?
4. How do anaerobic and aerobic treatment processes differ in their approach to wastewater purification, and what are the benefits of each method?
5. What role do microbial consortia play in enhancing the efficiency of wastewater treatment, and why are they often more effective than monoculture

treatments?
6. What challenges are associated with the biological treatment of wastewater, and what future directions are being explored to improve the effectiveness and sustainability of these methods?

ACKNOWLEDGEMENTS

The authors would like to acknowledge Ismael Manuel Urrutia Ortega for reviewing data curation, investigation, formal analysis, writing original draft preparation, and figure editing.

LIST OF ABBREVIATIONS

BAS	Bio-Stimulated Aerobic Reactors
BOD	Biochemical Oxygen Demand
BOD$_5$	5 Days Biochemical Oxygen Demand
COD	Chemical Oxygen Demand
DNA	Deoxyribonucleic Acid
EBPR	Enhanced Biological Phosphorus Removal
H$_2$S	Hydrogen Sulfide
N	Nitrogen
NH$_4^+$	Ammonium Ion
NO$_3^-$	Nitrate
P	Phosphorus
POAs	Advanced Oxidation Processes
PSB	Purple Sulfur Bacteria
WW	Wastewater

REFERENCES

[1]   Yang Z, Zhou Q, Sun H, Jia L, Zhao L, Wu W. Metagenomic analyses of microbial structure and metabolic pathway in solid-phase denitrification systems for advanced nitrogen removal of wastewater treatment plant effluent: A pilot-scale study. Water Res 2021; 196: 117067.
[http://dx.doi.org/10.1016/j.watres.2021.117067] [PMID: 33773452]

[2]   Wu G, Yin Q. Microbial niche nexus sustaining biological wastewater treatment. npj Clean Water. 2020; 3(1): 33.
[http://dx.doi.org/10.1038/s41545-020-00080-4]

[3]   Ren Y, Hao Ngo H, Guo W, *et al.* New perspectives on microbial communities and biological nitrogen removal processes in wastewater treatment systems. Bioresour Technol 2020; 297: 122491.
[http://dx.doi.org/10.1016/j.biortech.2019.122491] [PMID: 31810739]

[4]   Zhang C, Li S, Ho SH. Converting nitrogen and phosphorus wastewater into bioenergy using microalgae-bacteria consortia: A critical review. Bioresour Technol 2021; 342: 126056.
[http://dx.doi.org/10.1016/j.biortech.2021.126056] [PMID: 34601027]

[5]   Ahmed SF, Mofijur M, Nuzhat S, *et al*. Recent developments in physical, biological, chemical, and hybrid treatment techniques for removing emerging contaminants from wastewater. J Hazard Mater 2021; 416: 125912.
[http://dx.doi.org/10.1016/j.jhazmat.2021.125912] [PMID: 34492846]

[6]   Sanghamitra P, Mazumder D, Mukherjee S. Treatment of wastewater containing oil and grease by biological method- a review. J Environ Sci Health Part A Tox Hazard Subst Environ Eng 2021; 56(4): 394-412.
[http://dx.doi.org/10.1080/10934529.2021.1884468] [PMID: 33573477]

[7]   Rahimi S, Modin O, Mijakovic I. Technologies for biological removal and recovery of nitrogen from wastewater. Biotechnol Adv 2020; 43: 107570.
[http://dx.doi.org/10.1016/j.biotechadv.2020.107570] [PMID: 32531318]

[8]   Saleh IA, Zouari N, Al-Ghouti MA. Removal of pesticides from water and wastewater: Chemical, physical and biological treatment approaches. Environ Technol Innov 2020; 19: 101026.
[http://dx.doi.org/10.1016/j.eti.2020.101026]

[9]   Maja MM, Ayano SF. The impact of population growth on natural resources and farmers' capacity to adapt to climate change in low-income countries. Earth Syst Environ 2021; 5(2): 271-83.
[http://dx.doi.org/10.1007/s41748-021-00209-6]

[10]   Sadigov R. Rapid growth of the world population and its socioeconomic results. ScientificWorldJournal 2022; 2022: 1-8.
[http://dx.doi.org/10.1155/2022/8110229] [PMID: 35370481]

[11]   Ambaye TG, Vaccari M, van Hullebusch ED, Amrane A, Rtimi S. Mechanisms and adsorption capacities of biochar for the removal of organic and inorganic pollutants from industrial wastewater. Int J Environ Sci Technol 2021; 18(10): 3273-94.
[http://dx.doi.org/10.1007/s13762-020-03060-w]

[12]   Tahir MB, Nawaz T, Nabi G, Sagir M, Khan MI, Malik N. Role of nanophotocatalysts for the treatment of hazardous organic and inorganic pollutants in wastewater. Int J Environ Anal Chem 2022; 102(2): 491-515.
[http://dx.doi.org/10.1080/03067319.2020.1723570]

[13]   de Carluccio M, Fiorentino A, Rizzo L. Multi-barrier treatment of mature landfill leachate: effect of Fenton oxidation and air stripping on activated sludge process and cost analysis. J Environ Chem Eng 2020; 8(5): 104444.
[http://dx.doi.org/10.1016/j.jece.2020.104444]

[14]   Widajatno RL, Kardena E, Arifianingsih NN, Helmy Q. Activated sludge: conventional dye treatment technique.Biological Approaches in Dye-Containing Wastewater. Singapore: Springer 2022; pp. 119-53.
[http://dx.doi.org/10.1007/978-981-19-0545-2_5]

[15]   Wang K, Chen X, Yan D, Xu Z, Hu P, Li H. Petrochemical and municipal wastewater treatment plants activated sludge each own distinct core bacteria driven by their specific incoming wastewater. Sci Total Environ 2022; 826: 153962.
[http://dx.doi.org/10.1016/j.scitotenv.2022.153962] [PMID: 35189240]

[16]   Heitkamp MA, Cerniglia CE. Mineralization of polycyclic aromatic hydrocarbons by a bacterium isolated from sediment below an oil field. Appl Environ Microbiol 1988; 54(6): 1612-4.
[http://dx.doi.org/10.1128/aem.54.6.1612-1614.1988] [PMID: 3415226]

[17]   Akagi JM. Dissimilatory sulfate reduction, mechanistic aspects.Biology of Inorganic Nitrogen and Sulfur. Berlin, Heidelberg: Springer 1981; pp. 178-87.
[http://dx.doi.org/10.1007/978-3-642-67919-3_13]

[18]   Thauer RK, Stackebrandt E, Hamilton WA. Energy metabolism and phylogenetic diversity of sulphate-reducing bacteria.Sulphate-Reducing Bacteria: Environmental and Engineered Systems.

Cambridge: Cambridge University Press 2007; pp. 1-38.
[http://dx.doi.org/10.1017/CBO9780511541490.002]

[19]   Pfennig N, Widdel F, Postgate JR, Postgate JR, Kelly DP. The bacteria of the sulphur cycle. Philos Trans R Soc Lond B Biol Sci 1982; 298(1093): 433-41.
[http://dx.doi.org/10.1098/rstb.1982.0090] [PMID: 6127734]

[20]   Ahmad F, Zhu D, Sun J. Bacterial chemotaxis: a way forward to aromatic compounds biodegradation. Environ Sci Eur 2020; 32(1): 52.
[http://dx.doi.org/10.1186/s12302-020-00329-2]

[21]   Hao DC, Li XJ, Xiao PG, Wang LF. The utility of electrochemical systems in microbial degradation of polycyclic aromatic hydrocarbons: discourse, diversity and design. Front Microbiol 2020; 11: 557400.
[http://dx.doi.org/10.3389/fmicb.2020.557400] [PMID: 33193139]

[22]   Ho L, Goethals PLM. Municipal wastewater treatment with pond technology: Historical review and future outlook. Ecol Eng 2020; 148: 105791-1.
[http://dx.doi.org/10.1016/j.ecoleng.2020.105791]

[23]   Charles KJ, Ashbolt NJ, Roser DJ, McGuinness R, Deere DA. Effluent quality from 200 on-site sewage systems: design values for guidelines. Water Sci Technol 2005; 51(10): 163-9.
[http://dx.doi.org/10.2166/wst.2005.0363] [PMID: 16104418]

[24]   Zeng Y, de Guardia A, Daumoin M, Benoist JC. Characterizing the transformation and transfer of nitrogen during the aerobic treatment of organic wastes and digestates. Waste Manag 2012; 32(12): 2239-47.
[http://dx.doi.org/10.1016/j.wasman.2012.07.006] [PMID: 22863068]

[25]   Maxfield M, Daniell WE, Treser CD, VanDerslice J. Aerobic residential onsite sewage systems: an evaluation of treated-effluent quality [Internet]. Journal of Environmental Health. 2003; 66(3): 14-22.
https://link.gale.com/apps/doc/A109569149/HRCA?u=anon~bd71f29f&sid=googleScholar&xid=c1464209

[26]   Ceretta MB, Nercessian D, Wolski EA. Current trends on role of biological treatment in integrated treatment technologies of textile wastewater. Front Microbiol 2021; 12: 651025.
[http://dx.doi.org/10.3389/fmicb.2021.651025] [PMID: 33841377]

[27]   Abdelfattah A, Hossain MI, Cheng L. High-strength wastewater treatment using microbial biofilm reactor: a critical review. World J Microbiol Biotechnol 2020; 36(5): 75.
[http://dx.doi.org/10.1007/s11274-020-02853-y] [PMID: 32390104]

[28]   Duan S, Zhang Y, Zheng S. Heterotrophic nitrifying bacteria in wastewater biological nitrogen removal systems: A review. Crit Rev Environ Sci Technol 2022; 52(13): 2302-38.
[http://dx.doi.org/10.1080/10643389.2021.1877976]

[29]   Deng M, Zhao X, Senbati Y, Song K, He X. Nitrogen removal by heterotrophic nitrifying and aerobic denitrifying bacterium *Pseudomonas sp*. DM02: Removal performance, mechanism and immobilized application for real aquaculture wastewater treatment. Bioresour Technol 2021; 322: 124555.
[http://dx.doi.org/10.1016/j.biortech.2020.124555] [PMID: 33352391]

[30]   Ibrahim S, El-Liethy MA, Elwakeel KZ, Hasan MAEG, Al Zanaty AM, Kamel MM. Role of identified bacterial consortium in treatment of Quhafa Wastewater Treatment Plant influent in Fayuom, Egypt. Environ Monit Assess 2020; 192(3): 161.
[http://dx.doi.org/10.1007/s10661-020-8105-9] [PMID: 32020301]

[31]   Diken-Gür S, Bakhshpour M, Denizli A. Applications of microbes in bioremediation of water pollutants. In: Inamuddin, Ahamed MI, Prasad R (eds) Recent Advances in Microbial Degradation. Environmental and Microbial Biotechnology. Springer: Singapore 2021; pp 465–83.
[http://dx.doi.org/10.1007/978-981-16-0518-5_19]

[32]   Khalid FE, Lim ZS, Sabri S, Gomez-Fuentes C, Zulkharnain A, Ahmad SA. Bioremediation of diesel

contaminated marine water by bacteria: a review and bibliometric analysis. J Mar Sci Eng 2021; 9(2): 155.
[http://dx.doi.org/10.3390/jmse9020155]

[33]   Dong D, Sun H, Qi Z, Liu X. Improving microbial bioremediation efficiency of intensive aquacultural wastewater based on bacterial pollutant metabolism kinetics analysis. Chemosphere 2021; 265: 129151.
[http://dx.doi.org/10.1016/j.chemosphere.2020.129151] [PMID: 33302206]

[34]   Kyrikou I, Briassoulis D. Biodegradation of agricultural plastic films: a critical review. J Polym Environ 2007; 15(2): 125-50.
[http://dx.doi.org/10.1007/s10924-007-0053-8]

[35]   de Carvalho CCCR. Adaptation of *Rhodococcus erythropolis* cells for growth and bioremediation under extreme conditions. Res Microbiol 2012; 163(2): 125-36.
[http://dx.doi.org/10.1016/j.resmic.2011.11.003] [PMID: 22146587]

[36]   Harms H, Schlosser D, Wick LY. Untapped potential: exploiting fungi in bioremediation of hazardous chemicals. Nat Rev Microbiol 2011; 9(3): 177-92.
[http://dx.doi.org/10.1038/nrmicro2519] [PMID: 21297669]

[37]   Hussain MM, Wang J, Bibi I, *et al.* Arsenic speciation and biotransformation pathways in the aquatic ecosystem: The significance of algae. J Hazard Mater 2021; 403: 124027.
[http://dx.doi.org/10.1016/j.jhazmat.2020.124027] [PMID: 33265048]

[38]   Touliabah HES, El-Sheekh MM, Ismail MM, El-Kassas H. A review of microalgae- and cyanobacteria-based biodegradation of organic pollutants. Molecules 2022; 27(3): 1141.
[http://dx.doi.org/10.3390/molecules27031141] [PMID: 35164405]

[39]   Xiong JQ, Kurade MB, Jeon BH. Can microalgae remove pharmaceutical contaminants from water? Trends Biotechnol 2018; 36(1): 30-44.
[http://dx.doi.org/10.1016/j.tibtech.2017.09.003] [PMID: 28993012]

[40]   Zeraatkar AK, Ahmadzadeh H, Talebi AF, Moheimani NR, McHenry MP. Potential use of algae for heavy metal bioremediation, a critical review. J Environ Manage 2016; 181: 817-31.
[http://dx.doi.org/10.1016/j.jenvman.2016.06.059] [PMID: 27397844]

[41]   Singh R, Ryu J, Kim SW. Microbial consortia including methanotrophs: some benefits of living together. J Microbiol 2019; 57(11): 939-52.
[http://dx.doi.org/10.1007/s12275-019-9328-8] [PMID: 31659683]

[42]   Martignier A, Filella M, Pollok K, *et al.* Marine and freshwater micropearls: biomineralization producing strontium-rich amorphous calcium carbonate inclusions is widespread in the genus *Tetraselmis* (Chlorophyta). Biogeosciences 2018; 15(21): 6591-605.
[http://dx.doi.org/10.5194/bg-15-6591-2018]

[43]   Krzmarzick MJ, Taylor DK, Fu X, McCutchan AL. Diversity and niche of archaea in bioremediation. Archaea 2018; 2018: 1-17.
[http://dx.doi.org/10.1155/2018/3194108] [PMID: 30254509]

[44]   Furmanczyk EM, Kaminski MA, Lipiński L, Dziembowski A, Sobczak A. *Pseudomonas laurylsulfatovorans* sp. nov., sodium dodecyl sulfate degrading bacteria, isolated from the peaty soil of a wastewater treatment plant. Syst Appl Microbiol 2018; 41(4): 348-54.
[http://dx.doi.org/10.1016/j.syapm.2018.03.009] [PMID: 29752019]

[45]   Tyagi B, Kumar N. Chapter 1 - Bioremediation: principles and applications in environmental management. En: Saxena G, Kumar V, Shah MP, editores. Bioremediation for Environmental Sustainability [Internet]. Elsevier; 2021. p. 3–28.
[http://dx.doi.org/10.1016/B978-0-12-820524-2.00001-8]

[46]   Verma S, Kuila A. Bioremediation of heavy metals by microbial process. Environ Technol Innov 2019; 14: 100369.

[http://dx.doi.org/10.1016/j.eti.2019.100369]

[47]   Yong YC, Wu XY, Sun JZ, Cao YX, Song H. Engineering quorum sensing signaling of *Pseudomonas* for enhanced wastewater treatment and electricity harvest: A review. Chemosphere 2015; 140: 18-25.
[http://dx.doi.org/10.1016/j.chemosphere.2014.10.020] [PMID: 25455678]

[48]   Cox MM, Battista JR. *Deinococcus radiodurans* — the consummate survivor. Nat Rev Microbiol 2005; 3(11): 882-92.
[http://dx.doi.org/10.1038/nrmicro1264] [PMID: 16261171]

[49]   Manobala T, Shukla SK, Subba Rao T, Dharmendira Kumar M. A new uranium bioremediation approach using radio-tolerant *Deinococcus radiodurans* biofilm. J Biosci 2019; 44(5): 122.
[http://dx.doi.org/10.1007/s12038-019-9942-y] [PMID: 31719231]

[50]   Brim H, McFarlan SC, Fredrickson JK, *et al.* Engineering *Deinococcus radiodurans* for metal remediation in radioactive mixed waste environments. Nat Biotechnol 2000; 18(1): 85-90.
[http://dx.doi.org/10.1038/71986] [PMID: 10625398]

[51]   Choi MH, Jeong SW, Shim HE, Yun SJ, Mushtaq S, Choi DS, et al. Efficient bioremediation of radioactive iodine using biogenic gold nanomaterial-containing radiation-resistant bacterium, Deinococcus radiodurans R1. Chem Commun. 2017; 53(28): 3937–40.
[http://dx.doi.org/10.1039/C7CC00720E] [PMID: 28317956]

[52]   Hau HH, Gilbert A, Coursolle D, Gralnick JA. Mechanism and Consequences of anaerobic respiration of cobalt by *Shewanella oneidensis* strain MR-1. Appl Environ Microbiol 2008; 74(22): 6880-6.
[http://dx.doi.org/10.1128/AEM.00840-08] [PMID: 18836009]

[53]   Heidelberg JF, Paulsen IT, Nelson KE, *et al.* Genome sequence of the dissimilatory metal ion–reducing bacterium *Shewanella oneidensis.* Nat Biotechnol 2002; 20(11): 1118-23.
[http://dx.doi.org/10.1038/nbt749] [PMID: 12368813]

[54]   Cameron MD, Timofeevski S, Aust SD. Enzymology of *Phanerochaete chrysosporium* with respect to the degradation of recalcitrant compounds and xenobiotics. Appl Microbiol Biotechnol 2000; 54(6): 751-8.
[http://dx.doi.org/10.1007/s002530000459] [PMID: 11152065]

[55]   Fernández PM, Viñarta SC, Bernal AR, de Figueroa LIC. Chapter 3 - Advances in bioremediation of hexavalent chromium: cytotoxicity, genotoxicity, and microbial alleviation strategies for environmental safety. En: Saxena G, Kumar V, Shah MP, editores. Bioremediation for Environmental Sustainability [Internet]. Elsevier; 2021. p. 55–72.
[http://dx.doi.org/10.1016/B978-0-12-820524-2.00003-1]

[56]   Harms H, Schlosser D, Wick LY. Untapped potential: exploiting fungi in bioremediation of hazardous chemicals. Nat Rev Microbiol 2011; 9(3): 177-92.
[http://dx.doi.org/10.1038/nrmicro2519] [PMID: 21297669]

[57]   Lim SL, Chu WL, Phang SM. Use of *Chlorella vulgaris* for bioremediation of textile wastewater. Bioresour Technol 2010; 101(19): 7314-22.
[http://dx.doi.org/10.1016/j.biortech.2010.04.092] [PMID: 20547057]

[58]   Zainith S, Saxena G, Kishor R, Bharagava RN. Chapter 20 - Application of microalgae in industrial effluent treatment, contaminants removal, and biodiesel production: Opportunities, challenges, and future prospects. En: Saxena G, Kumar V, Shah MP, editores. Bioremediation for Environmental Sustainability [Internet]. Elsevier; 2021. p. 481–517.
[http://dx.doi.org/10.1016/B978-0-12-820524-2.00020-1]

[59]   Carr S, Buan NR. Insights into the biotechnology potential of *Methanosarcina.* Front Microbiol 2022; 13: 1034674.
[http://dx.doi.org/10.3389/fmicb.2022.1034674] [PMID: 36590411]

[60]   Ye Q, Liang C, Chen X, Fang T, Wang Y, Wang H. Molecular characterization of methanogenic microbial communities for degrading various types of polycyclic aromatic hydrocarbon. J Environ Sci

(China) 2019; 86: 97-106.
[http://dx.doi.org/10.1016/j.jes.2019.04.027] [PMID: 31787194]

[61]   Somee MR, Dastgheib SMM, Shavandi M, Zolfaghar M, Zamani N, Ventosa A, et al. Chapter 11 - Halophiles in bioremediation of petroleum contaminants: challenges and prospects. En: Saxena G, Kumar V, Shah MP, editores. Bioremediation for Environmental Sustainability [Internet]. Elsevier; 2021. p. 251–91.
[http://dx.doi.org/10.1016/B978-0-12-820524-2.00011-0]

[62]   Bhatt P, Bhatt K, Sharma A, Zhang W, Mishra S, Chen S. Biotechnological basis of microbial consortia for the removal of pesticides from the environment 2021; 41(3): 317-38.
[http://dx.doi.org/10.1080/07388551.2020.1853032]

[63]   Cao Z, Yan W, Ding M, Yuan Y. Construction of microbial consortia for microbial degradation of complex compounds. Front Bioeng Biotechnol 2022; 10: 1051233.
[http://dx.doi.org/10.3389/fbioe.2022.1051233] [PMID: 36561050]

[64]   Chen W, Brühlmann F, Richins R, Mulchandani A. Engineering of improved microbes and enzymes for bioremediation. Curr Opin Biotechnol 1999; 10(2): 137-41.
[http://dx.doi.org/10.1016/S0958-1669(99)80023-8] [PMID: 10209138]

[65]   Grémy O. Medical countermeasures against radionuclide contamination: An overview. 2019; 14: 06001.
[http://dx.doi.org/10.1051/bioconf/20191406001]

[66]   Ojuederie OB, Amoo AE, Owonubi SJ, Ayangbenro AS. Nanoparticles-assisted phytoremediation: advances and applications.Assisted Phytoremediation. Elsevier 2022; pp. 155-78.
[http://dx.doi.org/10.1016/B978-0-12-822893-7.00011-2]

[67]   Fuhrmann M, Lasat MM, Ebbs SD, Kochian LV, Cornish J. Uptake of cesium-137 and strontium-90 from contaminated soil by three plant species; application to phytoremediation. J Environ Qual 2002; 31(3): 904-9.
[http://dx.doi.org/10.2134/jeq2002.9040] [PMID: 12026094]

[68]   Willey N, Hall S, Mudigantia A. Assessing the potential of phytoremediation at a site in the U.K. contaminated with [137]Cs. Int J Phytoremediation 2001; 3(3): 321-33.
[http://dx.doi.org/10.1080/15226510108500062]

[69]   Alisawi HAO. Performance of wastewater treatment during variable temperature. Appl Water Sci 2020; 10(4): 89.
[http://dx.doi.org/10.1007/s13201-020-1171-x]

[70]   Cardoso BJ, Rodrigues E, Rodrigues Gaspar A, et al. Energy performance factors in wastewater treatment plants: A review. Journal of Cleaner Production 2021; 322: 129107.
[http://dx.doi.org/10.1016/j.jclepro.2021.129107]

[71]   Donkadokula NY, Kola AK, Naz I, Saroj D. A review on advanced physico-chemical and biological textile dye wastewater treatment techniques. Rev Environ Sci Biotechnol 2020; 19(3): 543-60.
[http://dx.doi.org/10.1007/s11157-020-09543-z]

[72]   Dutta D, Arya S, Kumar S. Industrial wastewater treatment: Current trends, bottlenecks, and best practices. Chemosphere 2021; 285: 131245.
[http://dx.doi.org/10.1016/j.chemosphere.2021.131245] [PMID: 34246094]

[73]   Pham TL, Bui MH. Removal of nutrients from fertilizer plant wastewater using *Scenedesmus* sp.: formation of bioflocculation and enhancement of removal efficiency. J Chem 2020; 2020: 1-9.
[http://dx.doi.org/10.1155/2020/8094272]

[74]   Rout PR, Shahid MK, Dash RR, *et al.* Nutrient removal from domestic wastewater: A comprehensive review on conventional and advanced technologies. J Environ Manage 2021; 296: 113246.
[http://dx.doi.org/10.1016/j.jenvman.2021.113246] [PMID: 34271353]

[75]   Egerland Bueno B, Américo Soares L, Quispe-Arpasi D, *et al.* Anaerobic digestion of aqueous phase

from hydrothermal liquefaction of *Spirulina* using biostimulated sludge. Bioresour Technol 2020; 312: 123552.
[http://dx.doi.org/10.1016/j.biortech.2020.123552] [PMID: 32502889]

[76]   Wu Y, Liu Y, Kamyab H, *et al.* Physico-chemical and biological remediation techniques for the elimination of endocrine-disrupting hazardous chemicals. Environ Res 2023; 232: 116363.
[http://dx.doi.org/10.1016/j.envres.2023.116363] [PMID: 37295587]

CHAPTER 10

Bioremediation of Contaminated Waters: Strategies and Success Cases

Nichdaly Ortiz Chacón[1,*], Aliana Zacaria Vital[1] and Israel Valencia Quiroz[1]

[1] *Phytochemistry Laboratory, UBIPRO, Superior Studies Faculty (FES)-Iztacala, National Autonomous University of Mexico (UNAM), Tlalnepantla de Baz, Mexico State, Mexico*

Abstract: Bioremediation of contaminated waters is an essential strategy to address pollution from various sources, such as industry, agriculture, and urban activities. This approach employs biological agents, including plants, microorganisms, and their enzymes, to detoxify and remove pollutants from aquatic environments. Bioaugmentation, involving the introduction of specialized cleanup microorganisms, is a significant technique, often requiring genetic engineering and extensive testing to ensure the microorganisms can survive and perform effectively in the target environment. Phytoremediation, where plants are used to absorb and degrade contaminants, is another crucial strategy. Contaminants affecting water bodies include oil, heavy metals, persistent organic pollutants (POPs), and agricultural chemicals, originating from point sources like factories and wastewater treatment plants, as well as non-point sources such as urban runoff and atmospheric deposition. The negative impacts of these contaminants range from aesthetic concerns to severe threats to human health and ecosystems. Bioremediation harnesses the natural detoxifying abilities of microorganisms and plants. Bacteria and fungi play a crucial role in transforming and detoxifying a broad spectrum of pollutants. Techniques like biostimulation enhance the activity of native microorganisms by adding nutrients or biosurfactants, facilitating the degradation of hydrocarbons and other contaminants. Phytoremediation utilizes plants to extract, stabilize, and degrade pollutants, providing a cost-effective and environmentally friendly solution. Success cases of bioremediation, such as the treatment of the Exxon Valdez oil spill and the recovery of Lake Washington from sewage pollution, demonstrate the effectiveness of these strategies. Challenges remain, including optimizing treatment efficiency and addressing emerging contaminants. However, ongoing research and technological advancements continue to improve the sustainability and applicability of bioremediation for large-scale environmental cleanup efforts.

Keywords: Bioremediation, Bioaugmentation, Detoxification, Environmental cleanup, Heavy metals, Microorganisms, Pollutants, Phytoremediation, Persistent organic pollutants, Water contamination.

---

* **Corresponding author Nichdaly Ortiz Chacón:** Phytochemistry Laboratory, UBIPRO, Superior Studies Faculty (FES)-Iztacala, National Autonomous University of Mexico (UNAM), Tlalnepantla de Baz, Mexico State, Mexico; Tel: +525556231137; E-mail: ortizchaconnichdaly@gmail.com

Israel Valencia Quiroz (Ed.)
All rights reserved-© 2025 Bentham Science Publishers

INTRODUCTION

Bioaugmentation usually requires large-scale propagation or the preparation of specialized cleanup microorganisms that have been genetically engineered. This process often involves significant investment and careful regulation. When considering the use of naturally occurring microorganisms for bioaugmentation purposes, it is important to perform extensive testing to determine the ability of the candidate microorganisms to survive in the receiving environment and perform the desired pollutant degradation processes. Reliable candidates for bioaugmentation can often be found or constructed by specialized methods, such as rational design or directed evolution. The process in which plants act in the reclamation of nutrients or pollutant levels in contaminated water or sediments through their natural growth and nutrient uptake is generally regarded as phytoextraction or phytoremediation (Fig. **1**) [1, 2].

Fig. (1). Main sources and types of contaminants affecting water bodies. Contaminants include products from human activities such as oil, heavy metals, POPs, detergents, and agricultural chemicals. The figure highlights both point sources, such as factories and wastewater treatment plants, and non-point sources, including agriculture and urban runoff, that contribute to the pollution of rivers, lakes, and seas. Figure made with DALL-E.

Despite all the efforts, water bodies continue to be affected by different sources of pollution. Industry, agriculture, services, and tourist activities are frequently responsible for water contamination, making the restoration of affected aquatic ecosystems a pressing priority. For that, one of the available technologies is bioremediation, which relies on the use of biological agents such as plants, microorganisms, or their enzymes to clean up pollution. When this biological approach is specifically facilitated by using microorganisms, it is often called "bioaugmentation" [3].

Sources and Types of Contaminants

The contamination of rivers, lakes, and sea waters with different types of chemicals is a well-known environmental problem. In general, the main contaminants in these waters are products from human activities such as oil, heavy metals, radioactive elements, detergents, agricultural chemicals (pesticides, herbicides, and fertilizers), BOD, and VOCs [4]. The effects caused by these contaminants can range from slight aesthetic problems to severe harm to human health and the environment. The sources of these aquatic pollutants are varied, and the contamination can occur from point sources (such as factories, refineries, sewage treatment plants, and vessels) as well as from non-point sources (agriculture, urban runoff, atmospheric deposition, and natural earth leaching) [5].

Numerous hydrocarbon pollutants are discharged into aquatic environments each year. These pollutants mainly originate from oil spills (accidental release of oil-related products into the environment) and discharges of oil from vessels, as well as from illegal land-based disposal of oil. POPs are another major group of contaminants in water. These chemicals are heavily used in manufacturing processes, and their accidental release or improper disposal can cause severe contamination. In addition, the heavy metals Hg, Cd, Pb, and As are among the most toxic contaminants in the aquatic environment due to their high toxicity and long-term effects on biota and human health. Although the concentrations of these toxic elements in the aquatic environment are generally very low, they tend to bioaccumulate within the food chain. As a result of the harmful accumulation of these substances in the marine environment, more and more restrictive measures are being implemented to protect marine ecosystems and human health [6 - 8].

Bioremediation: an Overview

Bioremediation is an innovative use of living (micro)organisms to clean up environmental pollution. These innocent, invisible helpers form a natural army that can detoxify and eliminate a wide range of environmental contaminants. It has long been appreciated that microorganisms such as bacteria and fungi have the unique ability to transform and detoxify a broad spectrum of organic compounds. In the past decades, bioremediation has matured as an accepted technology. When properly designed and implemented, it often offers a cost-effective and environmentally friendly solution for the cleanup of soil and water contaminants. With water scarcity becoming an ever more pressing global issue, there is a growing interest in the use of reclaimed water. The recent development of drought-resistant GM plants will facilitate the purification of water for this purpose [9, 10].

The contamination of water bodies with hazardous chemicals is a common problem that requires rapid and efficient countermeasures. Bioremediation, or the use of biological agents to remove or detoxify pollutants, is an ecologically friendly and often cost-effective solution. Bacteria, algae, plants and their inherent detoxification and degradation pathways can help to purify and detoxify contaminated water bodies. The present review focuses on water bioremediation using bacterial, algal, and plant systems. Successful examples from recent literature in which native or genetically modified organisms have been deployed to clean up contaminated waters are reported throughout. The contaminants addressed range from heavy metals to organic pollutants such as polychlorinated biphenyls, pesticides, and pharmaceutical residues. The biological systems described feature different modes of water treatment, for example, surface runoff treatment, groundwater treatment, or purification of water in large reservoirs or urban canals. The potential and limitations of biological water treatment are critically discussed [11, 12].

Definition and Principles

The bioremediation of water is generally conducted in one of two approaches: the pollution-affected water is either removed or treated above the ground in a tank or a reactor. It is retained in the original location, such as a pond, lake, or lagoon, and treated in the original location. In the first case, the water is usually pumped through a treatment system, passing over or through a biological support material such as activated carbon, compost, or peat. In the second case, the microorganisms are supplied with the required nutrients and electron acceptors and allowed to naturally biodegrade the pollutants. Various bioremediation treatments have been used to clean surface waters, including marsh, wetland, and lagoon systems. These systems promote the growth of natural bacteria, algae, and other aquatic organisms to enhance the degradation of contaminants. Among the different biological agents, bacteria play an important role in the cleanup of contaminated waters. Their natural metabolic activities can transform and detoxify many organic pollutants. Fungi, on the other hand, are primarily involved in the degradation of more complex hydrocarbons [3, 13].

Bioremediation accelerates the process of the natural biological restoration of a site by the destruction or removal of pollutants, thus preventing or moderating long-term environmental damage. Any technology that utilizes biology to restore a polluted environment is potentially a form of bioremediation. This could involve the cleansing of contaminants from groundwater, surface water, or soil. The general principle of bioremediation is the use of biological agents such as bacteria, algae, fungi, and other plants to degrade harmful substances and convert them into less toxic or non-toxic compounds that do not threaten the water, soil, or wildlife [10].

Microbial Bioremediation Techniques

The bioremediation of contaminated waters by microorganisms is usually a simple and low-cost technique. In distributive applications, especially if the pollutant has a high BOD or is otherwise toxic to aquatic life, the microbial approach should be preferred over phyto- or myco-remediation technologies. When the bioremediation strategy is based on the use of microorganisms, several techniques can be used alone or in combination. While small molecular weight products from microbial activity are being consumed by fast-growing algae, the primary aim of the addition of higher plant (phyto-remediation) or fungal (myco-remediation) species in the cleansing of a water body is not the degradation of the pollutant but rather biomass accumulation and concentration of the pollutant within the harvestable biomass [14, 15].

Bioremediation of contaminated water or soil by microorganisms can be achieved through various strategies. Immobilized cells in inert matrices like alginate can be added to waters, allowing extended pollutant degradation. The addition of selected cultivable pollutant-degrading microbes in the form of a bioaugmentation cocktail is another commercial possibility for adding efficiency to water bioremediation. Often unanticipated but always contributing to successful bioremediation is the enhancement of the diverse, mostly uncultivable, degrading microbial populations existing in the polluted environment through biostimulation. Nutrient addition, surfactant, or other elicitors of degradation may be added to allow and enhance pollutant biodegradation in the water bodies [16].

Biostimulation

It should be noted that nitrogen- and phosphorous-containing nutrients can also stimulate the biodegradation of hydrocarbons. Under nitrogen- and/or phosphorous-limiting conditions, some hydrocarbon-degrading microorganisms are known to produce biosurfactants. Biosurfactants lower the surface and interfacial tensions of water, thus facilitating the uptake and utilization of hydrocarbons by microorganisms. Since the biodegradation of contaminants via biosurfactant-producing microorganisms does not necessarily require the addition of nitrogen and phosphorus-containing nutrients [17].

The term "biostimulation" has been used in the past to describe an increase in the activity of naturally occurring microorganisms that are capable of biodegrading environmental contaminants but which exist in insufficient numbers. Biostimulation increases the ability of microorganisms to degrade contaminants by providing them with the required additional nutrients. The nutrients added during biostimulation do not have to be growth-limiting nutrients for indigenous microorganisms, as is the case with the addition of nitrogen and phosphorous into many polluted surface and ground waters. Furthermore, biostimulation also does not necessarily involve a change in redox conditions in the contaminated aquifer, although a decrease in redox potential can be achieved in combination with biostimulation [18].

Plant-based Bioremediation Techniques

Phytoremediation techniques can also be coupled with other remediation methods, such as soil amendments, and have been used successfully in real-case scenarios. Plants are the most attractive natural degraders and pioneer colonizers of contaminated sites. The plant-based phytoremediation technique is the most eco-friendly of the current remediation technologies and can be applied at the original location. As a consequence, this technique is the least invasive approach and would cost much less than cleaning construction. However, there is a disadvantage to the application of plant-based phytoremediation: the clean-up process may take a long time, and the efficiency of the removal of contaminants can be quite low, especially in large-scale contaminated sites [19, 20].

Bioremediation of contaminated water and soils can be performed using several techniques. Some of the most interesting ones involve plants that do not require a large energy input, and they contribute to the recovery of contaminated landscapes. Inside the plants, several processes occur that help alleviate the contamination: phytoextraction, where metals are extracted from the ground and accumulate in the shoots; phytodegradation, which involves the plants degrading organic compounds through their roots or other aerial parts; phytostabilization in which the plants affect the mobility of the contaminants, often by immobilizing them in the root zone; and phytovolatilization, where plants facilitate the evaporation of contaminants [21, 22].

Phytoremediation

Phytoremediation is an attractive, cost-effective, and environmentally sound technology for the remediation of metals from contaminated water. Over the past decade, commercial use and development of the technology have expanded rapidly. A wide variety of water bodies have been treated, including industrial effluents, acid mine drainage, and storm runoff. Several aquatic plant species have been utilized, and it is known that hydrophytes play two key roles in metal remediation: they create a microenvironment that enhances biological activities in the sediment, and they take up, accumulate, and store metals from the surrounding water. These accumulated metals can be harvested and recycled. The use of macrophytes for metal remediation is an attractive aspect of phytoremediation as it simplifies the metal recovery process compared to other treatment systems, which require either disposal of metal-laden materials accumulated by microorganisms or further processing to concentrate the metals [19, 23].

Within the response to escalating environmental pollution, there is an ever-increasing demand for the development of efficient ecotechnology-based clean-up programs. These are programs that work in concert with nature to mitigate environmental contaminants. Over the past several years, several diverse strategies for bioremediation of contaminated waters have been developed and reported. These range from the use of superadherent microbial systems in rotating biological contactors for the removal of heavy metals from domestic and industrial wastewater to enzymes and whole cells immobilized in biospecific sorbents. Plant systems, including cultivated wetland species, are particularly well adapted for the removal of a wide spectrum of organic chemicals and heavy metals from contaminated water. This property of plants is called phytoremediation. Simple plants, predominantly algae, are also used to clean up nutrient-rich wastewater as well as wastewater containing different organic chemicals [14, 24].

Successful Cases of Bioremediation

Bioremediation refers to any process that uses microorganisms, fungi, plants, or their enzymes to restore the environments altered by contaminants to their original condition. Despite numerous successes in the bioremediation of contaminated soils, it is rather slow to get consent for applying bioremediation to contaminated waters. Many localized spills of organic chemicals in water, such as fuels, solvents, or chlorinated hydrocarbons, have been amenable to treatment by bioremediation. Municipal wastewater treatment has long relied on bioremediation to clean up our unhealthy waste. The bioremediation of large rivers, lakes, or coastal marine areas impacted by organics is of interest to ecological health. The time when the pollution health risks for affected populations have fallen below acceptable levels is usually the endpoint of bioremediation [25, 26].

Bioremediation is permitted in some countries as a technique for cleaning up contaminated environments. Some successful cases of bioremediation have become accepted, where the health risks of organic contaminant pollution in the environment have subsided. Large volumes of oil have impacted coastal ecosystems after shipping accidents. Oils are essentially natural products, but the longer-chain hydrocarbons and associated chemicals cause inherent toxicity. Oil affects birds through smothering and loss of thermal insulation, as well as hydrophobic closure of air passages. The response to major oil spills has usually been to implement a cleanup involving the physical removal of the oil from affected areas and, in recent times, directed burning. This ensures that the hydrocarbon contaminants do not enter the food chain and impact larger mammals and humans. Natural weathering removes about 50% of the oil within a few days after the spill. The remainder is subject to photodegradation, evaporation, biodegradation, emulsification, and transport in the environment [27, 28].

Case Study: Bioremediation in Various Contexts

Exxon Valdez Oil Spill

The first success story of bioremediation was achieved with the Exxon Valdez oil spill. After this, many oil spills have been treated using bioremediation techniques, mainly involving the addition of fertilizers to promote the growth of indigenous hydrocarbon-degrading microorganisms. These microorganisms use the oil compounds as sources of carbon and energy and are capable of degrading the hydrocarbons. This type of bioremediation is relatively easy to perform, cost-effective, and has been implemented on a large scale. As a result, the affected areas usually experience a speedy recovery, both ecologically and in terms of lost services [29, 30].

Several water bodies with diverse pollution problems have been treated with one or more of the bioremediation strategies discussed in this review, leading to improved conditions. Successes, failures, unanticipated improvements, and difficulties that were encountered by the practitioners with the different approaches are described with the hope that they can help others in designing optimized remediation strategies for their particular water contamination problems [31].

Polychlorinated Biphenyls (PCBs)

Bioremediation has been successfully applied to address water contamination, particularly concerning PCBs. One notable case involved the accumulation of PCBs in sediment, which posed a significant threat to aquatic life and human health. A bioremediation approach introduced a PCB-degrading bacterium,

Dehalococcoides, under anaerobic conditions, resulting in a significant reduction of PCB levels. This approach highlighted the importance of selecting the right microorganisms for specific contaminants. *Dehalococcoides* species have been identified as key organisms capable of extensively dechlorinating PCBs in various environments [32 - 34]. These bacteria can remove chlorines from different PCB congeners [32, 33]. The growth of *Dehalococcoides* is linked to PCB dechlorination, with cell numbers increasing during the process [33, 35]. Several reductive dehalogenase genes, such as rd14 and pcbA5, have been associated with PCB dechlorination [34, 35].

White-rot fungi, such as *Phanerochaete chrysosporium*, have been used due to their ability to degrade a wide range of persistent organic pollutants, including PCBs. These fungi produce lignin-degrading enzymes like laccases and peroxidases that can break down PCBs into less harmful compounds. Additionally, bioaugmentation with engineered strains of bacteria capable of enhanced PCB degradation has shown promising results in both laboratory and field settings. *P. chrysosporium* has demonstrated significant potential for the biodegradation of PCBs and other persistent environmental pollutants. Studies have shown that this fungus can mineralize various PCB congeners, including biphenyl and lightly chlorinated PCBs, to CO_2 [36, 37]. The fungus's ability to degrade PCBs is attributed to its extracellular lignin-degrading enzyme system [38]. *P. chrysosporium* has been found to effectively degrade PCB mixtures, with degradation efficiency decreasing as chlorination increases [39]. The fungus can also synergistically interact with indigenous soil microorganisms to enhance PCB removal from contaminated soil [40]. An example of this process is shown in Fig. (**2**).

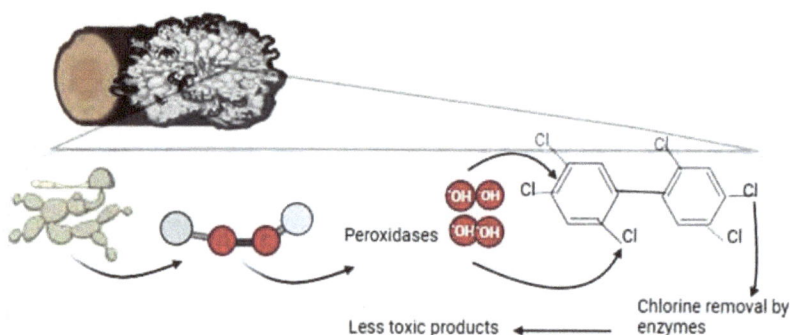

Fig. (2). The degradation of polychlorinated biphenyls carried out by PCBs generates H_2O_2 that activates the peroxidase enzymes for degradation, generating free radicals that oxidize the aromatic rings. In this process, the Cl present is eliminated by the same enzymes, generating less toxic products. Figure made with Biorender.

Oil Spill

Conventional methods were inadequate to address the scale of an oil spill, endangering marine life and local economies. Bioremediation strategies employ naturally occurring oil-degrading bacteria that effectively consume the oil, allowing for quicker recovery of the affected areas. This highlighted the efficacy of bioremediation in large-scale oil contamination and the importance of promptly utilizing native oil-degrading bacteria. Bioremediation using oil-degrading bacteria is an effective and eco-friendly approach for cleaning up petroleum-contaminated environments [41, 42]. Various bacterial genera, including *Pseudomonas*, *Bacillus*, and *Staphylococcus*, have been identified as capable of degrading petroleum hydrocarbons [42, 43]. These bacteria can utilize hydrocarbons as carbon and energy sources, breaking them down into harmless compounds [44, 45]. Factors such as pollutant concentration and environmental conditions can affect bacterial growth and degradation efficiency [42, 46]. Bioremediation techniques include adding exogenous bacterial populations or stimulating indigenous ones to enhance degradation rates [47]. Recent studies have demonstrated the potential of isolated bacterial strains to degrade crude oil effectively within 28 days [48]. This technology offers a cost-effective and sustainable solution for cleaning up oil-contaminated soil and water [44, 48].

Microbial enzymes are known to degrade various pollutants, restore and remediate the ecosystem, and regulate various biochemical processes. Extracellular enzymes like laccases, oxidases, and peroxidases secreted by white-rot fungi can degrade PAHs and lignins. Oxidoreductases can detoxify phenolic pollutants. Methane monooxygenase from methanotrophs can metabolize aromatic and heavy metals, making it a useful bioremediation tool [49]. Lipase-producing microorganisms are often exploited for the bioremediation of oil-polluted environments. The use of enzymes is more feasible and efficient than using whole cells, which requires proper environmental conditions and nutrition to survive. The large-scale production of metal-degrading enzymes from various microorganisms and their *in situ* application at polluted sites are strategies worth investigating further [49].

Heavy Metals

Rhizofiltration is an emerging phytoremediation technology that uses plant roots to remove heavy metals from contaminated water [50, 51]. Hyperaccumulator plants possess the remarkable ability to extract and concentrate extraordinary amounts of heavy metals in their aerial tissues without toxicity [52]. This ability is primarily due to the overexpression of metal transporter genes [52]. The bacterial rhizobiome plays a crucial role in enhancing metal solubility and uptake by hyperaccumulators [53]. Plant growth-promoting rhizobacteria (PGPR) can alleviate metal-induced stress and improve the efficiency of phytoremediation [54]. Combining hyperaccumulators with PGPR offers a promising, cost-effective approach for remediating metal-contaminated sites [54]. Phytoremediation using hyperaccumulators has gained attention as an environmentally friendly alternative to traditional remediation methods [55, 56], with potential applications in both soil cleanup and phytomining [52].

Biosurfactant Assisted Phytoremediation

Biosurfactants consist of both hydrophobic and hydrophilic moieties, making them amphipathic surface-active molecules. These compounds act at liquid interfaces (oil/water or water/oil), reducing surface tension and increasing the surface contact of the hydrophobic molecules, thereby enhancing their solubility and bioavailability. Biosurfactants, produced by various microorganisms, can enhance the bioavailability and biodegradation of HOCs like PAHs [57, 58]. As biosurfactants can enhance the mobility and biodegradation of organic compounds, they are suitable for bioremediation purposes. Low molecular weight compounds effectively reduce surface and interfacial tensions, while high molecular weight compounds can efficiently stabilize emulsions. Biosurfactant molecules possess both hydrophobic and hydrophilic domains, enabling them to lower surface tension and emulsify hydrophobic compounds [59]. The efficiency of biosurfactants is measured using critical micelle concentration [49].

Biosurfactants are less toxic, biodegradable, and highly selective, with the ability to detoxify pollutants and operate in a wide range of temperatures, pH, and saline conditions, making them favored over synthetic surfactants. However, the impact of biosurfactants on HOC bioavailability can vary depending on the specific biosurfactant and microbial strain involved [60]. These properties have gained much attention from scientists and environmentalists [49].

Biosurfactants form complexes with heavy metals at the rhizosphere-soil interface, inducing the desorption of metals and increasing their bioavailability in soil for better uptake by plant roots. Biosurfactants can enhance pollutant desorption from soil and facilitate microbial uptake [61, 62]. The use of biosurfactants can enhance the phytoremediation process, and many studies have been conducted to assess this new approach. There are two ways to achieve biosurfactant-assisted phytoremediation, namely, (1) application of biosurfactants to the contaminated site and (2) inoculation of biosurfactant-producing microorganisms to the contaminated site (Fig. **3**).

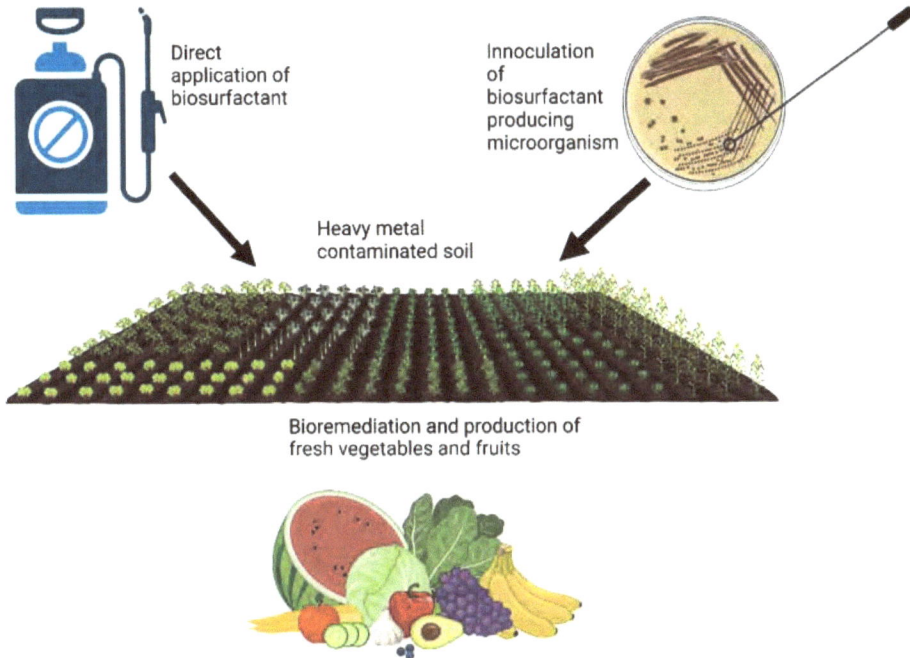

Fig. (3). Two approaches to biosurfactant-assisted phytoremediation: (1) direct application of biosurfactants to the contaminated site to enhance pollutant removal and (2) inoculation of biosurfactant-producing microorganisms, which generate biosurfactants *in situ* to aid in the remediation process and the recovery of soil for food production. Figure made with Biorender.

Bacteria *Pseudomonas aeruginosa* and *Bacillus subtilis,* which produce rhamnolipids and surfactin, respectively, have been extensively studied as large producers of biosurfactants [63]. Species of the genus *Candida* are commonly studied yeasts in biosurfactant production. Rhamnolipid is efficient in the bioremediation of oil-contaminated sites. Certain strains of *Bacillus subtilis* have been shown to produce rhamnolipids while degrading hydrocarbons. Additionally, a surfactin-producing strain of *B. subtilis* was identified and isolated, demonstrating its potential for removing contaminants in pharmaceuticals, environmental applications, cosmetics, and oil recovery [64]. Rhamnolipids affect cell surface hydrophobicity, influencing bacterial interactions with hydrophobic substrates [65]. They play diverse roles in bacterial physiology, including substrate uptake, virulence, and biofilm development [66]. Rhamnolipids can inhibit adhesion and disrupt pre-formed biofilms of marine bacteria like *B. pumilus* [67]. Their unique properties make rhamnolipids promising candidates for industrial and clinical applications [68].

Sophorolipids, lipopeptides, and other biosurfactants have also been evaluated for their bioremediation potential in hydrocarbon-contaminated marine and terrestrial locations. Rhamnolipid enhanced the uptake of PAHs by ryegrass (*Lolium multiflorum*). Rhamnolipids, biosurfactants produced by various bacteria, have diverse applications in plant and animal systems. They can enhance the uptake of polycyclic aromatic hydrocarbons by ryegrass roots [69] and improve the phytoextraction of heavy metals by *Lolium perenne* when combined with other amendments [70].

Rhamnolipids and their precursors activate plant innate immunity through distinct mechanisms [71] and have antimicrobial properties [72]. Using a bioaugmentation strategy in phytoremediation, the positive influence of biosurfactant-producing *Bacillus* strain on the biomass of tomatoes and its Cd uptake was evaluated. Several studies have investigated the effects of bacteria on cadmium (Cd) toxicity in tomato plants. Certain bacterial strains, including *Pseudomonas sp.*, *Bacillus sp.*, and *Burkholderia sp.*, can promote plant growth and reduce Cd uptake in tomatoes [73 - 75]. These bacteria produce growth-promoting substances and can increase soil Cd solubilization [74, 75]. The practicability of biosurfactant-producing microbial strains in the original location of bioremediation strategies in highly polluted environments is a novel and eco-feasible approach. However, little is known about this, as most research is done in laboratory conditions. Further efforts are needed to evaluate other potential applications of these surface-active molecules in the clean-up of toxic contaminants from the environment [49].

Sewage

Lake Washington's severe pollution due to untreated sewage discharges led to eutrophication (Fig. **4**). An ambitious effort to divert sewage to a wastewater treatment plant allowed the lake's natural ecosystem to recover, significantly improving the lake's water quality over several decades. This showcased the importance of holistic, long-term approaches to bioremediation.

Sewage sludge compost can be used as a soil conditioner because it enhances soil fertility properties such as carbon content, N, P, K, and water retention capacity while improving soil physical characteristics like bulk density and porosity. It has been suggested that sewage sludge compost can revive the fertility of reclaimed soil from landfills, providing a cost-effective solution to modern disposal problems. In various studies, sewage sludge application has been shown to improve soil fertility and crop yields. It increases soil organic matter, nutrients (N, P, K, Ca, Mg), and cation exchange capacity while reducing soil acidity [76 - 79]. Sewage sludge can enhance soil structure, decrease bulk density, and increase water-stable aggregates [76]. However, concerns exist regarding heavy metal accumulation in soils and crops [79, 80]. Alperujo compost (olive mill waste) and pig slurry have also been reported to enhance soil microbial biomass carbon and nitrogen compared to mineral fertilizer applications. Alperujo compost, derived from olive oil production waste, has shown promising results as a soil amendment in various studies. It can improve soil quality by increasing organic matter content, nutrient availability, and microbial activity [81 - 83]. Maximum microbial enzymatic activities (acid phosphatase, dehydrogenase and β-glucosidase, urease) were reported in soil amended with sewage sludge. The microbial community present in sewage sludge is tolerant of organic contaminants and can reduce them when employed discreetly in the soil [84].

Fig. (4). Eutrophication in a freshwater ecosystem. Excess nutrients, primarily nitrogen and phosphorus from agricultural runoff and wastewater stimulate excessive algae growth. As algae die and decompose, oxygen levels drop dramatically, creating hypoxic conditions that harm aquatic life, leading to 'dead zones' where biodiversity is severely reduced. Figure made with DALL-E.

Municipal sewage waste compost application in soils has been reported to enhance plant yield and diminish metal availability. Compost produced from crop residues applied to selenium-contaminated soils was found to decrease Se availability to *Brassica napus* plants. The soil fertility of a polluted site increased with the addition of alperujo compost, promoting the growth of *Paulownia fortunei* trees. The compost's application has been found to enhance soil fertility, particularly in degraded or contaminated soils, and can support plant growth and productivity [85]. Alperujo compost can also contribute to carbon sequestration in olive grove agroecosystems, although its effectiveness may vary compared to other organic amendments like biosolid compost [86]. The use of alperujo compost aligns with circular economy principles and sustainable waste management practices in olive oil production [87]. Similarly, poultry manure compost, farmyard manure, and plant leaves increased selenium volatilization and decreased uptake in *Zea mays* and *Vigna unguiculata* [84].

Biosolid compost applied to Cr, As, and Cu-contaminated soil improved the productivity of lettuce and carrots three to fivefold compared to unamended soil [84, 88] (Fig. **5**). *Rosmarinus officinalis* L. grown in compost-amended soil showed increased plant nutrient content and biomass [88]. An increment in root yield and shoot biomass of *Lupinus albus* L. was reported when grown in soil amended with compost comprising sewage sludge, olive, and vegetable waste. Biosolids application on topsoil increased the dry plant yield of *Trifolium subterraneum* and *Lolium rigidum* compared to clay/sand [88]. Sewage sludge as an amendment in contaminated soil enhanced the physiological parameters and biomass of three tree species (*Liriodendron tulipifera*, *Quercus acutissima*, and *Betula schmidtii*), with no substantial accumulation of heavy metals in the leaves [84].

Biochar and Sewage Sludge Compost for Soil Improvement

Biochar is generally alkaline, with pH values ranging from 7.95 to 10.6, due to the decomposition of organic substances and ash formation during pyrolysis, as well as the presence of organic functional groups, carbonates, and inorganic alkalis on its surface. Biochar, produced through the pyrolysis of biomass, exhibits alkaline properties that can significantly impact soil characteristics. Higher pyrolysis temperatures generally increase biochar alkalinity and pH [89, 90]. Biochar alkalinity comprises organic structural, other organic, carbonate, and inorganic components, all contributing to its total alkalinity [91]. Biochar's alkaline pH is particularly beneficial when applied to acidic mine soils. However, some biochars can have a neutral or acidic pH, depending on the feedstock and pyrolysis temperature [92].

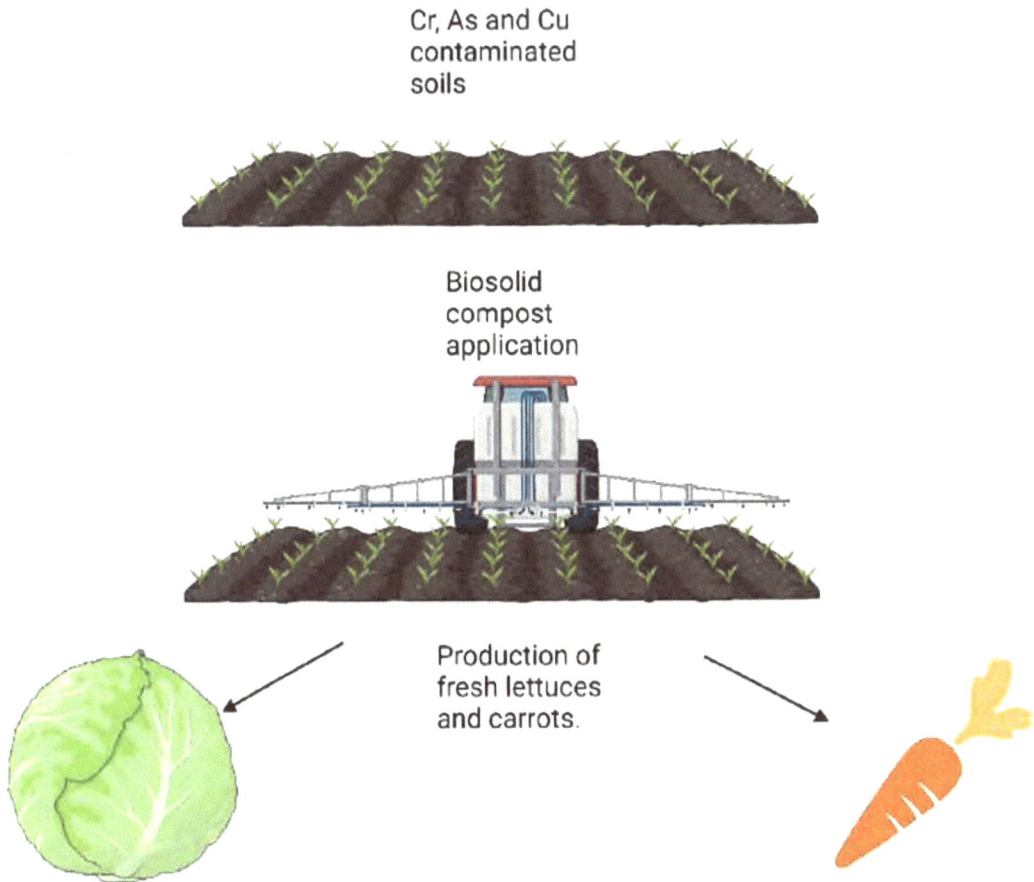

Fig. (5). Effect of biosolid compost on crop productivity in contaminated soils. Lettuce and carrots grown in Cr, As, and Cu-contaminated soil treated with biosolid compost showed a marked improvement in yield, with a three to fivefold increase compared to those grown in untreated contaminated soil. Figure made with Biorender.

Procedure Steps: Implementing Microbial Bioremediation

The implementation of microbial bioremediation to recover contaminated waters involves several critical steps:

1. Identification of Contaminants: analyze the type and concentration of contaminants present in the water bodies.
2. Selection of Microorganisms: identify and select appropriate microorganisms, including bacteria, fungi, yeasts, and algae, capable of biosorbing toxic ions.
3. Laboratory Testing: evaluate the efficacy of the selected microorganisms in removing contaminants under controlled conditions.
4. Site Preparation: prepare the contaminated water bodies for treatment, which may involve adjusting environmental conditions to optimize the activity of the

selected microorganisms.
5. Bioremediation Process: implement the bioremediation process using living or dead biomass for biosorption and bioaccumulation of contaminants.
6. Monitoring and Assessment: regularly monitor the water bodies to assess the progress of the remediation process, including changes in contaminant concentrations and microbial population dynamics.
7. Post-Treatment Evaluation: conduct a final assessment to determine the effectiveness of the remediation and ensure that contaminant levels have been reduced to acceptable standards.

Challenges and Future Directions

Bioremediation of contaminated water using microorganisms represents one of the most sustainable solutions for environmental protection. However, several factors can limit its efficiency, such as the low transfer of contaminants from the aqueous phase to the gaseous phase; the toxicity of the pollutants to the degrading microorganisms; the low bioavailability of the contaminants when dissolved in water; and the constant input of the pollutants. To overcome these limitations, several strategies have been developed, such as the selection and engineering of more resistant degrading microorganisms as well as the combination of bioremediation with physical processes or with other biological processes. In this chapter, we describe in detail these challenges and present some successful cases of strategies used to alleviate the limits of bioremediation of organic chemicals in water. Based on the lessons learned from these cases, we also discuss possible future directions to increase the efficiency of bioremediation in large-scale applications [12, 93].

Key Challenges in Bioremediation

Bioremediation has a high potential for decontaminating waters since it employs the ecological power of self-regeneration and self-purification of aquatic systems instead of attempting to create these characteristics artificially. However, the application of bioremediation technologies is far from a straightforward task, and several challenges must be addressed to make it effective and reliable. Optimization and validation of treatment efficiency are key tasks to be accomplished to allow the large-scale application of bioremediation to thick freshwater contaminants. Additionally, the development of bioremediation for emerging contaminants, the combination of bioremediation with other techniques or the use of nanoparticles, and the promotion of international regulations to address liability issues are identified as further challenges to the success of bioremediation. Overcoming these challenges would help to make bioremediation a commonly used approach for cleaning contaminated waters at different scales with proven efficiency [94].

CONCLUDING REMARKS

Bioremediation of contaminated waters stands as a pivotal approach to addressing environmental pollution. By leveraging the natural capabilities of microorganisms and plants, bioremediation offers an effective, sustainable, and often cost-efficient method for detoxifying polluted aquatic environments. Throughout this chapter, we have explored various strategies, including bioaugmentation and phytoremediation, which utilize specialized microorganisms and plants to degrade and remove contaminants. These methods have shown significant success in treating pollutants such as heavy metals, POPs, and hydrocarbons.

Success stories, such as the Exxon Valdez oil spill and the recovery of Lake Washington, underscore the potential of bioremediation to restore polluted water bodies. These examples highlight the importance of selecting appropriate biological agents and optimizing conditions to enhance the effectiveness of the remediation process. Additionally, the use of biosurfactants and other innovative techniques has demonstrated the potential to improve bioremediation efficiency, particularly in challenging environments. Despite its promise, bioremediation faces several challenges, including the variability of environmental conditions, the complexity of contaminant mixtures, and the need for ongoing research to address emerging pollutants. Advances in genetic engineering, microbial consortia, and combined remediation approaches are paving the way for more robust and scalable bioremediation solutions. In conclusion, bioremediation represents a vital tool in the global effort to combat water pollution. Continued research, technological innovation, and collaborative efforts are essential to overcoming existing challenges and expanding the application of bioremediation techniques. By harnessing the natural detoxifying abilities of biological agents, we can achieve significant progress in the sustainable management and restoration of our precious water resources.

AUTHORS' CONTRIBUTION

The authors confirm their contribution to the chapter as follows:

Conceptualization, methodology, investigation, data curation, writing-original draft preparation, writing review and editing, supervision, figure editing: Nichdaly Ortiz Chacón; Methodology, investigation, data curation, formal analysis, writing-original draft preparation, writing review, and editing: Aliana Zacaria Vital; Conceptualization, validation, writing review and editing, supervision: Israel Valencia Quiroz.

QUESTIONS RELATED TO THE TEXT

1.- What is bioremediation, and what are the main techniques used in the remediation of contaminated water bodies?

2.- What are the primary sources and types of contaminants in water bodies that can be treated through bioremediation?

3.- How does the technique of bioaugmentation work, and what challenges are associated with its implementation in water bioremediation?

4.- What role do plants play in the phytoremediation of contaminated waters, and what are the advantages and disadvantages of this technique?

5.- Mention and explain a success case in the bioremediation of contaminated waters, highlighting the methods used and the outcomes achieved.

6.- What are the main future challenges for the effective application of bioremediation in cleaning contaminated waters?

LIST OF ABBREVIATIONS

BOD   Biochemical Oxygen Demand

HOCs Hydrophobic Organic Compounds

PAHs Polycyclic Aromatic Hydrocarbons

PGPR Plant Growth-Promoting Rhizobacteria

POPS Persistent Organic Pollutants

VOCs Volatile Organic Compounds

REFERENCES

[1]   Kurniawan SB, Ramli NN, Said NSM, *et al.* Practical limitations of bioaugmentation in treating heavy metal contaminated soil and role of plant growth promoting bacteria in phytoremediation as a promising alternative approach. Heliyon 2022; 8(4): e08995.
[http://dx.doi.org/10.1016/j.heliyon.2022.e08995] [PMID: 35399376]

[2]   Tondera K, Chazarenc F, Chagnon PL, Brisson J. Bioaugmentation of treatment wetlands – A review. Sci Total Environ 2021; 775: 145820.
[http://dx.doi.org/10.1016/j.scitotenv.2021.145820] [PMID: 33618303]

[3]   Saeed MU, Hussain N, Sumrin A, *et al.* Microbial bioremediation strategies with wastewater treatment potentialities – A review. Sci Total Environ 2022; 818: 151754.
[http://dx.doi.org/10.1016/j.scitotenv.2021.151754] [PMID: 34800451]

[4]   Häder DP, Banaszak AT, Villafañe VE, Narvarte MA, González RA, Helbling EW. Anthropogenic pollution of aquatic ecosystems: Emerging problems with global implications. Sci Total Environ 2020; 713: 136586.
[http://dx.doi.org/10.1016/j.scitotenv.2020.136586] [PMID: 31955090]

[5]   Schwarzenbach RP, Escher BI, Fenner K, *et al.* The challenge of micropollutants in aquatic systems. Science 2006; 313(5790): 1072-7.
[http://dx.doi.org/10.1126/science.1127291] [PMID: 16931750]

[6]   Alengebawy A, Abdelkhalek ST, Qureshi SR, Wang MQ. Heavy metals and pesticides toxicity in agricultural soil and plants: Ecological risks and human health implications. Environ Sci Pollut Res Int

2022; 29: 53934-53.

[7] Srivastav AL. Chemical fertilizers and pesticides: role in groundwater contamination. Environ Sci Pollut Res Int 2020; 27: 4073-84.

[8] Pal AK, Singh J, Soni R, *et al.* The role of microorganism in bioremediation for sustainable environment management. In Bioremediation of Pollutants. 2020; 227–49.

[9] Pande V, Pandey SC, Sati D, Pande V, Samant M. Bioremediation: an emerging effective approach towards environment restoration. Environmental Sustainability 2020; 3(1): 91-103.
[http://dx.doi.org/10.1007/s42398-020-00099-w]

[10] Duodu MG, Singh B, Christina E. Waste management through bioremediation technology: An eco-friendly and sustainable solution. Waste Manag Res 2022; 2: 205-34.

[11] Zahed MA, Matinvafa MA, Azari A, Mohajeri L. Biosurfactant, a green and effective solution for bioremediation of petroleum hydrocarbons in the aquatic environment. Discover Water. 2022; 2(1): 5.

[12] Simon M, Joshi H. A review on green technologies for the rejuvenation of polluted surface water bodies: Field-scale feasibility, challenges, and future perspectives. J Environ Chem Eng. 2021; 9(4): 105763.

[13] Berillo D, Al-Jwaid A, Caplin J. Polymeric materials used for immobilisation of bacteria for the bioremediation of contaminants in water. Polymers (Basel) 2021; 13(7): 1073.
[http://dx.doi.org/10.3390/polym13071073] [PMID: 33805360]

[14] Catania V, Lopresti F, Cappello S, Scaffaro R, Quatrini P. Innovative, ecofriendly biosorbent-biodegrading biofilms for bioremediation of oil- contaminated water. N Biotechnol 2020; 58: 25-31.
[http://dx.doi.org/10.1016/j.nbt.2020.04.001] [PMID: 32485241]

[15] Sánchez-Castro I, Martínez-Rodríguez P, Abad MM, Descostes M, Merroun ML. Uranium removal from complex mining waters by alginate beads doped with cells of *Stenotrophomonas* sp. Br8: Novel perspectives for metal bioremediation. J Environ Manage 2021; 296: 113411.
[http://dx.doi.org/10.1016/j.jenvman.2021.113411] [PMID: 34351286]

[16] Sayed K, Baloo L, Sharma NK. Bioremediation of total petroleum hydrocarbons (TPH) by bioaugmentation and biostimulation in water with floating oil spill containment booms as bioreactor basin. Int J Environ Res Public Health 2021; 18(5): 2226.
[http://dx.doi.org/10.3390/ijerph18052226] [PMID: 33668225]

[17] Juárez-Maldonado A, Tortella G, Rubilar O, Fincheira P, Benavides-Mendoza A. Biostimulation and toxicity: The magnitude of the impact of nanomaterials in microorganisms and plants. J Adv Res 2021; 31: 113-26.
[http://dx.doi.org/10.1016/j.jare.2020.12.011] [PMID: 34194836]

[18] Ansari AA, Naeem M, Gill SS, AlZuaibr FM. Phytoremediation of contaminated waters: An eco-friendly technology based on aquatic macrophytes application. Egypt J Aquat Res 2020; 46(4): 371-6.
[http://dx.doi.org/10.1016/j.ejar.2020.03.002]

[19] Priya AK, Muruganandam M, Ali SS, Kornaros M. Clean-up of heavy metals from contaminated soil by phytoremediation: A multidisciplinary and eco-friendly approach. Environ Sci Pollut Res Int 2020.

[20] Khan S, Masoodi TH, Pala NA, *et al.* Phytoremediation prospects for restoration of contamination in the natural ecosystems. Water 2023; 15(8): 1498.
[http://dx.doi.org/10.3390/w15081498]

[21] Sophia S, Shetty Kodialbail V. Phytoremediation of Soil for Metal and Organic Pollutant Removal. In: Jerold M, Arockiasamy S, Sivasubramanian V, editores. Bioprocess Engineering for Bioremediation: Valorization and Management Techniques [Internet]. Cham: Springer International Publishing; 2020. 45–66.

[22] Ali S, Abbas Z, Rizwan M, *et al.* Application of Floating Aquatic Plants in Phytoremediation of Heavy Metals Polluted Water: A Review. Sustainability [Internet]. 2020; 12(5).

[23]   Rempel A, Gutkoski JP, Nazari MT, *et al.* Current advances in microalgae-based bioremediation and other technologies for emerging contaminants treatment. Sci Total Environ 2021; 772: 144918.
[http://dx.doi.org/10.1016/j.scitotenv.2020.144918] [PMID: 33578141]

[24]   Khalid FE, Lim ZS, Sabri S, Gomez-Fuentes C, Zulkharnain A, Ahmad SA. Bioremediation of diesel contaminated marine water by bacteria: A review and bibliometric analysis. J Mar Sci Eng 2021; 9(2): 155.
[http://dx.doi.org/10.3390/jmse9020155]

[25]   Molalign Medfu Tarekegn FZS, Ishetu AI. Microbes used as a tool for bioremediation of heavy metal from the environment. Yildiz F, editor. Cogent Food & Agriculture. 2020; 6(1): 1783174.

[26]   Bala S, Garg D, Thirumalesh BV, *et al.* Recent strategies for bioremediation of emerging pollutants: a review for a green and sustainable environment. Toxics 2022; 10(8): 484.
[http://dx.doi.org/10.3390/toxics10080484] [PMID: 36006163]

[27]   Solanki P, Dotaniya ML, Khanna N, *et al.* Recent advances in bioremediation for clean-up of inorganic pollutant-contaminated soils. Frontiers in Soil and Environmental Microbiology 2020; 3: 299-310.
[http://dx.doi.org/10.1201/9780429485794-31]

[28]   Luo Q, Hou D, Jiang D, Chen W. Bioremediation of marine oil spills by immobilized oil-degrading bacteria and nutrition emulsion. Biodegradation 2021; 32(2): 165-77.
[http://dx.doi.org/10.1007/s10532-021-09930-5] [PMID: 33683578]

[29]   Mafiana MO, Bashiru MD, Erhunmwunsee F, Dirisu CG, Li SW. An insight into the current oil spills and on-site bioremediation approaches to contaminated sites in Nigeria. Environ Sci Pollut Res Int 2021; 28(4): 4073-94.
[http://dx.doi.org/10.1007/s11356-020-11533-1] [PMID: 33188631]

[30]   Haripriyan U, Gopinath KP, Arun J, Govarthanan M. Bioremediation of organic pollutants: a mini review on current and critical strategies for wastewater treatment. Arch Microbiol 2022; 204(5): 286.
[http://dx.doi.org/10.1007/s00203-022-02907-9] [PMID: 35478273]

[31]   Adrian L, Dudková V, Demnerová K, Bedard DL. *"Dehalococcoides"* sp. strain CBDB1 extensively dechlorinates the commercial polychlorinated biphenyl mixture aroclor 1260. Appl Environ Microbiol 2009; 75(13): 4516-24.
[http://dx.doi.org/10.1128/AEM.00102-09] [PMID: 19429555]

[32]   Bedard DL, Ritalahti KM, Löffler FE. The Dehalococcoides population in sediment-free mixed cultures metabolically dechlorinates the commercial polychlorinated biphenyl mixture aroclor 1260. Appl Environ Microbiol 2007; 73(8): 2513-21.
[http://dx.doi.org/10.1128/AEM.02909-06] [PMID: 17308182]

[33]   Wang S, Chng KR, Wilm A, *et al.* Genomic characterization of three unique Dehalococcoides that respire on persistent polychlorinated biphenyls. Proceedings of the National Academy of Sciences of the United States of America.
[http://dx.doi.org/10.1073/pnas.1404845111]

[34]   Ewald JM, Humes SV, Martinez A, Schnoor JL, Mattes TE. Growth of *Dehalococcoides* spp. and increased abundance of reductive dehalogenase genes in anaerobic PCB-contaminated sediment microcosms. Environ Sci Pollut Res Int 2020; 27(9): 8846-58.
[http://dx.doi.org/10.1007/s11356-019-05571-7] [PMID: 31209752]

[35]   Dietrich D, Hickey WJ, Lamar R. Degradation of 4,4'-dichlorobiphenyl, 3,3',4,4'-tetrachlorobiphenyl, and 2,2',4,4',5,5'-hexachlorobiphenyl by the white rot fungus *Phanerochaete chrysosporium.* Appl Environ Microbiol 1995; 61(11): 3904-9.
[http://dx.doi.org/10.1128/aem.61.11.3904-3909.1995] [PMID: 8526503]

[36]   Thomas DR, Carswell KS, Georgiou G. Mineralization of biphenyl and PCBs by the white rot fungus *Phanerochaete chrysosporium.* Biotechnol Bioeng 1992; 40(11): 1395-402.

[http://dx.doi.org/10.1002/bit.260401114] [PMID: 18601096]

[37] Bumpus JA, Tien M, Wright D, Aust SD. Oxidation of persistent environmental pollutants by a white rot fungus. Science 1985; 228(4706): 1434-6.
[http://dx.doi.org/10.1126/science.3925550] [PMID: 3925550]

[38] Yadav JS, Quensen JF III, Tiedje JM, Reddy CA. Degradation of polychlorinated biphenyl mixtures (Aroclors 1242, 1254, and 1260) by the white rot fungus *Phanerochaete chrysosporium* as evidenced by congener-specific analysis. Appl Environ Microbiol 1995; 61(7): 2560-5.
[http://dx.doi.org/10.1128/aem.61.7.2560-2565.1995] [PMID: 7618867]

[39] Fernández-Sánchez JM, Rodríguez-Vázquez R, Ruiz-Aguilar G, Alvarez PJJ. PCB biodegradation in aged contaminated soil: interactions between exogenous *Phanerochaete chrysosporium* and indigenous microorganisms. J Environ Sci Health Part A Tox Hazard Subst Environ Eng 2001; 36(7): 1145-62.
[http://dx.doi.org/10.1081/ESE-100104869] [PMID: 11545344]

[40] Shahaby AF, Alharthi AA, Tarras AE. Bioremediation of Petroleum Oil by Potential Biosurfactant-Producing Bacteria using Gravimetric Assay. Int. J. Curr. Microbiol. App. Sci 2015; 4(5); 390-403.

[41] Marinescu M, Lacatusu A, Gament E, Plopeanu G, Carabulea V. Bioremediation Potential of Native Hydrocarbons Degrading Bacteria in Crude Oil Polluted Soil. Bulletin of University of Agricultural Sciences and Veterinary Medicine Cluj-Napoca Agriculture, 2017; 74(1): 19–25.

[42] Islam TH, Ghosh B, Magnet MMH, Fatema K, Akter S, Khan MAR, Datta S. Isolation and identification of petroleum degrading bacteria from oil contaminated soil & water and assessment of their potentiality in bioremediation. IOSR J Environ Sci Toxicol Food Technol. 2013; 5(2): 55-58.

[43] Agrawal I. Oil degrading bacteria: remediation of environmental pollution resulting from petroleum hydrocarbons. Biotechnol Kiosk. 2020; 2(10) :5-10.
[http://dx.doi.org/10.37756/bk.20.2.10.1]

[44] Xu X, Liu W, Tian S, *et al.* Petroleum Hydrocarbon-Degrading Bacteria for the Remediation of Oil Pollution Under Aerobic Conditions: A Perspective Analysis. Front Microbiol 2018; 9: 2885.
[http://dx.doi.org/10.3389/fmicb.2018.02885] [PMID: 30559725]

[45] Ganesan M, Mani R, Sai S, *et al.* Bioremediation by oil degrading marine bacteria: An overview of supplements and pathways in key processes. Chemosphere 2022; 303(Pt 1): 134956.
[http://dx.doi.org/10.1016/j.chemosphere.2022.134956] [PMID: 35588873]

[46] Ramesh MA, Somashekar P. Bioremediation of oil spill: an invasion by bacteria to a safe environment. Indian J Appl Res. 2014; 4(9).
[http://dx.doi.org/10.36106/ijar]

[47] Madan A, Kumar S, Waheed SM. Isolation & Characterization of Hydrocarbon Degrading Bacteria: A Bio-remedial Approach to Clean-up Oil Spills. Adv Mater Lett 2023; 14(4): 2304-1736.
[http://dx.doi.org/10.5185/amlett.2023.041736]

[48] Malik G, Hooda S, Majeed S, Pandey VC. Understanding assisted phytoremediation: Potential tools to enhance plant performance.Assisted Phytoremediation. Elsevier 2022; pp. 1-24. [Internet] https://www.sciencedirect.com/science/article/pii/B978012822893700015X
[http://dx.doi.org/10.1016/B978-0-12-822893-7.00015-X]

[49] Rawat K, Fulekar MH, Pathak B. Rhizofiltration: a green technology for remediation of heavy metals. Int J Innov Bio-Sci. 2012; 2(4): 193-199. Available at: http://www.parees.co.in/ijibs.htm

[50] Dushenkov V, Kumar PBAN, Motto H, Raskin I. Rhizofiltration: the use of plants to remove heavy metals from aqueous streams. Environ Sci Technol 1995; 29(5): 1239-45.
[http://dx.doi.org/10.1021/es00005a015] [PMID: 22192017]

[51] Rascio N, Navari-Izzo F. Heavy metal hyperaccumulating plants: How and why do they do it? And what makes them so interesting? Plant Sci 2011; 180(2): 169-81.
[http://dx.doi.org/10.1016/j.plantsci.2010.08.016] [PMID: 21421358]

[52]   Visioli G, D'Egidio S, Sanangelantoni AM. The bacterial rhizobiome of hyperaccumulators: future perspectives based on omics analysis and advanced microscopy. Front Plant Sci 2015; 5: 752.
[http://dx.doi.org/10.3389/fpls.2014.00752] [PMID: 25709609]

[53]   Asad SA, Farooq M, Afzal A, West H. Integrated phytobial heavy metal remediation strategies for a sustainable clean environment - A review. Chemosphere 2019; 217: 925-41.
[http://dx.doi.org/10.1016/j.chemosphere.2018.11.021] [PMID: 30586789]

[54]   Tongbin WC. Hyperaccumulators and phytoremediation of heavy metal contaminated soil: a review of studies in China and aborad. Acta Ecologica Sinica. 2001; 21(7): 1196-1203.

[55]   Peer W, Baxter I, Richards EL, Freeman J, Murphy A. Phytoremediation and hyperaccumulator plants. International Journal of Phytoremediation. 2005.

[56]   Das P, Mukherjee S, Sen R. Improved bioavailability and biodegradation of a model polyaromatic hydrocarbon by a biosurfactant producing bacterium of marine origin. Chemosphere 2008; 72(9): 1229-34.
[http://dx.doi.org/10.1016/j.chemosphere.2008.05.015] [PMID: 18565569]

[57]   Tecon R, van der Meer JR. Effect of two types of biosurfactants on phenanthrene availability to the bacterial bioreporter Burkholderia sartisoli strain RP037. Appl Microbiol Biotechnol 2010; 85(4): 1131-9.
[PMID: 19730847]

[58]   Cameotra SS, Makkar RS, Kaur J, Mehta SK. Synthesis of biosurfactants and their advantages to microorganisms and mankind. Adv Exp Med Biol 2010; 672: 261-80.
[http://dx.doi.org/10.1007/978-1-4419-5979-9_20] [PMID: 20545289]

[59]   Liu Y, Zeng G, Zhong H, *et al.* Effect of rhamnolipid solubilization on hexadecane bioavailability: enhancement or reduction? J Hazard Mater 2017; 322(Pt B): 394-401.
[http://dx.doi.org/10.1016/j.jhazmat.2016.10.025] [PMID: 27773441]

[60]   Kaczorek E, Pacholak A, Zdarta A, Smułek W. The Impact of Biosurfactants on Microbial Cell Properties Leading to Hydrocarbon Bioavailability Increase. Colloids and Interfaces [Internet]. 2018; 2(3).
[http://dx.doi.org/10.3390/colloids2030035]

[61]   Mata-Sandoval JC, Karns J, Torrents A. La influencia de surfactantes y biosurfactantes en la biodisponibilidad de contaminantes orgánicos hidrofóbicos en ambientes subterráneos. Rev Int Contam Ambient [Internet]. 2012; 16(4): 193-20. Available at: https://www.revistascca.unam.mx/rica/index.php/rica/article/view/32584

[62]   Zhang Y, Maier WJ, Miller RM. Effect of Rhamnolipids on the Dissolution, Bioavailability, and Biodegradation of Phenanthrene. Environ Sci Technol 1997; 31(8): 2211-7.
[http://dx.doi.org/10.1021/es960687g]

[63]   Christova N, Tuleva B, Nikolova-Damyanova B. Enhanced hydrocarbon biodegradation by a newly isolated *Bacillus subtilis* strain. Z Naturforsch C J Biosci 2004; 59(3-4): 205-8.
[http://dx.doi.org/10.1515/znc-2004-3-414] [PMID: 15241927]

[64]   Zhao Z, Selvam A, Wong JWC. Effects of rhamnolipids on cell surface hydrophobicity of PAH degrading bacteria and the biodegradation of phenanthrene. Bioresour Technol 2011; 102(5): 3999-4007.
[http://dx.doi.org/10.1016/j.biortech.2010.11.088] [PMID: 21208798]

[65]   Abdel-Mawgoud AM, Lépine F, Déziel E. Rhamnolipids: diversity of structures, microbial origins and roles. Appl Microbiol Biotechnol 2010; 86(5): 1323-36.
[http://dx.doi.org/10.1007/s00253-010-2498-2] [PMID: 20336292]

[66]   Dusane DH, Nancharaiah YV, Zinjarde SS, Venugopalan VP. Rhamnolipid mediated disruption of marine *Bacillus pumilus* biofilms. Colloids Surf B Biointerfaces 2010; 81(1): 242-8.
[http://dx.doi.org/10.1016/j.colsurfb.2010.07.013] [PMID: 20688490]

[67] Henkel M, Geissler M, Weggenmann F, Hausmann R. Production of microbial biosurfactants: Status quo of rhamnolipid and surfactin towards large☐scale production. Biotechnol J 2017; 12(7): 1600561.
[http://dx.doi.org/10.1002/biot.201600561] [PMID: 28544628]

[68] Zhu L, Zhang M. Effect of rhamnolipids on the uptake of PAHs by ryegrass. Environ Pollut 2008; 156(1): 46-52.
[http://dx.doi.org/10.1016/j.envpol.2008.01.004] [PMID: 18281132]

[69] Gunawardana B, Singhal N, Johnson A. Effects of amendments on copper, cadmium, and lead phytoextraction by *Lolium perenne* from multiple-metal contaminated solution. Int J Phytoremediation 2011; 13(3): 215-32.
[http://dx.doi.org/10.1080/15226510903567448] [PMID: 21598788]

[70] Schellenberger R, Crouzet J, Nickzad A, *et al.* Bacterial rhamnolipids and their 3-hydroxyalkanoate precursors activate *Arabidopsis* innate immunity through two independent mechanisms. Proc Natl Acad Sci USA 2021; 118(39): e2101366118.
[http://dx.doi.org/10.1073/pnas.2101366118] [PMID: 34561304]

[71] Vatsa P, Sanchez L, Clement C, Baillieul F, Dorey S. Rhamnolipid biosurfactants as new players in animal and plant defense against microbes. Int J Mol Sci 2010; 11(12): 5095-108.
[http://dx.doi.org/10.3390/ijms11125095] [PMID: 21614194]

[72] Zhang J, Xiao Q, Wang P. Phosphate-solubilizing bacterium *Burkholderia* sp. strain N3 facilitates the regulation of gene expression and improves tomato seedling growth under cadmium stress. Ecotoxicol Environ Saf 2021; 217: 112268.
[http://dx.doi.org/10.1016/j.ecoenv.2021.112268] [PMID: 33930768]

[73] He LY, Chen ZJ, Ren GD, Zhang YF, Qian M, Sheng XF. Increased cadmium and lead uptake of a cadmium hyperaccumulator tomato by cadmium-resistant bacteria. Ecotoxicol Environ Saf 2009; 72(5): 1343-8.
[http://dx.doi.org/10.1016/j.ecoenv.2009.03.006] [PMID: 19368973]

[74] Dourado MN, Martins PF, Quecine MC, Piotto FA, Souza LA, Franco M, *et al. Burkholderia* sp. SCMS54 reduces cadmium toxicity and promotes growth in tomato. J Hazard Mater 2013.

[75] Börjesson G, Kätterer T. Soil fertility effects of repeated application of sewage sludge in two 30-yea-old field experiments. Nutr Cycl Agroecosyst 2018; 112(3): 369-85.
[http://dx.doi.org/10.1007/s10705-018-9952-4]

[76] Colodro G, Espíndola CR. Alterações na fertilidade de um latossolo degradado em resposta à aplicação de lodo de esgoto. Acta Scientiarum Agronomy. 2006; 28: 1-5.

[77] Barbosa JZ, Poggere GC, Dalpisol M, Serrat BM, Bittencourt S, Motta ACV. Alkalinized sewage sludge application improves fertility of acid soils. Ciênc agrotec 2017; 41: 483-93.

[78] Zhou L, Hu A, Ge N. Study on utilization of municipal sewage sludge in farmland and forest land. Acta Ecol Sin. 2012; 19(2): 185-93.

[79] Moreira RS, Mincato RL, Santos BR. Heavy metals availability and soil fertility after land application of sewage sludge on dystroferric Red Latosol. Ciência e Agrotecnologia 2013; 37(6): 512-20.

[80] Fornes F, García-de-la-Fuente R, Belda RM, Abad M. 'Alperujo' compost amendment of contaminated calcareous and acidic soils: Effects on growth and trace element uptake by five *Brassica* species. Bioresour Technol 2009; 100(17): 3982-90.
[http://dx.doi.org/10.1016/j.biortech.2009.03.050] [PMID: 19369067]

[81] Alburquerque JA, de la Fuente C, Bernal MP. Improvement of soil quality after "alperujo" compost application to two contaminated soils characterised by differing heavy metal solubility. J Environ Manage 2011; 92(3): 733-41.
[http://dx.doi.org/10.1016/j.jenvman.2010.10.018] [PMID: 21035939]

[82] Monetta P, Bueno L, Cornejo V, Babelis G. Short-term dynamics of soil chemical parameters after

application of alperujo in high-density drip-irrigated olive groves in Argentina. Int J Environ Stud 2012; 69(4): 578-88.

[83] Pandey J, Sarkar S, Pandey VC. Compost-assisted phytoremediation.Assisted Phytoremediation. Elsevier 2022; pp. 243-64.https://www.sciencedirect.com/science/article/pii/B978012822893700001X [Internet]
[http://dx.doi.org/10.1016/B978-0-12-822893-7.00001-X]

[84] Madejón P, Alaejos J, García-Álbala J, Fernández M, Madejón E. Three-year study of fast-growing trees in degraded soils amended with composts: Effects on soil fertility and productivity. J Environ Manage 2016; 169: 18-26.
[http://dx.doi.org/10.1016/j.jenvman.2015.11.050] [PMID: 26716572]

[85] Panettieri M, Moreno B, de Sosa LL, Benítez E, Madejón E. Soil management and compost amendment are the main drivers of carbon sequestration in rainfed olive trees agroecosystems: An evaluation of chemical and biological markers. Catena 2022; 214: 106258.
[http://dx.doi.org/10.1016/j.catena.2022.106258]

[86] Liebert RJ. Compost. In: Psycurity [Internet]. Routledge; 2018; pp. 115-29.
[http://dx.doi.org/10.4324/9781315203874-6]

[87] Cala V, Cases MA, Walter I. Biomass production and heavy metal content of *Rosmarinus officinalis* grown on organic waste-amended soil. J Arid Environ 2005; 62(3): 401-12.
[http://dx.doi.org/10.1016/j.jaridenv.2005.01.007]

[88] Yuan JH, Xu RK, Zhang H. The forms of alkalis in the biochar produced from crop residues at different temperatures. Bioresour Technol 2011; 102(3): 3488-97.
[http://dx.doi.org/10.1016/j.biortech.2010.11.018] [PMID: 21112777]

[89] Chen W, Li K, Chen Z, Xia M, Chen Y, Yang H, *et al.* A new insight into chemical reactions between biomass and alkaline additives during pyrolysis process. Proc Combust Inst 2021; 38(3): 3881-90.
[http://dx.doi.org/10.1016/j.proci.2020.06.023]

[90] Fidel RB, Laird DA, Thompson ML, Lawrinenko M. Characterization and quantification of biochar alkalinity. Chemosphere 2017; 167: 367-73.
[http://dx.doi.org/10.1016/j.chemosphere.2016.09.151] [PMID: 27743533]

[91] Lebrun M, Nandillon R, Miard F, Bourgerie S, Morabito D. Biochar assisted phytoremediation for metal(loid) contaminated soils. Assisted Phytoremediation. Elsevier 2022; pp. 101-30. [Internet]
https://www.sciencedirect.com/science/article/pii/B9780128228937000100
[http://dx.doi.org/10.1016/B978-0-12-822893-7.00010-0]

[92] Demarco CF, Quadro MS, Selau Carlos F, Pieniz S, Morselli LBGA, Andreazza R. Bioremediation of aquatic environments contaminated with heavy metals: A review of mechanisms, solutions and perspectives. Sustainability (Basel) 2023; 15(2): 1411.
[http://dx.doi.org/10.3390/su15021411]

[93] Kumar R, Patil SA. Removal of heavy metals using bioelectrochemical systems.In: Integrated Microbial Fue[94] Kumar R, Patil SA Removal of heavy metals using bioelectrochemical systems In: Integrated Microbial Fuel Cells for Wastewater Treatment 2020; p 49-711 Cells for Wastewater Treatment. 2020; pp. 49-71.

Luminescent (Bio)sensors for Pesticide Detection: An Innovative Tool for Water Monitoring

María K. Salomón-Flores[1,*], Iván J. Bazany-Rodríguez[2], Helen Paola Toledo-Jaldin[3,4], Juan Pablo León-Gómez[1] and Alejandro Dorazco-González[1,*]

[1] *Institute of Chemistry, National Autonomous University of Mexico (UNAM), Mexico City, Mexico*

[2] *Faculty of Chemistry, National Autonomous University of Mexico (UNAM), Mexico City, Mexico*

[3] *Institute of Metallurgy, Autonomous University of San Luis Potosi, San Luis Potosi, Mexico*

[4] *National Technological of Mexico, Technological of Superior Studies of Tianguistenco, Mechanical Engineering Division, Tenango-La Marquesa Km22, Santiago Tilapa, Santiago Tianguistenco, Mexico*

Abstract: This chapter provides an overview of new fluorescent (bio)sensors designed for commonly used pesticides with a focus on molecular design and applications in real samples. Organophosphate and organochlorine pesticides have many applications in silviculture, public health, pest control, the food industry, and agriculture. Chronic exposure and acute contact with these agrochemicals result in toxic levels in animals, plants, humans, and ecosystems in general. Due to the toxicological, biochemical, and environmental effects of the accumulation of toxic agrochemicals in soil, food, and natural water resources, there is an imperative need to achieve analytical tools capable of working with real samples. In the last decade, research has explored the structural, reactivity, and detection aspects of sensory systems, ranging from the small organic molecules to more complex networks coordinated to metal centers involving transition metal or lanthanide ions, as well as biological nano-systems such as biosensors. The primary goal of (bio)sensors is to develop affordable, easy-to-process, efficient, economical, and stable methods for the accurate and reliable quantification of pesticides in real samples by simple visual detection. The challenge to achieve this goal starts with the (bio)synthesis strategies and their functionality in aqueous media to get efficient environmental monitoring. This chapter describes the relevant features for the development of (bio)sensors based on metal-free organic luminophores, luminescent metal-complexes, metal complexes, metal-organic frameworks, and fluorescent biosensors containing enzymes. These (bio)sensors can be used to quantitatively detect common pesticides and agrochemicals in soil, fruits, vegetables, and water. Key features include molecular strategies, luminescence detection mechanisms, scientific methodology, sensitivity, and analytical precision.

---

[*] **Corresponding authors María K. Salomón-Flores and Alejandro Dorazco-González:** Institute of Chemistry, National Autonomous University of Mexico (UNAM), Mexico City, Mexico; Tel: +525556231137; E-mail: ortizchaconnichdaly@gmail.com

Keywords: Acetylcholinesterase, azinphos-methyl, Biosensors, Chemosensors, Chemodetection, Detection, Enzymatic biosensor, Fluorescence, Glyphosate, Metal complexes, Malathion, MOF, Organic luminophores, Organophosphates, Organochlorines, Parathion-methyl, Pesticides, Paraoxon, Quantification, Soil, Water.

INTRODUCTION

Efficient monitoring of water quality is a scientific topic of global interest. It is very challenging due to the enormous increase in water consumption in agriculture, the food industry, the textile industry, and mining, among other human activities. The monitoring includes the quantitative detection of pollutant chemical species such as pesticides to know if a water source is suitable and safe for human and animal consumption, as well as for the aquatic ecosystem in the world [1]. Pesticide monitoring water is essential to evaluate the efficacy of mitigation measures aimed at curbing pollution [2]. The development of novel analytical methodologies for selective fast sensing of target pesticides capable of operating in aqueous phase or real samples is an active and relevant field of modern analytical chemistry that impacts biological chemistry and environmental sciences [3].

In recent decades, the production of food with longer shelf life and supply chains has been considered key and challenging objectives due to the growing demand of the world's population. In this context, the growing of vegetables, crops, and fruits has been challenging without the use of agrochemicals. It is well known that pesticides commonly enter groundwater bodies, rivers, lakes, and seas through surface runoff, often from an agricultural field, industrial activities, or neighborhoods where they are applied, causing irreversible damage and posing significant health risks to humans [4]. Considering the health damage that pesticides cause in living organisms, it is urgent to develop analytical tools that detect these chemicals in food and irrigation water [5].

To date, very few analytical methodologies can detect agrochemicals, such as herbicides or pesticides, with considerable sensitivity in aqueous media. Instrumental techniques such as chromatographic-mass spectrometry and some spectroscopies (*e.g.*, Raman) are difficult to perform as they often require pre-treatment of the samples and specialized labs [6, 7]. Among the available analytical techniques, luminescence stands out due to simpler detection techniques, high signal to noise ratio, real-time responses, and efficiency in quantifying harmful chemicals such as pesticides [8].

This chapter highlights recent examples of fluorescent sensors that efficiently detect pesticides and herbicides in water and real samples, such as fruits and vegetables. The selected examples are based on organic fluorophores, metal complexes, metal-organic frameworks, and biosensors involving enzymes.

In all cases, the optical sensing of specific agrochemicals is achieved by monitoring the changes in their photoluminescence features.

The examples presented in this book chapter focus on the fluorescent detection of the most widely used pesticides worldwide based on organophosphate and organochlorine molecules. Fig. (**1**) compiles the chemical structure of the pesticides addressed herein.

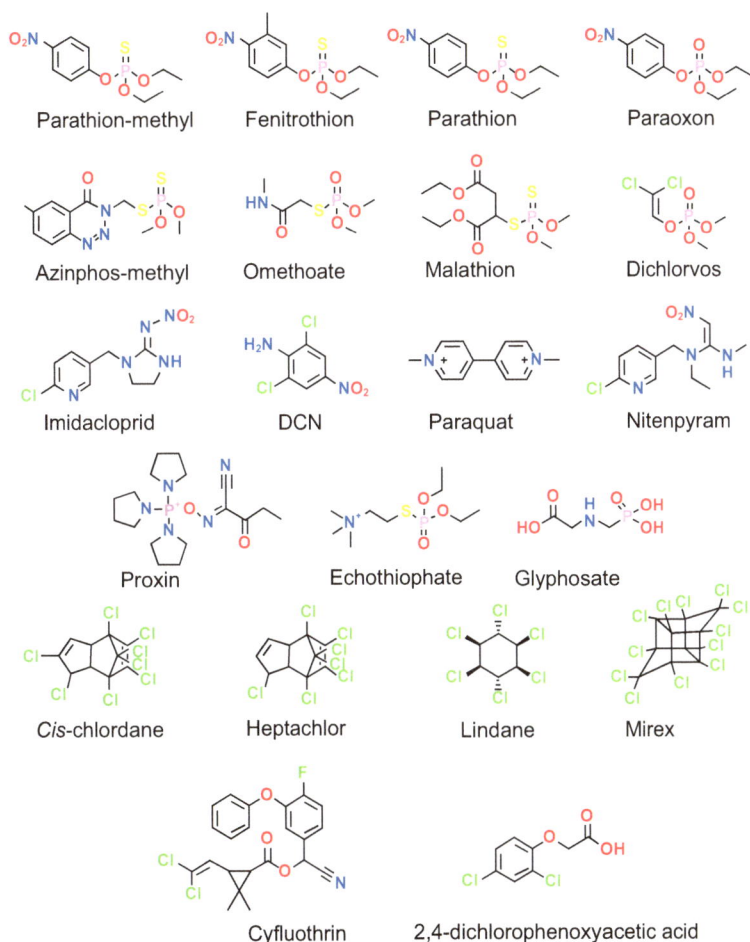

Fig. (1). Chemical structure of pesticides addressed in this work.

Fluorescent Metal-free Organic Sensors for Pesticides in Real Samples

The development of molecular sensors based on organic fluorophores with specific recognition sites for pesticides/herbicides is one of the recent strategies for detecting these pollutant agrochemicals in real samples.

This section highlights recent examples and general characteristics of synthetic fluorescent sensors derived from small organic compounds. The main reported metal-free species used as optical sensors in water and real samples (*e.g.*, soil and food) include macrocycles, acyclic sensors, and polymers. General concepts and experimental quantification methodologies are addressed below.

Macrocycles

A macrocycle is a cyclic macromolecule made up of units that repeat uniformly and form host-guest complexes in their cavities [9]. Among the macrocycles, the cyclodextrins, cucurbiturils, calixarenes and pillararenes are relevant examples because they can selectively capture and detect pesticides due to their cavity with a defined size and convergent binding sites. Fig. (**2**) illustrates the structures of these macrocycles.

Cyclodextrins

They are built from D-glucopyranoside units *via* α-1,4-glycosidic bonds. They are non-toxic and have good water solubility. The diameters of the hydrophobic cavities of the most common cyclodextrins measure from 0.5 to 0.8 nm [10].

Calixarenes

They are formed by substituted phenols linked by methylene bridges; their cavities rich in π electrons adopt various three-dimensional conformations [11].

Curcubiturils

The name "cucurbituril" is derived from the pumpkin-like shape of these molecules. They are built by glycoluril units linked together by two methylene lines, forming a cavity with a negative charge density [12].

Pillararenes

They are formed by hydroquinone groups linked by methylene bridges. In 2008, Ogoshi reported the first 5-membered pillararene synthesized in the laboratory [13, 14].

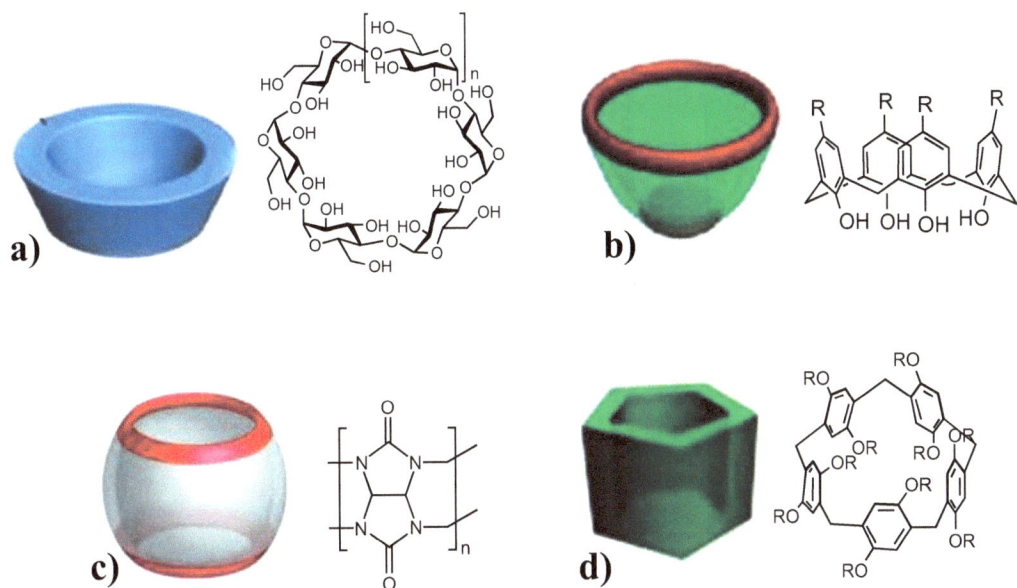

Fig. (2). Molecular structures and their schematic representations of the four types of macrocycles: **a**) cyclodextrins; **b**) calixarenes; **c**) curcubiturils; **d**) pillararenes.

According to current literature, the detection of pesticides by macrocycles through fluorescent emission occurs in two ways: either a non-fluorescent macrocycle acts as a scaffold attached to a signaling system, or a fluorescent macrocycle respond to the presence of the pesticide by changing its emission properties [15].

Relevant Examples of Macrocycle-based Sensors for Pesticides in Real Samples

Cyclodextrins encapsulating pesticides have been used in soil and water remediation procedures [16]. Mindy Levine's research group has reported on the use of cyclodextrins for the detection of organic contaminants [15]. In 2017, Levine's group reported Me-β-CD cyclodextrin for the detection of four organochlorine pesticides (cis-chlordane, heptachlor, lindane, and mirex (see Fig. 3) in freshwater and saltwater sources in the U.S.A. (Rhode Island). This system was found to be sensitive, with a detection limit at the sub-ppm level and showed excellent selectivity [17].

Described below is an example of the methodology for fluorescence modulation and sensing using the combination of Me-β-CD cyclodextrin and BODIPY fluorophore.

A mixture of 1.25 mL of a 10 mM Me-β-CD cyclodextrin solution in phosphate-buffered water (PBS) and 1.25 mL of the water sample was prepared in a quartz cuvette. 100 µL of fluorophore BODIPY (Fig. **3**) [of a 0.1 mg/mL in tetrahydrofuran (THF) solution] was added, and the emission spectrum was taken at an excitation wavelength of 460 nm. Next, 20 µL the corresponding pesticide 1-4 (Fig. **3**) (of a 1.0 mg/mL solution in THF) was then added to the cuvette, and the emission spectrum was taken again (λ_{ex} 460 nm). The fluorescence modulation was measured by the ratio of the integrated emission in the presence of the analyte to the integrated emission of the fluorophore in the absence of the analyte, as shown in equation (1). The geographical location of the water samples influenced the experimental results, as shown in Table **1**, due to the different levels of salinity and pH [17].

Calixarenes are versatile molecules that can be used to recognize organophosphate pesticides, organochlorine herbicides, and carbamate fungicides [11]. In 2018, Haibing Li presented a coumarin-appended calix [4]arene (C4C2) assembled on a silicon surface (C4C2-SAMs), which showed a response to phoxim (organophosphate pesticide) with a detection limit of 1×10^{-6} M. Phoxim was analyzed in local rivers of China and found to be free of contamination by this pesticide; C4C2-SAM changes from hydrophobic to hydrophilic in the presence of phoxim and no change was observed in river water samples [18].

Curcubiturils: In 2021, Jing-Xin Liu *et al.* investigated the binding properties of cucurbituril (Q [10]) to protonated acridine (AD, fluorophore). The AD was encapsulated in the cavity of Q [10], forming the inclusion complex AD2@Q [10], which resulted in the strong quenching of the fluorescent emission. Upon the introduction of the pesticide dodine DD, the fluorophore AD was displaced, and the emission was recovered. The authors suggest that this assembly has the potential for detecting DD residue in agricultural products [19].

Pillararenes: In 2024, Xin Xiao *et al.* developed a fluorescence sensor, AC@PyP5, through the supramolecular interaction between pyridine pillar [5]arene (PyP5) and acridine derivatives (AC) in water, which demonstrated selectivity towards the pesticides hymexazol and ethiofencarb [20] (Fig. **4**).

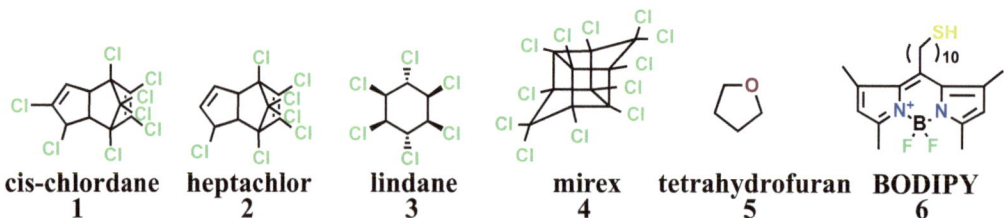

cis-chlordane	heptachlor	lindane	mirex	tetrahydrofuran	BODIPY
1	**2**	**3**	**4**	**5**	**6**

Fig. (3). Chemical structure of analytes **1-5** and fluorophore **6** [17].

$$Fluorescence\ modulation = \frac{Fl_{analyte}}{Fl_{blank}} \qquad (1)$$

Table 1. Selected fluorescence modulation results comparing sampling location.

Water Sample	Cyclodextrin	Arcadia Lake	Narragansett Bay	Ocean	Providence River
1	Me-β-CD	1.17 ± 0.01	1.26±0.02	1.03±0.01	1.05±0.0003
	PBS	1.21 ± 0.02	1.14 ±0.01	1.15±0.01	1.20±0.01
2	Me-β-CD	1.08 ± 0.01	1.23 ± 0.02	1.17±0.01	1.05±0.01
	PBS	1.28 ± 0.01	1.14 ± 0.01	1.14±0.01	1.18±0.02

Relevant Examples of Organic Acyclic Sensors for Pesticides in Real Samples

Two relevant examples of organic molecules for pesticide detection in real samples and their scientific methodology are presented here.

In 2022, Ziya Aydin *et al.* reported an optical sensor for malathion (Mal) and glyphosate (Glyp). The sensor, N,N-dimethyl-4-(3-(pyrazin-2-ilimino)p-op-1-en-1-yl)aniline CNP (Fig. **5**), presented changes in its photophysical properties (fluorescence, absorbance, and color) due to its interaction with Glyp through hydrogen bonds and with Mal through coordination bonds. The detection limits for Glyp and Mal were 0.375 µM and 0.066 µM, respectively [21]. The methodology that was carried out is described below.

Apple sample preparation: Samples of 10 g of apple (previously washed with distilled water and cut) were weighed and placed in 10 mL of diethyl ether for 2 hours, then filtered with a 0.45 mm microporous membrane. The diethyl ether was evaporated, the residue dissolved in 2 mL of ethanol, and it was then diluted with the working solvent system [21].

Soil sample preparation: 20 g of soil was dispersed in 20 mL of acetonitrile/acetate buffer solution for 30 min under ultrasound. It was filtered with a 0.22 µm microporous membrane. The resulting filtrate was diluted four times more with an acetonitrile/acetate buffer solution [21].

Potato sample preparation: 20 g of potatoes were ground in a mixer and centrifuged for 10 min at 5000 rpm. After filtering through a 0.45 µm membrane, potato juice was diluted with acetonitrile/acetate buffer [21].

Different concentrations of Glyp or Mal (1.0, 3.0, and 5.0 ppm) were added to real samples (apple, soil, potato, and tap water) and analyzed with CNP under the same conditions as shown in the selectivity studies. Table **2** shows the results, and the authors propose that the CNP sensor can quantify Glyp and Mal in real samples [21].

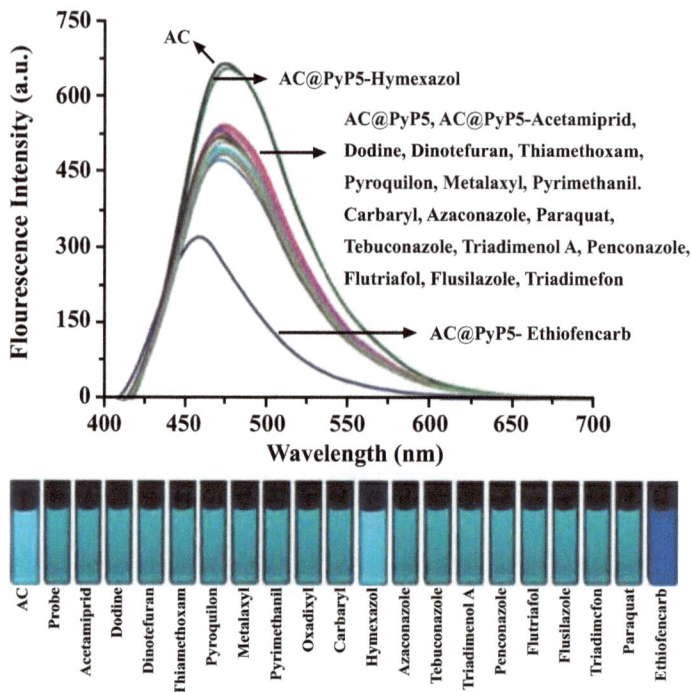

Fig. (4). (Top) Fluorescent emission change of AC@PyP5 (20 µM) upon addition of pesticides; (Bottom) Photograph of AC@PyP5 with pesticides under UV light (365 nm) [20].

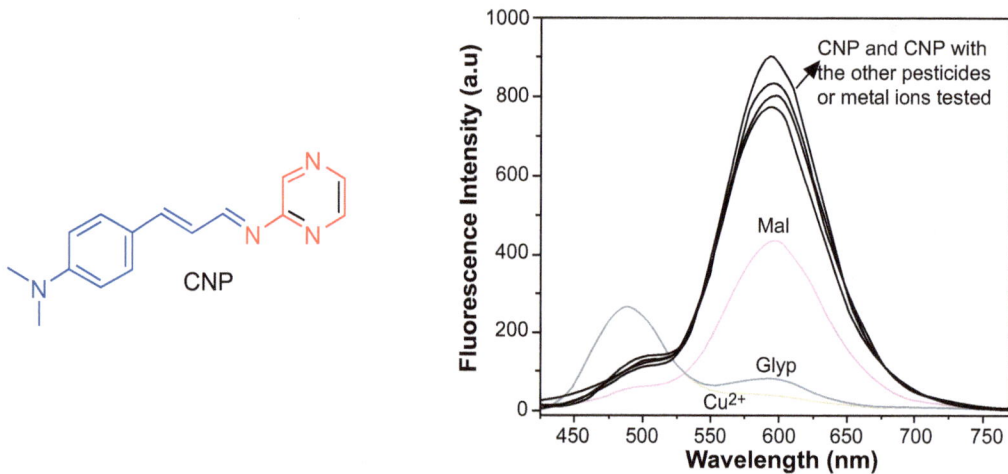

Fig. (5). Fluorescent emission change of CNP (100µM) after the addition of pesticides (Mal, Glyp, Fena, PQ) and metal in acetonitrile/acetate buffer [21].

In 2018, Zipin Zhang *et al.* used the fluorophore pyranine for fluorescent quenching detection of paraquat (PQ), which is a highly toxic herbicide, through the non-covalent pyranine-PQ interaction, as shown in Fig. (**6**). Tap water and reservoir water samples were analyzed as real samples [22].

Relevant Examples of Organic Polymers for Pesticides in Real Samples

Polymers are high-molecular-weight chemicals made of repeating units called monomers, which are linked by covalent bonds [23]. Recently, fluorescent polymers have been reported for pesticide detection by quenching fluorescent emission [15].

In 2020, Zhonggang Wang *et al.* presented three polymers carbazole-based monomers (N-benzylcarbazole PAN-C, N-benzyl dibromo-carbazole PAN-C–Br, and N-benzyl dimethoxy-carbazole PAN-C–OCH3). These polymers exhibited fluorescent emission changes to common pesticides (trifluralin, isopropalin, glyphosate, fenitrothion, imidacloprid, cyfluothrin) due to the varied porosity of the polymers and molecular sizes of the pesticides [24]. Fig. (**7**) shows the bright cyan fluorescent emission under a 365 nm UV lamp of a PAN-C polymer test paper. After dipping the test paper into six aqueous solutions of pesticides at the same concentration, it exhibits different fluorescence changes for each pesticide [25].

Table 2. Determination of malathion and glyphosate in real samples.

Pesticide	Sample	Pesticide added (ppm)	Pesticides found (ppm)	Recovery (%)	RSD (n=3)
Malathion	Apple	0	-	-	-
	-	1	0.87	87.0	7.32
	-	3	2.83	94.2	1.64
	-	5	4.65	93.0	1.40
	Tap water	0	-	-	-
	-	1	0.84	83.7	2.45
	-	3	2.81	93.8	2.14
	-	5	4.69	93.8	3.09
Glyphosate	Potato	0	-	-	-
	-	1	0.81	80.7	4.78
	-	3	2.88	95.9	1.61
	-	5	4.41	88.2	1.58
	Soil	0	-	-	-

(Table 2) cont.....

Pesticide	Sample	Pesticide added (ppm)	Pesticides found (ppm)	Recovery (%)	RSD (n=3)
	-	1	0.83	83.0	5.47
	-	3	2.90	96.7	3.25
	-	5	4.59	91.8	3.15
	Tap water	0	-	-	-
	-	1	0.88	87.6	5.42
	-	3	2.86	95.4	2.07
	-	5	4.64	92.8	1.32

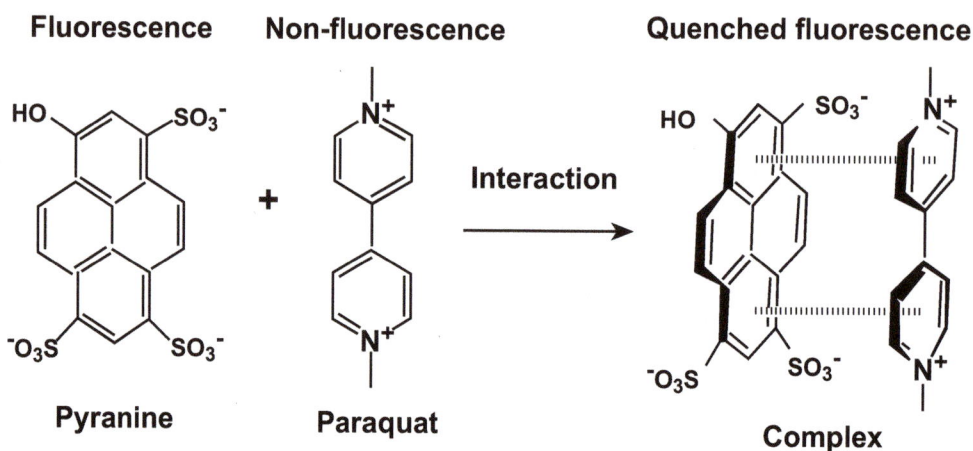

Fig. (6). Schematic representation of paraquat detection by pyranine [22].

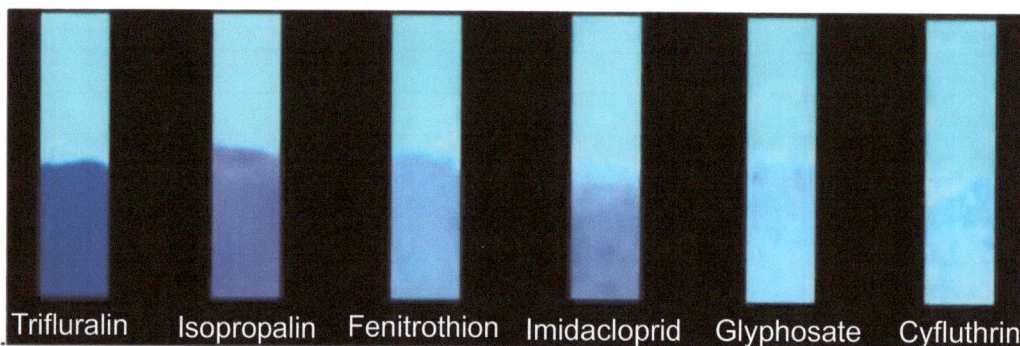

Fig. (7). PAN-C test paper of six pesticides in a water medium [25].

Luminescent Metal Molecular Complexes for Chemodetection of Organophosphorus Pesticides

Design strategies for luminescent chemodetection of organophosphorus pesticides by metal molecular complexes include the indicator-displacement assay (IDA) and direct coordination of the organophosphorus pesticide with the metal ion. These chemodetection strategies involve the use of artificial receptors that can act as chemosensors or chemodosimeters [26]. The difference between both types of artificial receptors lies in the reversibility of the detection system: chemosensors can detect an analyte reversibly, which is suitable for real-time sensing, while chemodosimeters are employed for irreversible and cumulative sensing [27]. Furthermore, both artificial receptors are composed of a binding unit (recognition site) and a signaling unit (signal transduction site) united by a covalent bond and/or supramolecular interaction (Fig. **8**) [28].

If the signaling site is a luminophore, in the presence of an analyte such as an organophosphorus pesticide, the signal transducer produces a luminescent change caused by the interaction of the analyte with the recognition unit of the receptor.

This luminescence change can be either ratiometric (analyte-induced luminescence intensity changes from two or more emission bands at different wavelengths) or luminescence enhancement/quenching.

Indicator-Displacement Assay (IDA)

Direct Coordination with the Metal Ion

$M^{n\pm}$ **Metal Ion**

(Binding Site)
Recognition Unit
Receptor

(Signaling Site)
Signal Transduction Unit
Indicator

Analyte

Fig. (8). Schematic representation of the common strategies used to generate chemosensors/chemodosimeters based on metal complexes for the luminescent detection of organophosphorus pesticides.

Indicator-displacement assays (IDA) frequently rely on the presence of an analyte to separate supramolecular/covalent interactions from the receptor-luminophore assembly. This leads to the formation of a new receptor-analyte adduct and the release of a luminophore for reversible or irreversible sensing. In the context of luminescent chemodetection, the fundamental mechanism of the IDA includes (a) the initial association of a luminescent indicator (luminophore) to a receptor, generating a "receptor-indicator assembly", (b) the addition of a competitive analyte into the system that causes the separation of the "receptor-indicator assembly" and the activation of a luminescent signal from the release of the indicator, and (c) the generation of a new "receptor-analyte adduct" [26].

Based on the fundamental mechanism of the IDA, Chow reported the chemodetection and catalytic augmentation properties of the assembly [Mn^{III}(TPP)(NCS)]–[Pd^{II}(DMSO)Cl$_2$] (TPP= 5,10,15,20-tetraphenyl-21H,2-H-porphine), where [Mn^{III}(TPP)(NCS)] is a colorimetric indicator that changes color from yellowish-brown to green, and [Pd^{II}(DMSO)Cl$_2$] is a receptor for binding the organophosphate pesticide dimethoate in acetone (Fig. **9**). Fitting the UV-Vis spectroscopic responses with the binding model reveals that [Mn^{III}(TPP)(NCS)]–[Pd^{II}(DMSO)Cl$_2$] associates two dimethoate (pesticide) molecules with an average association constant (log K) of 4.67. Curiously, the free "receptor-analyte adduct" Pd^{II}(dimethoate)$_2$Cl$_2$ released works as a catalyst for the Heck coupling reaction of 5-diethylamino-2-iodo-phenyl ester, generating the highly luminescent 7-diethylaminocoumarin (λ_{em}= 450 nm). This amplifies the signal and improves the initial sensitivity by 25 times, with a limit of detection (LOD) of 0.9 µM in MeCN/H$_2$O [29].

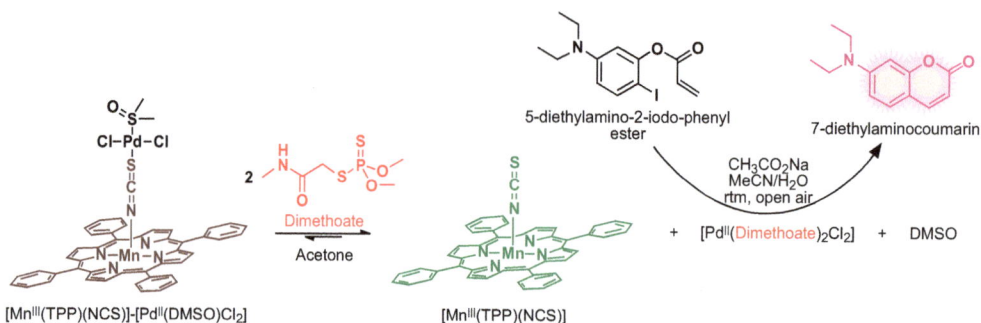

Fig. (9). Proposed tandem recognition, signaling, and amplification by [Mn^{III}(TPP)(NCS)]–[Pd^{II}(DMSO)Cl$_2$] for chemodetection dimethoate.

On the other hand, Chow also reported on the bimetallic donor-acceptor assembly [Re^{I}(tBu-bpy)(CO)$_3$(NCS)]–[Pt^{II}(DMSO)(Cl)$_2$] (tBu-bpy = 4,4′-di-tert-butyl-,2′-bipyridine), where [Re^{I}(tBu-bpy)(CO)$_3$(NCS)] is a yellow luminescent indicator (λ_{em}= 578 nm) that upon coordination with the [Pd^{II}(DMSO)Cl$_2$] acceptor shifts to 572 nm and is significantly quenched in CHCl$_3$. The acceptor complex [Pd^{II}(DMSO)Cl$_2$] is a receptor for binding the organophosphate pesticides with aliphatic mercapto functionality. This assembly showed luminescent chemodosimetric selectivity and sensibility for phorate with a LOD of 3.84 μM (1 ppm) and a formation constant (log K) of 3.43 (1:1 Benesi–Hildebrand model) (Fig. **10**) [30].

Zheng reported a tetranuclear bimetallic complex, [Ru^{II}(bpy)$_2$(CN)$_2$]$_2$ [Cu^{II}]$_2$(bpy=2,2'-bipyridine), integrating two catalytic-recognition units [Cu^{I}], and two luminescent indicators ([Ru^{II}(bpy)$_2$(CN)$_2$]) to simultaneously detect and degrade methyl parathion, a highly toxic organophosphate pesticide *via* the IDA. The catalytic-recognition units [Cu^{I}] are luminescent quenchers for [Ru^{II}(bpy)$_2$(CN)$_2$] and function simultaneously as catalysts to hydrolyze the phosphate ester group of methyl parathion and as strong receptors to bind phosphates ([Cu^{I}]-catalyzed hydrolysis products). Fig. (**11**) illustrates the recognition mechanism of assembly [Ru^{II}(bpy)$_2$(CN)$_2$]$_2$[Cu^{II}]$_2$, where the active [Cu^{I}] catalytic-recognition units in [Ru^{II}(bpy)$_2$(CN)$_2$]$_2$[Cu^{II}]$_2$ hydrolyzed the phosphate ester of methyl parathion to 4-nitrophenolate anion (detectable by UV/Vis; ΔA 400 nm) and O, O-dimethyl thiophosphate (DTP) anion, the DTP interacted with the [Cu^{I}] centers in the assembly, displacing two [Ru^{II}(bpy)$_2$(CN)$_2$] and inducing a red luminescence enhancement (λ_{em}= 621 nm).

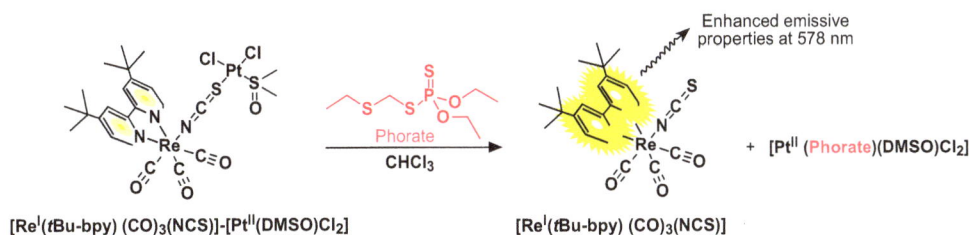

Fig. (10). Proposed chemodosimetric mechanism of [Re^{I}(tBu-bpy)(CO)$_3$(NCS)]–[Pt^{II}(DMSO)(Cl)$_2$] with phorate.

The formation of the more stable [Cu^{I}(DTP)] complex (ΔG= −20.3 kJ mol^{-1}; *log K*= 3.56) concerning the Ru-complex (ΔG= −20.2 kJ mol^{-1}; *log K*= 3.54) was the driving force for this displacement. The LOD of [Ru^{II}(bpy)$_2$(CN)$_2$]$_2$[Cu^{II}]$_2$ toward methyl parathion was determined to be 125 μM (33 ppm) and 30 μM (8 ppm) for the UV and luminescent modes of detection, respectively [31].

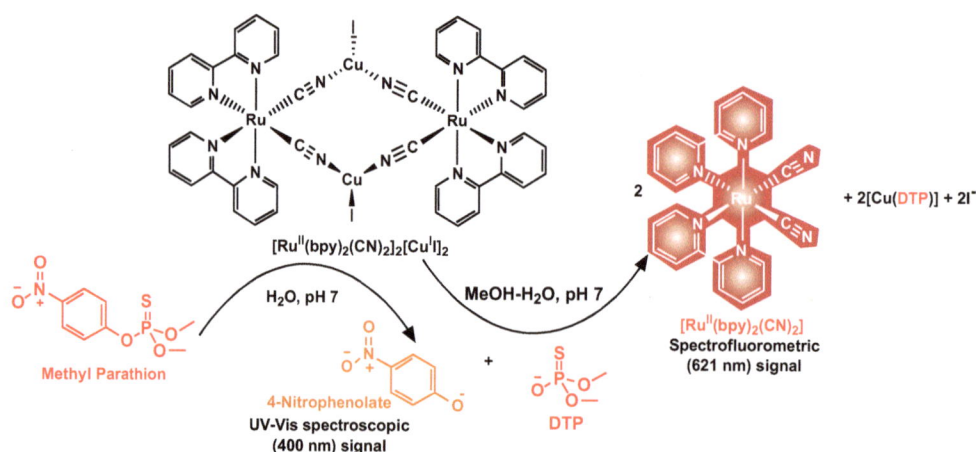

Fig. (11). Proposed mechanism for the chemodetection of methyl parathion by complex $[Ru^{II}(bpy)_2(CN)_2]_2[Cu^{II}]_2$.

Glyphosate is an organophosphate herbicide widely used in agriculture despite causing several human diseases. As a cost-effective, quick, and simple alternative for the detection and quantification of glyphosate in environmental samples, recently, reports have increased for the detection of glyphosate through optical chemosensors or chemodosimeters that require an instrument such as fluorescence spectroscopy. In this sense, several research groups have designed molecular chemodetection systems based on d-block and f-block metal complexes, mainly Cu (II), Zn(II), Fe(III), and Eu(III), which are metals with good affinity and selectivity to coordinate glyphosate in water.

Based on IDA, the assemblies Cu^{II}-Luminophore, Zn^{II}-Luminophore, and Fe^{III}-Luminophore are presented in Fig. (**12**). Table **3** compiles the most outstanding sensor systems based on functional transition metal complexes in aqueous media. Another strategy for luminescent chemodetection of organophosphorus pesticides such as glyphosate/parathion is to use direct coordination of the organophosphorus pesticide with the metal ion. Some examples reported lately are illustrated and summarized in Fig. (**13**) and Table **4**, respectively.

Table 3. Chemosensors based on IDA for luminescent detection of glyphosate pesticide.

Assembly	Signal	LOD	Solvent	Ref.
7	Off-On	0.63 µM	DMSO/H$_2$O (1:4, v/v) pH = 6.0	[32]
8	Off-On	0.41 µM	DMSO/H$_2$O (1:4, v/v) pH = 6.0	[32]
9	Off-On	36 nM	DMSO/H$_2$O (9:1, v/v) pH = 7.0	[33]
10	Off-On	4.1 nM	H$_2$O/MeOH (8:2, v/v)	[34]

(Table 3) cont.....

Assembly	Signal	LOD	Solvent	Ref.
11	Off-On	25 nM	H_2O pH 7.4	[35]
12	On-Off	49.3 nM	DMSO	[36]
13	Off-On	52 nM	DMSO/H_2O (95:5, v/v) pH 7.4	[37]
14	Off-On	2.48 nM	EtOH/H_2O solution (9/1, v/v) pH 7.4	[38]
15	Off-On	18.77 nM	DMSO/H_2O (9/1, v/v) pH 7.4	[39]
16	Off-On	11.26 nM	EtOH/H_2O (9/1, v/v) pH 7.4	[40]
17	Off-On	13.16 nM	H_2O pH 7.4	[41]
18	Off-On	1.87 µM	MeCN	[42]
19	Off-On	10.3 nM	H_2O pH 7.0	[43]
20	Off-On	71.4 nM	H_2O pH 7.0	[44]
21	Off-On	0.11 µM	EtOH/H_2O solution (9/1, v/v) pH 6.0	[45]

Table 4. Chemosensors based on direct coordination of pesticide with the metal ion for luminescent detection of glyphosate and parathion.

Chemosensor	Pesticide_	Signal_	Log K_	LOD_	Solvent_	Ref.
22	glyphosate_	On-Off	4.67	-a	H_2O pH 8.0	[46]
23	glyphosate_	Off-On	5.24	-a	H_2O pH 8.0	[46]
24	parathion	On-Off	5.03	2.05 µM	H_2O pH 7.0	[47]
25	glyphosate_	Off-On	5.36	-a	H_2O pH 5.9	[48]

a⁻ not reported.

Luminescent Metal-organic Frameworks (LMOF$_S$) for the Sensing of Pesticides in Real Samples

In this section, we discuss the synthesis, characteristics, and sensing mechanisms of different metal-organic frameworks (MOFs) for pesticide detection, including their application in real samples.

The combination of organic linkers with metallic ions generates materials with unique properties called MOFs. These properties are sought and designed according to the intended applications of the material. Their flexible design allows one to obtain materials of various shapes and sizes (Fig. **14**), which will be conditioned by the synthesis methods. The applications of MFOs are diverse, from medical [49] to environmental applications such as contaminant sensing [50, 51].

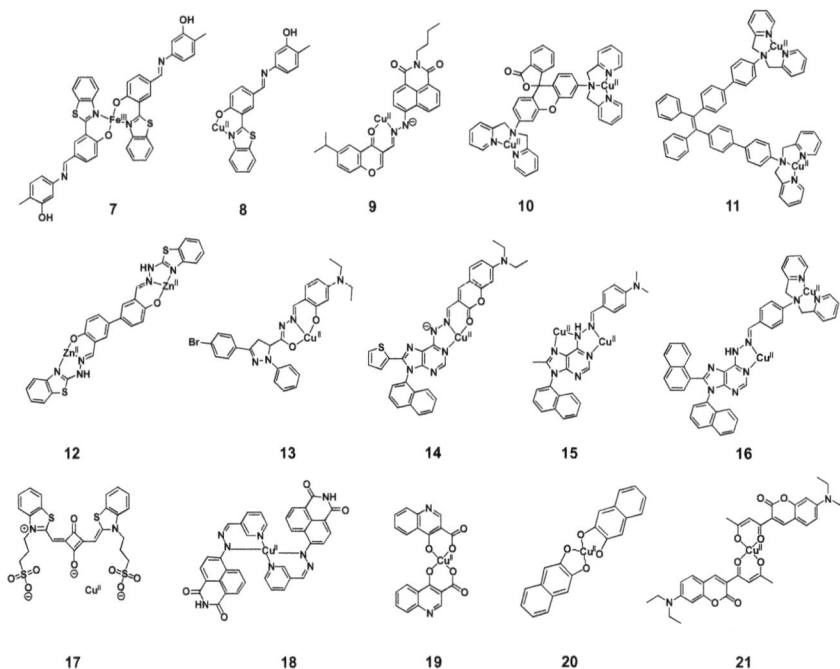

Fig. (12). Metal-complexes-based chemosensors on IDA for luminescent detection of glyphosate pesticide.

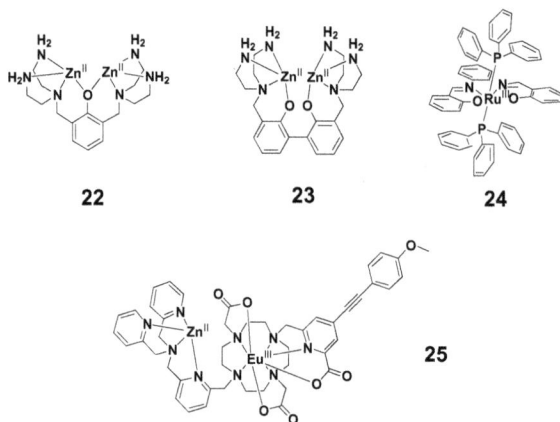

Fig. (13). Chemosensors based on direct coordination with metal for luminescent detection of parathion.

MOFs-based sensors work differently depending on their optical, mechanical, catalytic, conductivity, or other properties that can be converted into detectable signals for identifying various targets. Optical MOFs-based sensors can be colorimetric [52], luminescent [53], electrochemiluminescent [54], and surface-enhanced Raman spectroscopy [55]. They can be analyzed based on their construction strategies, sensing mechanisms, and real sample application.

A luminescent MOF emits light after irradiation with a certain wavelength, which can be fluorescence or phosphorescence. After light absorption, the molecule jumps to an excited state while coming down to the ground; a radiative transition is free, such as fluorescence or phosphorescence. The difference is that the fluorescence is faster than phosphorescence; the first occurs among energy levels of equal spin multiplicity ($S_1 \rightarrow S_0$), and the second is between different multiplicity spin states ($T_1 \rightarrow S_0$). Phosphorescence emission wavelength occurs in microseconds to seconds, contrary to fluorescence [56]. Luminescence can result from different sources, such as π conjugated organic ligands alone; lanthanide metal ions (Ln) can produce this response by an antenna effect (Fig. 15). As shown in Fig. (16), one mechanism may be when UV light excites the ligand, and the MOF emits light by ligand-based luminescence. It is important to determine if the analyte can absorb part of the UV light to determine the mechanisms correctly. The analyte can also receive energy from the ligand or by electron transfer. Among luminescent MOFs constructed with divalent transition metals, Zn(II) is widely used because it is less expensive and toxic than others, as well as a very versatile metal that provides thermally stable MOFs with optical properties [57].

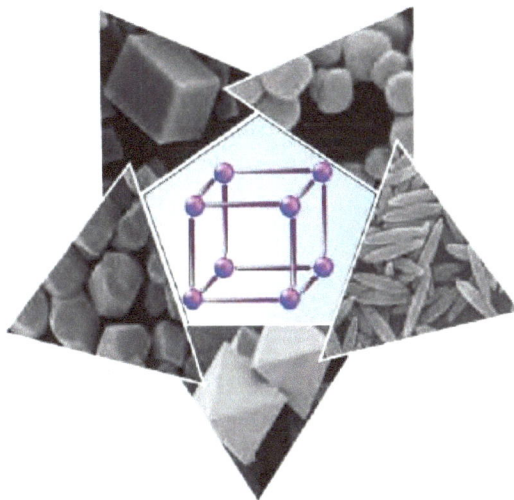

Fig. (14). MOF different structures characterized by SEM.

Table **5** compiles outstanding examples of Zn-MOF-based luminescent sensors with analytical response type and detection limit, as well as the real sample tested. In general, these Zn-MOFs have a large surface area (> 278 m^2/g) and a very low detection limit at nanomolar levels. These Zn-MOFs were used to detect pesticides in real samples. In irrigation water, methyl-parathion and parathion were detected using the standard addition method.

Luminescent MOFs based on diamagnetic d^{10} transition metals such as Cu(I) and Cd(II) have also been reported as efficient optical sensors for pesticides (Table **6**). For example, $[Cd_3(PDA)(tz)_3Cl(H_2O)_4]$ (PDA= 1,4-phenylenediacetate; tz = 1,2,4-triazolate) senses azinphos-methyl by combining static quenching, electron transfer, resonance energy transfer, and excitation energy absorption [58]. On the other hand, a Cu-MOF with the general formula $[Cu_2(H_4DET)(H_2O)_2]$ detects glyphosate by a quenching signal. The literature also features luminescent MOFs containing high-valent metal ions (Ti(IV), Zr(IV), Al(III), Fe(III), and Cr(III)) for pesticide sensing. Table **7** shows a few examples tested on real samples. He *et al.* reported a water-stable Zr-MOF able to sense parathion-methyl with LOD= 0.45 nM [59]. In this system, the $-NO_2$ group caused the transfer of electrons from the Zr-MOF to parathion-methyl. Satisfactory recoveries ranging from 78-100% were obtained for spiked food and environmental samples (tap water, rainwater, and lettuce).

Ghosh reported a Zr-MOF that can detect the herbicide 2,4-dichlorophenoxyacetic acid (2,4-D) in different fruits and vegetables with a LOD of 10.9 nM. The analytical response was elucidated as a transfer of electrons, which turned on the fluorescence intensity in the presence of 2,4-D [60].

Several lanthanide-MOFs have been designed for the luminescent sensing of pesticides. However, only a few have been tested in real samples, likely due to issues with hydrostability. The most common MOFs include Tb(III) and Eu(III). Table **8** lists some recent and relevant examples in this line.

Table 5. Luminescent Zn-MOFs for pesticide sensing in real samples.

MOFs	Signal	Pesticide	Sample	Sup. area m^2/g	LOD (M)	Ref.
$[Zn_2(TCPB)]$	Turn-off	parathion-methyl	irrigation water	4073.9	0.45×10^{-9}	[61]
$[Zn_4(TCPP)_2(TCPB)_2]$	Turn-off	Parathion	irrigation water	-a	6.60×10^{-9}	[62]
$[Zn_3(DDB)(DPE)]$	Turn-off	2,6-DCPNA	Fruits	-a	2.70×10^{-7}	[63]
$[Zn_4(CPTA)_2(OH)_2$ DMF)0.5]	Turn-off	Nitenpyram	Aquafarm	278.5	0.62×10^{-6}	[64]

a= no reported. 2,6-DCPNA= 2,6-dichloro-4-nitroamine; TCPB= 1,2,4,5-tetrakis(4-carboxyphenyl)benzene; TCP= tetrakis (4-carboxyphenyl) porphyrin; DDB= 3,5-di(2',4'-dicarboxylphenyl)benzoic acid; DPE= 1,2-di(4-pyridyl)ethylene; CPTA= [2-(2-carboxypyridin-4-yl)terephthalic acid]; DMF= N,N'-dimethylformamide.

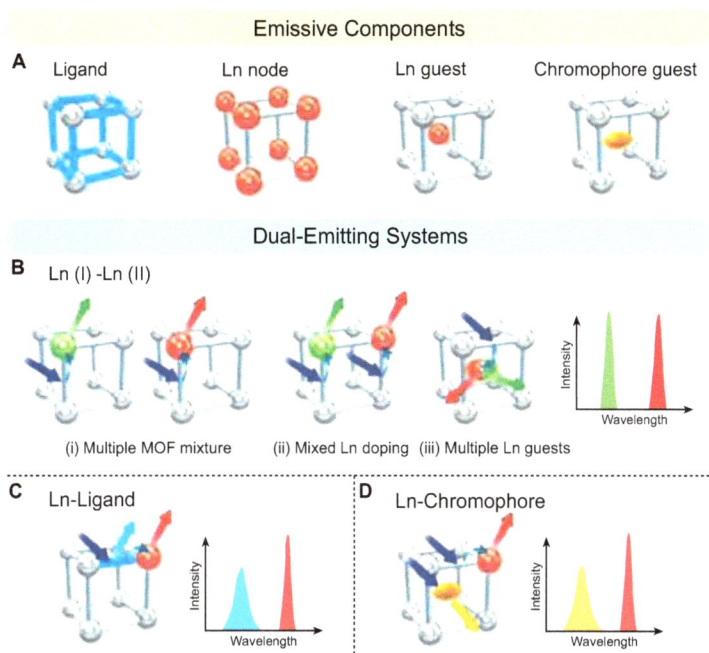

Fig. (15). Schematic of possible single emissive components (A), the dual-emitting system of fluorescent sensors in Ln MOFs (B–D) [56].

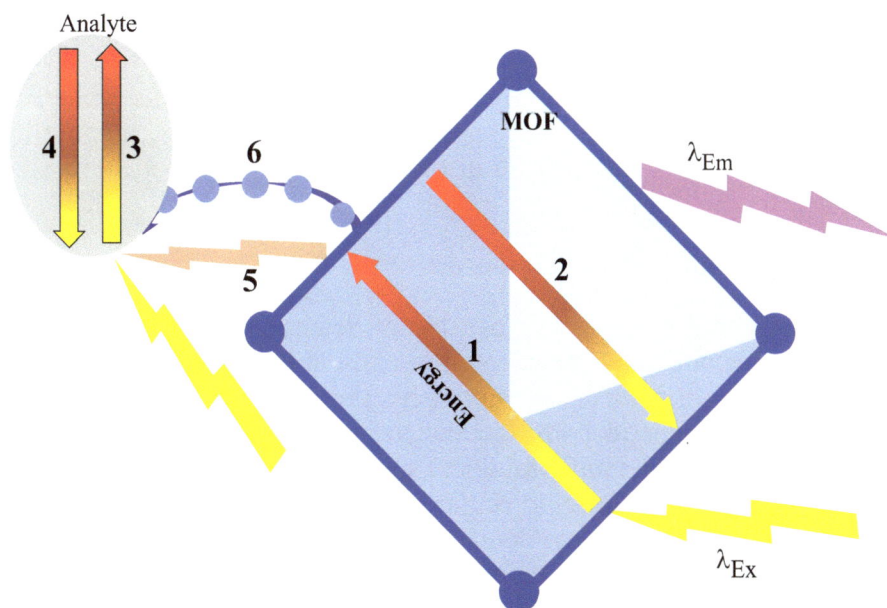

Fig. (16). Schematic of various energy/electron transfer processes in the presence of an analyte. UV light excites the ligand;(1) ligand-based luminescence by emission; (2) analyte absorbs a part of UV light by the analyte; (3) analyte-based luminescence; (4) energy transfer from ligand to the analyte; (5) electron transfer from ligand to the analyte; (5) electron transfer from ligand to the analyte (6).

Table 6. Luminescent Cd-MOFs, Cu-MOFs for pesticide sensing in real samples.

MOFs	Signal	Pesticide	Sample	LOD (M)	Ref.
$[Cd_3(PDA)(tz)_3Cl(H_2O)_4]$	Turn-off	azinphos-methyl	apples, tomatoes	2.50×10^{-8}	[58]
$[Cu_2(DET)(H_2O)_2]$	Turn-off	Glyphosate	apples, oranges	0.07×10^{-6}	[65]

PDA= 1,4-phenylenediacetate; tz = 1,2,4-triazolate; DET= 2,1',5,5'-diphenyl ether tetracarboxylic acid.

Table 7. Luminescent Zr-MOFs for pesticide sensing in real samples.

MOFs	Signal	Pesticide	Sample	LOD (M)	Ref.
$[Zr(TCBP)]$	Turn-off	parathion-methyl	Rainwater	0.43×10^{-9}	[59]
$[Zr_6O_4(OH)_4(BDC-NH_2)_6]$	Turn-on	2,4-D	Fruits	10.9×10^{-9}	[60]

TCBP= 1,2,4,5-tetrakis(4-carboxyphenyl)benzene; BDC-NH2= dianion of 2-aminobenzene dicarboxylic acid; 2,4-D= 2,4-dichlorophenoxyacetic acid.

Table 8. Luminescent lanthanide-MOFs for pesticide sensing in real samples.

MOFs	Signal	Pesticide	Sample	LOD (M)	Ref.
$Tb(TMA)(H_2O)_6$	Turn -on	Chlorpyrifos	Strawberries	0.11×10^{-9}	[66]
$[Tb_3(HDDB)(DDB)(H_2O)_6]$	Turn-off	DCN	Fruits	0.14×10^{-6}	[67]

TMA= trimesic acid; DDB= 3,5-di(2',4'-dicarboxylphenyl) benzoic acid; DCN= 2,6-dichloro-4 nitroaniline.

Luminescent Biosensors for Pesticides in Real Samples

In the last decade, efficient biosensors for pesticides have been developed using bio-macromolecules such as enzymes and DNA-aptamers, among others. Concepts and an overview description of outstanding enzymatic examples are shown below.

Fluorescent Enzymatic-based Biosensors

A *Biosensor* is a molecular device capable of detecting any target analyte. It is composed of two elements: a bio-recognition unit able to recognize the chemical, physical, or biological response produced by the target analyte and a transducer element that converts the response generated into a measurable signal. The bio-recognition element can be any biological species able to interact with the target analyte through specific supramolecular interactions. These include enzymes, antibodies, DNA aptamers, molecularly imprinted polymers, or nanozymes. However, novel bio-recognition elements, including nanobodies, peptides, modified enzymes, affibodies, and DNAzymes, have also been used [68].

Recently, some nanomaterials have emerged as low-cost alternatives to biomolecules for use in (bio)analytical assays. These materials cover a wide variety of chemical compounds, including porphyrins, transition metal complexes, metal oxides, nanocomposites, smart polymers, and carbonaceous nanomaterials [69].

The use of biomolecules for the detection of analytes is novel research which is rapidly growing over the past 20 years. As these biomaterials are helpful in the analytical measurement of any kind of analyte in both laboratory and field testing, they have wide utility in agricultural product safety, monitoring of environmental issues, safety measurements of food, clinical diagnostics, as biological and chemical warfare agents or biomedical research to process control [70]. Besides, biosensors designed for the sensing of pesticide substances turned out to be more reliable, specific, convenient, and reproducible as compared to other widely used analytical methods available, such as colorimetry, mass spectrometry, capillary electrophoresis, enzyme-linked, and chromatographic techniques [71].

Biosensors have a significant impact on food, agricultural, environmental, and biomedical applications. These systems provide many advantages in comparison to conventional analytical detection methods (*e.g.*, chromatographic techniques), such as minimal sample preparation and handling, faster time analysis, rapid detection of the target analytes, the use by non-skilled personnel, and portability for uses in field applications.

Pesticides can cause metabolic perturbation in pest species, including plants, fungi, or animals, by inhibiting specific enzymes. This inhibition characteristic allows enzymes to serve as biological recognition elements in biosensors, considering that an enzyme can be inhibited by several pesticides and other compounds, leading to specificity issues [72]. The detection mechanisms in such biosensors can involve two main types of responses: enzyme inhibition, where the analyte inhibits the enzymatic activity of enzymes such as alkaline phosphatase, tyrosinase, acetylcholinesterase, or butyrylcholinesterase, resulting in a measurable decrease in response; and catalytic activity, where enzymes like organophosphate hydrolase catalyze the hydrolysis of specific analytes, leading to an increase in response as the analyte concentration rises.

Based on the transducer used, enzyme inhibition-based biosensors can be classified into four types: 1) electrochemical, 2) thermal, 3) piezoelectric, and 4) optical (fluorescence systems). Among these, fluorescence analysis is considered a promising sensing technology due to its advantages of rapid response, high sensitivity, simplicity of operation, and low cost.

Based on the measurement of optical parameters like fluorescence, absorbance, or chemiluminescence, different kinds of biosensors have been fabricated with the help of optoelectronic transducers.

The underlying principle of the fluorescence-based biosensor is the specific interaction that occurs between the bio-recognition moiety and the target analyte. Because of this chemical interaction, an optical signal in the form of a change in absorption or emission band is easily detected by the fluorophore, and it is directly correlated with the concentration of the analyte.

Fluorescent biosensors based on acetylcholinesterase (AChE) are the most widely used in the development of new inhibition-based sensors. The first fluorescent sensor constructed from the synergy of a biological (enzyme) fragment with a luminescent substance was reported by Rogers in 1991 [73]. This system includes an immobilized fluorescein isothiocyanate (FITC) as a transducer element. When the acetylcholine (ACh) is hydrolyzed, a drop in the pH of the solution takes place, leading to a luminescence quenching of FITC-AChE. Thus, suppression of the enzyme activity by an inhibitor substrate will alter the pH-dependent fluorescent signal generated by FITC-AChE. The strong interaction and subsequent AChE inactivation by the carbamate insecticides bendiocarb and methomyl and the organophosphorus pesticides, such as echothiophate and paraoxon, thus permitted the detection of these compounds in the nanomolar to micromolar range. Since then, this research area has shown sustained growth, increasing mainly since the 2000s. A novel system, reported by Liu [74], using AChE as bio-recognition element with a thiamine/MnO_2 system as a transducer element, was able to detect dichlorvos and chlorpyrifos with a detection limit (LOD) of 5.11 pM and 2.45 nM, respectively. Interestingly, Han presented an enzyme-inhibition-based system working on the electrostatic binding of gold nanoclusters with thiocoline, detecting paraoxon with a LOD of 33 fM [75], showing the high performance and sensitivity of this advanced kind of chemosensors.

To date, more complex biosensor devices have been constructed based on multi-target inhibitor design. This approach involves combining or merging distinct frameworks into a single chemical entity with cleavable or non-cleavable linkers. An interesting example of this kind of device was given by Yan [76] developing a multienzyme-targeted fluorescent probe design, which merges five different enzymes (acetylcholinesterase, butyrylcholinesterase, chymotrypsin, and carboxylesterases) into a complex molecular sensor. This sensor was capable of detecting dichlorvos, carabaryl and deltamethrin, making it the first biosensor recognizing pyrethroid pesticides.

Another successful strategy for pesticide detection is based on the hydrolysis catalytic type enzymatic process, which has been mainly applied for the detection of OPs using organophosphate hydrolase (OPH). OPH is an enzyme encoded from the "opd" gene by many soil bacteria species with the capability to hydrolyze organophosphorus pesticides [77]. One example is an OPH-based optical biosensor, where sulfo-N-hydroxy succinimide modified-gold nanoparticles are covalently linked to OPH *via* its lysine residue. The change in the distance between fluorophore and gold nanoparticle causes a variation in the fluorescence, allowing to directly detect paraoxon with high sensitivity [78]. Melo enhanced this sensitivity with a biosensor integrating polyethyleneimine (PEI) functionalized silica nanoparticles (Si NP) with Sphingomonas sp. cells. The biohybrid was optimized for the hydrolysis of methyl parathion (MP) into p-nitrophenol achieving a LOD value of 1.0 ppm [79]. Table **9** compiles recent examples of enzymatic optical biosensors for common pesticides with application in real samples.

Table 9. Enzymatic fluorescent biosensors for the detection of organophosphate pesticides (OPs).

Enzyme	Transducer element	Detection mechanism	Pesticide	LOD	Ref.
BuChE	Carbon quantum dots (CQDs)	Enzyme-inhibition (turn-off)	Paraoxon	0.18 nM	[80]
Multienzyme AChE BuChE Chy CE1, CE2	quinazolinone derivative	Enzyme-inhibition (turn-off)	-dichlorvos - carbaryl	25 nM 17 µM	[76]
AChE	CdTe QD	Enzyme-inhibition (turn-off)	-paraoxon -dichlorvos -malathion	1.62 fM 75.3 fM 0.23 nM	[81]
AChE	Gold nanoclusters	Enzyme-inhibition (turn-on)	-parathion-methyl	0.53 nM	[82]
ACP	Ce(IV)-based coordination polymer	Enzyme-inhibition (turn-off)	-malathion	0.14 µM	[83]

AChE= acetil cholinesterase; BuChE= butyrylcholinesterase; ACP= acid phosphatase; CE= carboxilesterases.

CONCLUDING REMARKS

The development of innovative analytical tools for sensing pesticides and herbicides in real-time and at low cost is a current and ongoing challenge caused by the indiscriminate usage and high toxicity of these agrochemicals.

Synthetic fluorescent sensors have proven to be successful in the last decade for the identification and quantification of pesticides/herbicides in water and real samples. However, the creation of an optical sensor with an affinity for a target pesticide is not a trivial task and requires a molecular strategy, a synthetic design for chemical affinity, and a good analytical response.

Among luminescent sensors, organic fluorophores, and metal complexes with binding sites available for pesticides are a subgroup with proven and successful applications, mainly due to the high affinity that can be achieved for the analyte and its detailed study at the molecular level.

On the other hand, luminescent MOFs containing diamagnetic transition metals, mainly Zn(II) or lanthanides, are at the forepart as optical sensors due to their simple preparation and their highly modifiable structures. By chemically functionalizing these structures, they can generate a large range of luminescent mechanisms for selective detection of target pesticide.

Finally, the combination of enzyme-based biosensors (*e.g.*, acetylcholinesterase, butyrylcholinesterase, and acid phosphatase) with synthetic fluorophores has been used for quantitative sensing of organophosphate pesticides with a high-selectivity and specificity.

Among these systems, biosensors constitute a highly efficient strategy, particularly due to their low detection limits, which are highly desired in real samples, such as water bodies, lakes, rivers, or irrigation water. In contrast, biosensors still have some drawbacks, such as their expensive purification, as well as poor thermal stability and efficiency, which are highly dependent on pH and temperature.

These drawbacks are not present in molecular sensors and MOFs, but improving the detection efficiency, mainly their detection limit of pesticides, remains an ongoing challenge.

QUESTIONS RELATED TO THE TEXT

1.- What are the main types of pesticides discussed in the chapter that are detected using luminescent (bio)sensors, and why is their detection important?

2.- Describe the key advantages of using fluorescent organic sensors and metal-organic frameworks (MOFs) for pesticide detection in real samples.

3.- How do luminescent biosensors based on enzymes, such as acetylcholinesterase, function in the detection of organophosphate pesticides?

4.- What are the challenges associated with detecting pesticides in aqueous environments, and how do luminescent (bio)sensors address these challenges?

5.- Explain the role of macrocycle-based sensors in detecting pesticides in water samples. How do their structural properties contribute to their effectiveness?

6.- What are the potential drawbacks of using biosensors for pesticide detection in real samples, as mentioned in the chapter, and how might these be mitigated?

AUTHORS' CONTRIBUTION

María K. Salomón-Flores led the development and writing of sections on "Fluorescent Metal-Free Organic Sensors for Pesticides in Real Samples," including detailed examples on "Macrocycle-Based Sensors", "Organic Acyclic Sensors", and "Organic Polymers for Pesticide Detection". She also contributed significantly to data analysis and interpretation related to these topics.

Iván J. Bazany-Rodríguez was responsible for the section on "Luminescent Metal Molecular Complexes for Chemodetection of Organophosphorus Pesticides" and conducted the literature review and synthesis for this area, highlighting recent advances and practical applications in pesticide detection.

Helen Paola Toledo-Jaldin authored the section on "Luminescent Metal-Organic Frameworks (LMOFs) for Sensing of Pesticides in Real Samples" and focused on the design and characterization of LMOFs, as well as their efficacy in detecting various pesticide types.

Juan Pablo León-Gómez contributed to the sections on "Luminescent Biosensors of Pesticides in Real Samples" and "Fluorescent Enzymatic-Based Biosensors" and provided insights into the biochemical mechanisms and technological innovations related to biosensor development.

Alejandro Dorazco-González coordinated the overall structure and flow of the chapter, including drafting the "Introduction", "Abstract", "Keywords", and "Conclusion" sections and ensured coherence and consistency across all sections, integrating contributions from all authors.

ACKNOWLEDGMENTS

María K. Salomón-Flores, Iván J. Bazany-Rodríguez, Helen P. Toledo-Jaldin and Juan Pablo León-Gómez are grateful to CONAHCYT Mexico for scholarship CVU: 848759, 662748, 516659 and 623898, respectively.

LIST OF ABBREVIATIONS

ΔA	Differential Absorbance
AC	Acridine
ACh	Acetylcholine
AChE	Acetylcholinesterase
AD	Acridine
BODIPY	Boron-dipyrromethene

C4C2	Cumarin-appended calix [4]arene
CHCl₃	Chloroform
CNP	N, N-dimethyl-4-(3-(pyrazin-2-ilimino)prop-1-en-1-yl)aniline
DCN	2,6-dichloro-4-nitroaniline
DD	Dodine
DMF	N,N′-dimethylformamide
DMSO	Dimethyl Sulfoxide
DNA	Deoxyribonucleic Acid
DTP	O,O-dimethyl thiophosphate
EtOH	Ethanol
Fena	Fenamiphos
FITC	Fluorescein Isothiocyanate
fM	Femtomolar
Glyp	Glyphosate
IDA	Indicator Displacement Assay
K	Association Constant
λ_{em}	Emission Length
λ_{ex}	Excitation Length
LMOFS	Luminescent Metal Organic Frameworks
LOD	Limit of Detection
Mal	Malathion
MeCN	Acetonitrile
MeOH	Methanol
µM	Micromolar
mL	Milliliter
Mm	Millimeter
MOFs	Metal Organic Frameworks
MP	Methyl Parathion
Nm	Nanometers
nM	Nanomolar
OPH	Organophosphate Hydrolase
OPs	Organophosphorus Pesticides
PBS	Phosphate-Buffered Saline
PEI	Polyethyleneimine
pH	Potential Hydrogen

Ppm	Parts per Million
PQ	Paraquat
PyP5	Pyridine pillar [5]arene
Q [10]	Cucurbit [10]uril
Rpm	Revolutions per Minute
RSD	Relative Standard Deviation
S_0	Singlet Basal State
S_1	First Excited Singlet State
Si NP	Silica Nanoparticles
T_1	Triplet State
THF	Tetrahydrofuran
UV/Vis	Ultraviolet–Visible

REFERENCES

[1] Chow R, Scheidegger R, Doppler T, Dietzel A, Fenicia F, Stamm C. A review of long-term pesticide monitoring studies to assess surface water quality trends. Water Res X 2020; 9: 100064.
[http://dx.doi.org/10.1016/j.wroa.2020.100064] [PMID: 32995734]

[2] Carazo-Rojas E, Pérez-Rojas G, Pérez-Villanueva M, *et al.* Pesticide monitoring and ecotoxicological risk assessment in surface water bodies and sediments of a tropical agro-ecosystem. Environ Pollut 2018; 241: 800-9.
[http://dx.doi.org/10.1016/j.envpol.2018.06.020] [PMID: 29909306]

[3] Rosales-Vázquez LD, Dorazco-González A, Sánchez-Mendieta V. Luminescent metal-organic frameworks for sensing of toxic organic pollutants in water and real samples. 2022.
[http://dx.doi.org/10.1016/B978-0-323-99425-5.00005-0]

[4] Pathak VM, Verma VK, Rawat BS, *et al.* Current status of pesticide effects on environment, human health and it's eco-friendly management as bioremediation: A comprehensive review. Front Microbiol 2022; 13: 962619.
[http://dx.doi.org/10.3389/fmicb.2022.962619] [PMID: 36060785]

[5] Toledo-Jaldin HP, Pinzón-Vanegas C, Blanco-Flores A, *et al.* Pesticides luminescent sensing by a Tb^{3+}-doped Zn metal-organic framework with selectivity towards parathion. Environ Pollut 2024; 343: 123195.
[http://dx.doi.org/10.1016/j.envpol.2023.123195] [PMID: 38142811]

[6] Vargas-Pérez M, Domínguez I, González FJE, Frenich AG. Application of full scan gas chromatography high resolution mass spectrometry data to quantify targeted-pesticide residues and to screen for additional substances of concern in fresh-food commodities. J Chromatogr A 2020; 1622: 461118.
[http://dx.doi.org/10.1016/j.chroma.2020.461118] [PMID: 32307105]

[7] Beneito-Cambra M, Gilbert-López B, Moreno-González D, *et al.* Ambient (desorption/ionization) mass spectrometry methods for pesticide testing in food: a review. Anal Methods 2020; 12(40): 4831-52.
[http://dx.doi.org/10.1039/D0AY01474E] [PMID: 33000770]

[8] Valdes-García J, Zamora-Moreno J, Salomón-Flores MK, *et al.* Fluorescence Sensing of Monosaccharides by Bis-boronic Acids Derived from Quinolinium Dicarboxamides: Structural and Spectroscopic Studies. J Org Chem 2023; 88(4): 2174-89.

[http://dx.doi.org/10.1021/acs.joc.2c02590] [PMID: 36735858]

[9]   Shan P, Hu J, Liu M, Tao Z, Xiao X, Redshaw C. Progress in host–guest macrocycle/pesticide research: Recognition, detection, release and application. Coord Chem Rev 2022; 467: 214580.
[http://dx.doi.org/10.1016/j.ccr.2022.214580]

[10]   Zhou J, Jia J, He J, Li J, Cai J. Cyclodextrin Inclusion Complexes and Their Application in Food Safety Analysis: Recent Developments and Future Prospects. Foods 2022; 11(23): 3871.
[http://dx.doi.org/10.3390/foods11233871] [PMID: 36496679]

[11]   Sanabria Español E, Maldonado M. Host–Guest Recognition of Pesticides by Calixarenes. Crit Rev Anal Chem 2019; 49(5): 383-94.
[http://dx.doi.org/10.1080/10408347.2018.1534200] [PMID: 30753109]

[12]   Kaifer AE. Portal Effects on the Stability of Cucurbituril Complexes. Isr J Chem 2018; 58(3-4): 244-9.
[http://dx.doi.org/10.1002/ijch.201700097]

[13]   Cao D, Kou Y, Liang J, Chen Z, Wang L, Meier H. A facile and efficient preparation of pillararenes and a pillarquinone. Angew Chem Int Ed 2009; 48(51): 9721-3.
[http://dx.doi.org/10.1002/anie.200904765] [PMID: 19924749]

[14]   Ogoshi T, Kanai S, Fujinami S, Yamagishi T, Nakamoto Y. para-Bridged symmetrical pillar[5]arenes: their Lewis acid catalyzed synthesis and host-guest property. J Am Chem Soc 2008; 130(15): 5022-3.
[http://dx.doi.org/10.1021/ja711260m] [PMID: 18357989]

[15]   Levine M. Fluorescence-Based Sensing of Pesticides Using Supramolecular Chemistry. Front Chem 2021; 9: 616815.
[http://dx.doi.org/10.3389/fchem.2021.616815] [PMID: 33937184]

[16]   Dragone M, Shitaye G, D'Abrosca G, *et al.* Inclusions of Pesticides by β-Cyclodextrin in Solution and Solid State: Chlorpropham, Monuron, and Propanil. Molecules 2023; 28(3): 1331.
[http://dx.doi.org/10.3390/molecules28031331] [PMID: 36771001]

[17]   DiScenza DJ, Lynch J, Miller J, Verderame M, Levine M. Detection of Organochlorine Pesticides in Contaminated Marine Environments *via* Cyclodextrin-Promoted Fluorescence Modulation. ACS Omega 2017; 2(12): 8591-9.
[http://dx.doi.org/10.1021/acsomega.7b00991] [PMID: 30023587]

[18]   Feng J, Yang G, Mei Y, *et al.* Macroscopic visual detection of phoxim by calix[4]arene-based host-guest chemistry. Sens Actuators B Chem 2018; 271: 264-70.
[http://dx.doi.org/10.1016/j.snb.2018.05.107]

[19]   Xu WT, Luo Y, Zhao WW, *et al.* Detecting Pesticide Dodine by Displacement of Fluorescent Acridine from Cucurbit[10]uril Macrocycle. J Agric Food Chem 2021; 69(1): 584-91.
[http://dx.doi.org/10.1021/acs.jafc.0c05577] [PMID: 33377764]

[20]   Zhu XY, Lu Y, Xu WT, Xiao X. A fluorescence sensor based on water-soluble pillar[5]arene and acridine hydrochloride for the detection of pesticide molecules. Microchem J 2024; 196: 109651.
[http://dx.doi.org/10.1016/j.microc.2023.109651]

[21]   Aydin Z, Keskinateş M, Akın Ş, Keleş H, Keleş M. A novel fluorescent sensor based on an enzyme-free system for highly selective and sensitive detection of glyphosate and malathion in real samples. J Photochem Photobiol Chem 2023; 435: 114340.
[http://dx.doi.org/10.1016/j.jphotochem.2022.114340]

[22]   Zhao Z, Zhang F, Zhang Z. A facile fluorescent "turn-off" method for sensing paraquat based on pyranine-paraquat interaction. Spectrochim Acta A Mol Biomol Spectrosc 2018; 199: 96-101.
[http://dx.doi.org/10.1016/j.saa.2018.03.042] [PMID: 29573700]

[23]   Bajpai P. Polymer Chemistry.Biermann's Handbook of Pulp and Paper. 3rd ed. Elsevier 2018; pp. 373-80.
[http://dx.doi.org/10.1016/B978-0-12-814238-7.00018-0]

[24] Zhang B, Yan J, Shang Y, Wang Z. Synthesis of Fluorescent Micro- and Mesoporous Polyaminals for Detection of Toxic Pesticides. Macromolecules 2018; 51(5): 1769-76.
[http://dx.doi.org/10.1021/acs.macromol.7b02669]

[25] Zhang B, Li B, Wang Z. Creation of Carbazole-Based Fluorescent Porous Polymers for Recognition and Detection of Various Pesticides in Water. ACS Sens 2020; 5(1): 162-70.
[http://dx.doi.org/10.1021/acssensors.9b01954] [PMID: 31927991]

[26] Chow CF, Zheng A, Huang M, Shen C. The power of dissociation: development of displacement assays for chemosensing and latent catalytic systems. Mater Chem Front 2020; 4(5): 1328-39.
[http://dx.doi.org/10.1039/C9QM00639G]

[27] Kaur K, Saini R, Kumar A, *et al.* Chemodosimeters: An approach for detection and estimation of biologically and medically relevant metal ions, anions and thiols. Coord Chem Rev 2012; 256(17-18): 1992-2028.
[http://dx.doi.org/10.1016/j.ccr.2012.04.013]

[28] Kolesnichenko IV, Anslyn EV. Practical applications of supramolecular chemistry. Chem Soc Rev 2017; 46(9): 2385-90.
[http://dx.doi.org/10.1039/C7CS00078B] [PMID: 28317053]

[29] Chow CF, Tang Q, Gong CB, Mung SWY. Indicator-catalyst displacement assay for tandem detection and signal amplification of dimethoate organophosphate pesticide. Sens Actuators B Chem 2024; 405: 135335.
[http://dx.doi.org/10.1016/j.snb.2024.135335]

[30] Chow CF, Ho KYF, Gong CB. Synthesis of a New Bimetallic Re(I)–NCS–Pt(II) Complex as Chemodosimetric Ensemble for the Selective Detection of Mercapto-Containing Pesticides. Anal Chem 2015; 87(12): 6112-8.
[http://dx.doi.org/10.1021/acs.analchem.5b00684] [PMID: 26039794]

[31] Zheng A, Shen C, Tang Q, Gong CB, Chow CF. Catalytic Chemosensing Assay for Selective Detection of Methyl Parathion Organophosphate Pesticide. Chemistry 2019; 25(41): 9643-9.
[http://dx.doi.org/10.1002/chem.201901656] [PMID: 31017704]

[32] Sun F, Yang L, Li S, *et al.* New Fluorescent Probes for the Sensitive Determination of Glyphosate in Food and Environmental Samples. J Agric Food Chem 2021; 69(43): 12661-73.
[http://dx.doi.org/10.1021/acs.jafc.1c05246] [PMID: 34672544]

[33] Liu Q, Li S, Wang Y, *et al.* Sensitive fluorescence assay for the detection of glyphosate with NAC Cu^{2+} complex. Sci Total Environ 2023; 882: 163548.
[http://dx.doi.org/10.1016/j.scitotenv.2023.163548] [PMID: 37080305]

[34] Guan J, Yang J, Zhang Y, *et al.* Employing a fluorescent and colorimetric picolyl-functionalized rhodamine for the detection of glyphosate pesticide. Talanta 2021; 224: 121834.
[http://dx.doi.org/10.1016/j.talanta.2020.121834] [PMID: 33379052]

[35] Kang Z, Zhang Z, Zhang Y, Chen S, Wang J, Yuan MS. Di-(2-picolyl)amine functionalized tetraphenylethylene as multifunctional chemosensor. Anal Chim Acta 2022; 1196: 339543.
[http://dx.doi.org/10.1016/j.aca.2022.339543] [PMID: 35151401]

[36] Liu Q, Yu Y, Wu M, Yan X, Wu W, You J. Synthesis and application of a dual-functional fluorescent probe for sequential recognition of Zn^{2+} and glyphosate. Spectrochim Acta A Mol Biomol Spectrosc 2023; 303: 123221.
[http://dx.doi.org/10.1016/j.saa.2023.123221] [PMID: 37544213]

[37] Cao X, You J, Liu Q, Liu B, Yu Y, Wu W. A dual-functional fluorescent sensor based on dihydropyrazole derivative for successive detection of Cu2+ and glyphosate and its applications. Mater Today Commun 2024; 38: 107975.
[http://dx.doi.org/10.1016/j.mtcomm.2023.107975]

[38] Shao Q, Jiang C, Chen X, *et al.* Sensing of organophosphorus pesticides by fluorescent complexes

based on purine-hydrazone receptor and copper (II) and its application in living-cells imaging. Spectrochim Acta A Mol Biomol Spectrosc 2023; 296: 122676.
[http://dx.doi.org/10.1016/j.saa.2023.122676] [PMID: 37031483]

[39] Chen X, Mao Y, Wang A, *et al.* Synthesis and application of purine-based fluorescence probe for continuous recognition of Cu^{2+} and glyphosate. Spectrochim Acta A Mol Biomol Spectrosc 2024; 304: 123291.
[http://dx.doi.org/10.1016/j.saa.2023.123291] [PMID: 37639808]

[40] Tao X, Mao Y, Alam S, *et al.* Sensitive fluorescence detection of glyphosate and glufosinate ammonium pesticides by purine-hydrazone-Cu^{2+} complex. Spectrochim Acta A Mol Biomol Spectrosc 2024; 314: 124226.
[http://dx.doi.org/10.1016/j.saa.2024.124226] [PMID: 38560950]

[41] Zhao S, Shi L, Zhang X, Sun X, Zhu W, Yu L. An on–off–on fluorescent probe for the detection of glyphosate based on a Cu^{2+}-assisted squaraine dye sensor. Anal Methods 2024; 16(9): 1341-6.
[http://dx.doi.org/10.1039/D3AY02128A] [PMID: 38334227]

[42] Sun F, Ye XL, Wang YB, *et al.* NPA-Cu^{2+} Complex as a Fluorescent Sensing Platform for the Selective and Sensitive Detection of Glyphosate. Int J Mol Sci 2021; 22(18): 9816.
[http://dx.doi.org/10.3390/ijms22189816] [PMID: 34575982]

[43] Che S, Zhuge Y, Shao X, Peng X, Fu H, She Y. A fluorescence ionic probe utilizing Cu^{2+} assisted competition for detecting glyphosate abused in green tea. Food Chem 2024; 447: 138859.
[http://dx.doi.org/10.1016/j.foodchem.2024.138859] [PMID: 38479145]

[44] Che S, Zhuge Y, Peng X, *et al.* An ion synergism fluorescence probe via Cu^{2+} triggered competition interaction to detect glyphosate. Food Chem 2024; 448: 139021.
[http://dx.doi.org/10.1016/j.foodchem.2024.139021] [PMID: 38574711]

[45] Wang X, Sakinati M, Yang Y, *et al.* The construction of a CND/Cu^{2+} fluorescence sensing system for the ultrasensitive detection of glyphosate. Anal Methods 2020; 12(4): 520-7.
[http://dx.doi.org/10.1039/C9AY02303H]

[46] Conti L, Flore N, Formica M, *et al.* Glyphosate and AMPA binding by two polyamino-phenolic ligands and their dinuclear Zn(II) complexes. Inorg Chim Acta 2021; 519: 120261.
[http://dx.doi.org/10.1016/j.ica.2021.120261]

[47] Bazany-Rodríguez IJ, Gómez-Vidales V, Bautista-Renedo JM, González-Rivas N, Dorazco-González A, Thangarasu P. Selective chemosensing of organophosphorus pesticide ethyl parathion explored by a luminescent Ru(III)-Salophen complex in water. Dyes Pigments 2023; 210: 110916.
[http://dx.doi.org/10.1016/j.dyepig.2022.110916]

[48] Jennings LB, Shuvaev S, Fox MA, Pal R, Parker D. Selective signalling of glyphosate in water using europium luminescence. Dalton Trans 2018; 47(45): 16145-54.
[http://dx.doi.org/10.1039/C8DT03823F] [PMID: 30378619]

[49] Valizadeh Harzand F, Mousavi Nejad SN, Babapoor A, *et al.* Recent advances in metal-organic framework (MOF) asymmetric membranes/composites for biomedical applications. Symmetry (Basel) 2023; 15(2): 403.
[http://dx.doi.org/10.3390/sym15020403]

[50] Rana A, Mir NUD, Banik A, Hazra A, Biswas S. Design of functionalized luminescent MOF sensor for the precise monitoring of tuberculosis drug and neonicotinoid pesticide from human body-fluids and food samples to protect health and environment. J Mater Chem C Mater Opt Electron Devices 2024; 12(3): 1030-9.
[http://dx.doi.org/10.1039/D3TC03712F]

[51] Li J, Liu M, Li J, Liu X. A MOF-on-MOF composite encapsulating sensitized Tb(III) as a built-in self-calibrating fluorescent platform for selective sensing of F ions. Talanta 2023; 259: 124521.
[http://dx.doi.org/10.1016/j.talanta.2023.124521] [PMID: 37058939]

[52]   Wang L, Hu Z, Wu S, Pan J, Xu X, Niu X. A peroxidase-mimicking Zr-based MOF colorimetric sensing array to quantify and discriminate phosphorylated proteins. Anal Chim Acta 2020; 1121: 26-34.
[http://dx.doi.org/10.1016/j.aca.2020.04.073] [PMID: 32493586]

[53]   Gan YL, Huang KR, Li YG, *et al.* Synthesis, structure and fluorescent sensing for nitrobenzene of a Zn-based MOF. J Mol Struct 2021; 1223: 129217.
[http://dx.doi.org/10.1016/j.molstruc.2020.129217]

[54]   Zhao L, Song X, Ren X, *et al.* Ultrasensitive near-infrared electrochemiluminescence biosensor derived from Eu-MOF with antenna effect and high efficiency catalysis of specific CoS_2 hollow triple shelled nanoboxes for procalcitonin. Biosens Bioelectron 2021; 191: 113409.
[http://dx.doi.org/10.1016/j.bios.2021.113409] [PMID: 34146971]

[55]   Cheng J, Li B, He J, Wang P. Zr-MOF-induced smart accumulation enables surface-enhanced Raman spectroscopic detection of dioxin at ppt level in food samples. ACS Sens 2023; 8(5): 2115-23.
[http://dx.doi.org/10.1021/acssensors.3c00639] [PMID: 37183968]

[56]   Pal TK. Metal–organic framework (MOF)-based fluorescence "turn-on" sensors. Mater Chem Front 2023; 7(3): 405-41.
[http://dx.doi.org/10.1039/D2QM01070D]

[57]   Rath BB, Vittal JJ. Water Stable Zn(II) Metal–Organic Framework as a Selective and Sensitive Luminescent Probe for Fe(III) and Chromate Ions. Inorg Chem 2020; 59(13): 8818-26.
[http://dx.doi.org/10.1021/acs.inorgchem.0c00545] [PMID: 32501007]

[58]   Singha DK, Majee P, Mandal S, Mondal SK, Mahata P. Detection of Pesticides in Aqueous Medium and in Fruit Extracts Using a Three-Dimensional Metal–Organic Framework: Experimental and Computational Study. Inorg Chem 2018; 57(19): 12155-65.
[http://dx.doi.org/10.1021/acs.inorgchem.8b01767] [PMID: 30221511]

[59]   He K, Li Z, Wang L, *et al.* A Water-Stable Luminescent Metal–Organic Framework for Rapid and Visible Sensing of Organophosphorus Pesticides. ACS Appl Mater Interfaces 2019; 11(29): 26250-60.
[http://dx.doi.org/10.1021/acsami.9b06151] [PMID: 31251555]

[60]   Ghosh S, Mal D, Mukherjee S, Biswas S. Sustainable Fabrication of an Eco-Friendly, Reusable Chitosan@Cotton@MOF Composite Sensor for 2,4-Dichlorophenoxyacetic Acid Herbicide and Nitroxoline Antibiotic. ACS Sustain Chem& Eng 2023; 11(35): 13179-86.
[http://dx.doi.org/10.1021/acssuschemeng.3c03517]

[61]   Xu X, Guo Y, Wang X, *et al.* Sensitive detection of pesticides by a highly luminescent metal-organic framework. Sens Actuators B Chem 2018; 260: 339-45.
[http://dx.doi.org/10.1016/j.snb.2018.01.075]

[62]   Wang L, He K, Quan H, Wang X, Wang Q, Xu X. A luminescent method for detection of parathion based on zinc incorporated metal-organic framework. Microchem J 2020; 153: 104441.
[http://dx.doi.org/10.1016/j.microc.2019.104441]

[63]   Wang XQ, Feng DD, Tang J, *et al.* A water-stable zinc(II)–organic framework as a multiresponsive luminescent sensor for toxic heavy metal cations, oxyanions and organochlorine pesticides in aqueous solution. Dalton Trans 2019; 48(44): 16776-85.
[http://dx.doi.org/10.1039/C9DT03195B] [PMID: 31674607]

[64]   Wang W, Yang F, Yang Y, Wang YY, Liu B. Rational synthesis of a stable rod MOF for ultrasensitive detection of nitenpyram and nitrofurazone in natural water systems. J Agric Food Chem 2022; 70(50): 15682-92.
[http://dx.doi.org/10.1021/acs.jafc.2c05780] [PMID: 36469812]

[65]   Li J, Zhu M, Zhang Y, Gao E, Wu S. A new visual and stable fluorescent Cu-MOF as a dual-function sensor for glyphosate and $Cr_2O_7{}^{2-}$. New J Chem 2022; 46(41): 19808-16.
[http://dx.doi.org/10.1039/D2NJ03186H]

[66] Zhang Z, Zhang L, Han P, Liu Q. A luminescent probe based on terbium-based metal–organic frameworks for organophosphorus pesticides detection. Mikrochim Acta 2022; 189(11): 438.
[http://dx.doi.org/10.1007/s00604-022-05508-x] [PMID: 36319758]

[67] Wang XQ, Ma X, Feng D, *et al.* Four Novel Lanthanide(III) Metal–Organic Frameworks: Tunable Light Emission and Multiresponsive Luminescence Sensors for Vitamin B $_6$ and Pesticides. Cryst Growth Des 2021; 21(5): 2889-97.
[http://dx.doi.org/10.1021/acs.cgd.1c00080]

[68] Bazin I, Tria SA, Hayat A, Marty JL. New biorecognition molecules in biosensors for the detection of toxins. Biosens Bioelectron 2017; 87: 285-98.
[http://dx.doi.org/10.1016/j.bios.2016.06.083] [PMID: 27568847]

[69] Nasir M, Nawaz MH, Latif U, Yaqub M, Hayat A, Rahim A. An overview on enzyme-mimicking nanomaterials for use in electrochemical and optical assays. Mikrochim Acta 2017; 184(2): 323-42.
[http://dx.doi.org/10.1007/s00604-016-2036-8]

[70] Bhatia D, Paul S, Acharjee T, Ramachairy SS. Biosensors and their widespread impact on human health. Sensors International 2024; 5: 100257.
[http://dx.doi.org/10.1016/j.sintl.2023.100257]

[71] Pundir CS, Malik A, Preety . Bio-sensing of organophosphorus pesticides: A review. Biosens Bioelectron 2019; 140: 111348.
[http://dx.doi.org/10.1016/j.bios.2019.111348] [PMID: 31153016]

[72] Octobre G, Delprat N, Doumèche B, Leca-Bouvier B. Herbicide detection: A review of enzyme- and cell-based biosensors. Environ Res 2024; 249: 118330.
[http://dx.doi.org/10.1016/j.envres.2024.118330] [PMID: 38341074]

[73] Rogers K, Cao CJ, Valdes JJ, Eldefrawi AT, Eldefrawi ME. Acetylcholinesterase fiber-optic biosensor for detection of anticholinesterases. Fundam Appl Toxicol 1991; 16(4): 810-20.
[http://dx.doi.org/10.1016/0272-0590(91)90166-2] [PMID: 1909249]

[74] Jiang W, Yang Z, Tong F, *et al.* Two birds with one stone: An enzyme-regulated ratiometric fluorescent and photothermal dual-mode probe for organophosphorus pesticide detection. Biosens Bioelectron 2023; 224: 115074.
[http://dx.doi.org/10.1016/j.bios.2023.115074] [PMID: 36638562]

[75] Liang B, Han L. Displaying of acetylcholinesterase mutants on surface of yeast for ultra-trace fluorescence detection of organophosphate pesticides with gold nanoclusters. Biosens Bioelectron 2020; 148(15): 111825.
[http://dx.doi.org/10.1016/j.bios.2019.111825] [PMID: 31677527]

[76] Guo WY, Fu YX, Liu SY, *et al.* Multienzyme-Targeted Fluorescent Probe as a Biosensing Platform for Broad Detection of Pesticide Residues. Anal Chem 2021; 93(18): 7079-85.
[http://dx.doi.org/10.1021/acs.analchem.1c00553] [PMID: 33906355]

[77] Singh BK. Organophosphorus-degrading bacteria: ecology and industrial applications. Nat Rev Microbiol 2009; 7(2): 156-64.
[http://dx.doi.org/10.1038/nrmicro2050] [PMID: 19098922]

[78] Simonian AL, Good TA, Wang SS, Wild JR. Nanoparticle-based optical biosensors for the direct detection of organophosphate chemical warfare agents and pesticides. Anal Chim Acta 2005; 534(1): 69-77.
[http://dx.doi.org/10.1016/j.aca.2004.06.056] [PMID: 17723332]

[79] Mishra A, Kumar J, Melo JS. An optical microplate biosensor for the detection of methyl parathion pesticide using a biohybrid of Sphingomonas sp. cells-silica nanoparticles. Biosens Bioelectron 2017; 87: 332-8.
[http://dx.doi.org/10.1016/j.bios.2016.08.048] [PMID: 27573300]

[80] Wu X, Song Y, Yan X, *et al.* Carbon quantum dots as fluorescence resonance energy transfer sensors

for organophosphate pesticides determination. Biosens Bioelectron 2017; 94: 292-7.
[http://dx.doi.org/10.1016/j.bios.2017.03.010] [PMID: 28315592]

[81] Korram J, Dewangan L, Karbhal I, *et al.* CdTe QD-based inhibition and reactivation assay of acetylcholinesterase for the detection of organophosphorus pesticides. RSC Advances 2020; 10(41): 24190-202.
[http://dx.doi.org/10.1039/D0RA03055D] [PMID: 35516221]

[82] Luo QJ, Li ZG, Lai JH, Li FQ, Qiu P, Wang XL. An on–off–on gold nanocluster-based fluorescent probe for sensitive detection of organophosphorus pesticides. RSC Advances 2017; 7(87): 55199-205.
[http://dx.doi.org/10.1039/C7RA11835J]

[83] Liu P, Zhao M, Zhu H, *et al.* Dual-mode fluorescence and colorimetric detection of pesticides realized by integrating stimulus-responsive luminescence with oxidase-mimetic activity into cerium-based coordination polymer nanoparticles. J Hazard Mater 2022; 423(A): 127077.
[http://dx.doi.org/10.1016/j.jhazmat.2021.127077]

CHAPTER 12

Nanotechnology for Water and Soil Conservation

María Alejandra Istúriz-Zapata[1,*] and **Alfredo Jiménez-Pérez**[1]

[1] *Centro de Desarrollos de Productos Bióticos, Instituto Politécnico Nacional (IPN), Yautepec, Morelos, Mexico*

Abstract: One of the great challenges of our time is to maintain and improve the quality of water, air, and soil so that the Earth can continue to support life. Human activity over the last two centuries has released large quantities of pollutants into the environment, damaging ecosystems as a result of the economic activities, including agriculture, that have enabled human development. These challenges include mitigating the effects of climate change, the unsustainable use of natural resources, and the excessive use of chemicals. Water scarcity is a global threat to life. Nanotechnology offers the potential to change agriculture and water management. Nanotechnology is the science of manipulating matter at the nanoscale, *i.e.*, the scale of 10^{-9} meters, by controlling its shape and size, which can be applied to the design, characterization, fabrication, and application of nanometric-sized structures, devices, and systems. This versatility spans from medicine to agriculture, offering efficient, flexible, and multifunctional processes. Nanoparticles can penetrate microscopic spaces, and their application in environmental sciences includes restoring soil quality and fertility and improving fertilizer efficiency. They, therefore, have great potential for sustainable agriculture. Nanomaterials can be used for water treatment because their physicochemical properties are entirely different from those of conventional-size materials. These properties enhance their efficiency and make methods such as adsorption/oxidation more powerful. The application of nanotechnology is expected to provide an efficient and economical means of supplying drinking water and removing contaminants from soils. This chapter discusses how nanotechnology can be used in water and soil conservation.

Keywords: Nanoparticles, Pollutants, Purification, Remediation, Soil quality, Water.

INTRODUCTION

Nanotechnology is the science of creating structures of a required shape at a scale of 10^{-9} meters by manipulating the physical and chemical properties of a substance at molecular levels [1 - 5].

* **Corresponding author María Alejandra Istúriz-Zapata:** Centro de Desarrollos de Productos Bióticos, Instituto Politécnico Nacional (IPN), Yautepec, Morelos, Mexico; Tel: +527353942020; E-mail: aljimenez@ipn.mx

Israel Valencia Quiroz (Ed.)
All rights reserved-© 2025 Bentham Science Publishers

Nanotechnology can be applied to design materials and equipment of minute size, conferring them greater reactivity contact surface. Nanomaterials have high reactivity, functionalization, a large specific surface area, and size-dependent properties [1 - 6].

Quantic effects and the increase in relative surface area are the two main characteristics of nanomaterials. Other important characteristics are a) morphology, b) appearance/size ratio, c) hydrophobicity, d) solubility/toxic species release ratio, d) surface area/rugosity ratio, e) induction of reactive species synthesis, f) structure/composition, g) binding sites, and h) dispersion/aggregation rate (Z-potential) [4, 7]. The physical properties of a redox process are important, but at the nanoparticle level, they are even more important [8, 9].

The adaptation of diverse practical methodologies for the synthesis of nanoparticles by modulating their shapes, sizes, and chemical compositions has yielded a multitude of multidimensional structures (1-D, 2-D, and 3-D), enabling their interaction with organic molecules and eliciting combined behaviours (Fig. 1). The two principal approaches to nanoparticle synthesis are the top-down and bottom-up approaches [10, 11].

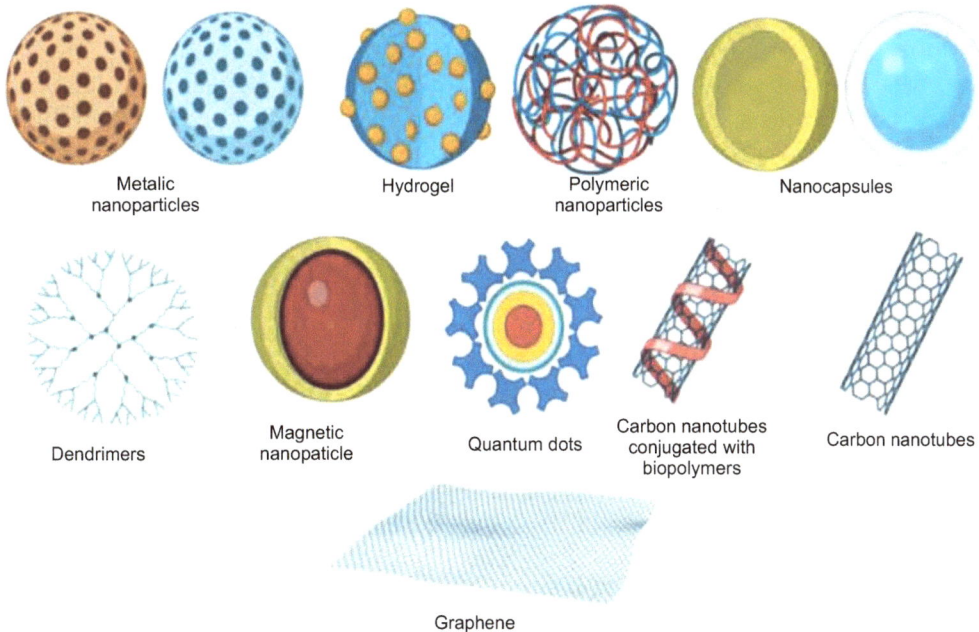

Fig. (1). Different nanoparticle types employed in water and soil conservation.

Nanotechnology processes are highly efficient, flexible, and multifunctional; they are a suitable option for retrofitting conventional infrastructures and developing high-performance, low-cost treatment solutions that are not dependent on large infrastructures [12, 13]. It can improve water purification processes and even reduce environmental problems. However, further research is needed to ensure it does not lead to additional pollution problems [5]. Nanotechnology has the potential to significantly change agriculture and water management [13 - 16].

This new technology can address long-standing issues such as hazardous waste landfills and is ideal for *in situ* applications [5, 14, 17]. Nanoparticles can penetrate small areas in the subsurface and remain suspended in groundwater due to their tiny size. They may be coated with different materials, such as biopolymers. Their size and coating allow them to travel farther and achieve a more homogeneous distribution. However, in practice, nanomaterials do not travel far from their point of injection [8, 18].

Applications of nanotechnology in environmental sciences include using nanoclays and zeolites to restore soil quality and fertility and improve the effectiveness of fertilizers. Another use is the development of nanopolymer-coated smart seeds programmed to germinate under favorable conditions [6, 17, 19, 20]. Crops are classified according to their nutrient needs, so the use of nanobiosensors with satellite systems allows a more precise supply of nutrients than with current methods [21]. This also enables the more efficient application of nanoherbicides and nanofertilizers, helping to avoid or reduce contamination issues [19]. Therefore, the application of nanotechnology in sustainable agriculture has great potential, especially in developing countries [22].

Nanotechnological methods have proven effective and are widely used in soil remediation. The recent use of nanoparticles for processes such as adsorption, redox reactions, precipitation, and co-precipitation processes has successfully removed contaminants from soil due to their enormous specific surface area [4, 17].

The treatment of water with nanomaterials is beneficial because the physicochemical properties of these materials are entirely different from those of conventional sizes [23]. The use of nanomaterials in conventional water treatment improves their efficiency. Methods such as adsorption, oxidation, and separation have been developed using nanoadsorbents, nanocatalysts, and bioactive nanoparticles, which are more effective due to their size and larger surface area/volume ratio [15, 16, 24].

Nanoparticles are easily added and dispersed in reactors to treat soils, sediments, and solid wastes. Nanoparticles can be immobilized in solid matrices, such as zeolites, carbon, or membranes to treat gaseous effluents [15]. Subsurface injection of Fe nanoparticles enhances the degradation of organochlorine contaminants such as trichloroethane, thereby reducing the toxicity of the pollutants in the soil [15].

Bimetallic Fe/Pd, Fe/Ag, or Zn/Pd nanoparticles are highly reductive and catalyze the transformation of various pollutants, including askarels and polychlorinated biphenyls ($C_2H_{10-n}Cl_n$, where n = 1-10, mainly 2-7), which are highly resistant to microbial degradation and are considered potent carcinogens. In addition, they can modify other halogenated compounds, such as solvents and pesticides, reducing most of the tested halogenates to hydrocarbons, which are degraded by the natural microflora of the site [25]. Fe-NPs are known to reduce a wide range of pollutant compounds, including perchlorates (ClO_4^-), nitrates (NO_3^-), dichromates ($Cr_2O_7^-$), Ni, Hg, and uranium dioxide (UO_2) [15].

The activity of heavy metals in soil is controlled by sorption-desorption reactions with other soil constituents [5]. A wide range of amendments have been used to manipulate the bioavailability of heavy metals and prevent their diffusion in soil by inducing various sorption processes: adsorption to mineral surfaces, formation of stable complexes with organic ligands, surface precipitation, and ion exchange processes [5, 26].

Water Conservation

One of the greatest challenges facing humanity in the coming centuries is to ensure that water is safe to drink [27]. Only 3% of the planet's total water is freshwater, and of this, only 0.5% is potable [28]. Water conservation refers to all those practices, activities, and techniques employed for the conscious and sustainable use of freshwater, as well as for the protection and preservation of all freshwater sources (rivers, lakes, aquifers, groundwater, and wetlands) and the natural ecosystems surrounding them, to ensure their long-term availability [29, 30].

The United Nations established Sustainable Development Goal 6 for water conservation, which states: "*Ensure availability and sustainable management of water and sanitation for all*" (2024) [31]. This goal calls for the application of four principles: 1) separate wastewater from drinking water; 2) provide universal access to drinking water and treat it to remove chemical and biological contaminants; 3) protect and restore freshwater ecosystems; and 4) protect access to and use of water. Therefore, urgent measures are needed for the decontamination, reuse, and conservation of water.

Contaminants in water directly or indirectly affect the health of the ecosystems near the water sources and prevent the supply of vital liquids to people and various economic activities. Therefore, water conservation and availability have become national security issues in some nations in the Americas, Europe, and Africa [24, 28]. Water may contain contaminants of different origins (natural or anthropogenic) and types (organic, inorganic, or microbiological), which may be persistent or recalcitrant [16]. A recalcitrant contaminant is not affected by microorganisms or any degradation process due to its chemical stability. These contaminants include organic and inorganic chemical compounds, pathogenic microorganisms (bacteria, fungi, parasites, and viruses), and water industrial wastes with heavy metals [32]. These pollutants alter the properties of water and inhibit its natural degradation when mixed with water [16, 24].

Water treatment systems vary depending on the type of water to treat. Wastewater treatment requires the implementation of biological, physical, and chemical technologies in four stages [24]. Biological treatment systems are usually the first choice because they use microorganisms to break down and consume certain organic molecules [33]. In other cases, when microorganisms cannot metabolize the contaminants, common physical processes are used for water purification: coagulation and flocculation, deionization, disinfection, dissolved air flotation, filtration, sedimentation, steam distillation, ion exchange, and reverse osmosis [16, 24, 33]. In recent years, plastic or metal screens have also been used to retain floating solids from the water surface [34]. However, conventional water treatment processes can be expensive and release secondary toxic contaminants into the environment [35, 36]. The problem of water pollution goes further because the generation of pollutants exceeds our purification capacity and the current demand for water treatment.

Current water shortages are the outcome of inadequate and disorganized water supply, prolonged droughts, contamination of natural sources, and increased water demand due to high population growth [37]. Contaminated water can be reused, but the lack of affordable cleaning/recycling technologies makes it expensive to use and makes clean water expensive [38, 39]. Nanotechnology enables water recovery, treatment, and conservation [40, 41]. Water treatment has two areas: drinking water treatment and wastewater treatment [42]. Nanotechnology is used in the following processes: a) purification/disinfection; b) decontamination of wastewater and industrial water; c) desalination; d) landfill leachate treatment; and e) water quality sensors (Fig. **2**).

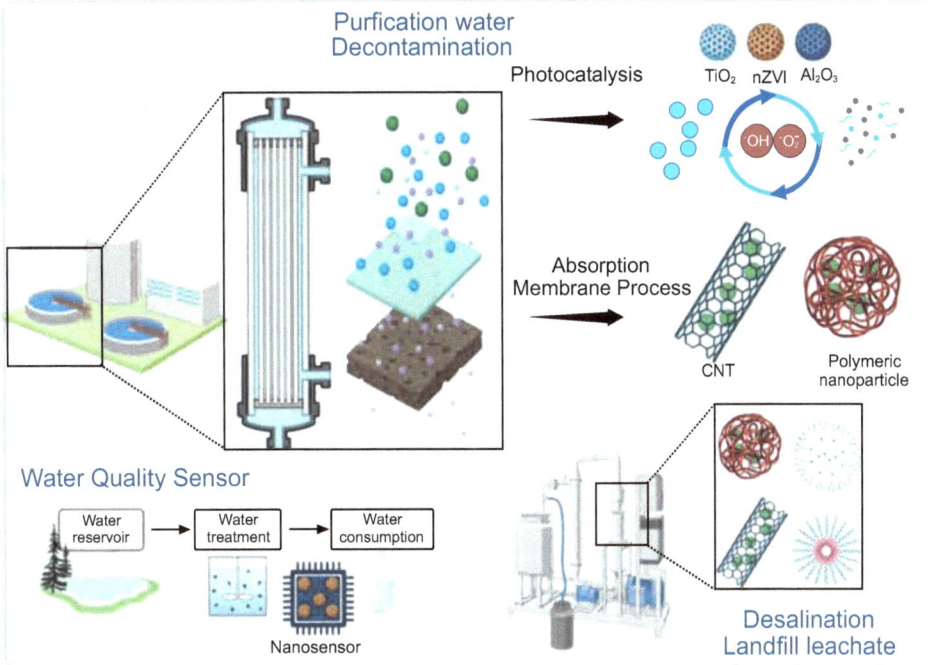

Fig. (2). Applications of nanotechnology using different types of nanoparticles for the treatment of drinking water and wastewater.

The use of nanotechnology has encouraged the development of nanosemiconductors for photocatalysis and photocatalytic nanoparticles [43, 44], nanobiopolymers [45], nanocrystalline zeolites [46], nanoreactive and/or nanofiltering membranes [47], nanoclays [48, 49], nanocatalysts [50, 51], magnetic nanoparticles [52, 53], nanosensors [54], nanoparticles [55, 56], carbon nanotubes [57, 58], dendrimers [59, 60], polymeric nanoparticles [60], nanoparticles with enzymes [61, 62], metallic nanoparticles (zerovalent iron or titanium), graphene, graphene oxide, quantum dots, carbon nanorods, carbon nano-onions, and reduced graphene [46], among others [25, 63 - 65]. Silver-derived materials have been used in water potabilization and disinfection because they enhance catalysis and oxidation processes [3, 42, 66, 67]. Magnetic adsorbents, in particular magnetite (Fe_3O_4), have emerged as a promising material for the purification of water. Its high surface area, supermagnetism, and ease of functionalization make it suitable for the adsorption of a wide range of contaminants. Recently, adsorbents used in water treatment have included carbon-based compounds and metal-organic frameworks (Fig. **3**) [5].

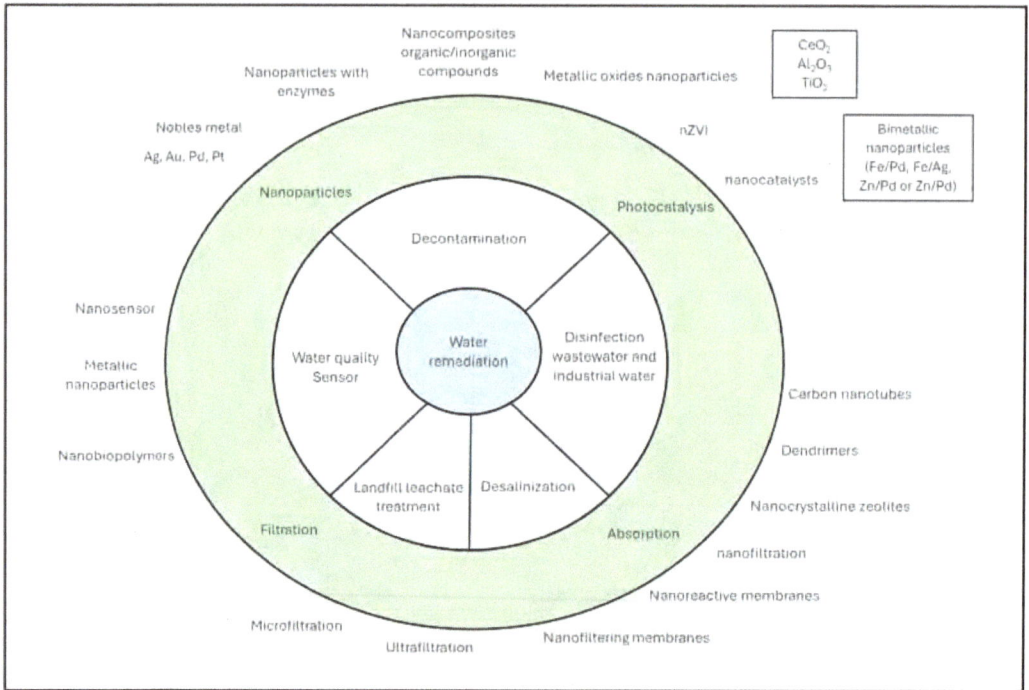

Fig. (3). Water Remediation. Nanotechnology used for water remediation is shown from the inside to the outside; different methods can be used in different areas.

Photocatalysis, an advanced oxidation process, is widely used for water decontamination as it degrades recalcitrant contaminants by forming highly reactive chemical species with low selectivity, such as hydroxyl radicals (-OH$^-$), superoxide (-O$_2^-$), and hydrogen peroxide (H$_2$O$_2$) [23, 68 - 70]. This method is particularly very effective when contaminants are present at very low concentrations [71, 72]. Advanced oxidation processes such as H$_2$O$_2$/UV process, Fenton and photo-Fenton process, ozone process, electrochemical oxidation process, wet air oxidation and supercritical water oxidation process, and photocatalytic process have been employed mainly to process wastewater [73]. The key advantage of nanotechnology is its ability to remove pollutants at low concentrations from wastewater, which can be challenging to eliminate using conventional methods [20].

Reverse photocatalysis efficiently removes organic, inorganic, and microbiological contaminants [23, 62]. It reduces the activation energy in a photochemical reaction using a photocatalyst in the solid state on a solid-liquid interface, acting as an agent capable of producing chemical transformations in the substrate, repeatedly and regenerating itself in each reaction cycle [71, 72, 74]. Therefore, a photocatalyst is a catalyst activated when irradiated by a determined wavelength, being the photocatalysts homogeneous or heterogeneous. Homogeneous photocatalysts include metal salts and organic or organometallic complexes and are difficult to extract from the reaction mixture for reuse [23]. The latter are insoluble solids, such as cerium oxide (CeO_2), aluminum oxide (Al_2O_3), or titanium oxide (TiO_2), which can be easily separated and reused at the end of the photocatalytic process [70, 71].

Metal nanoparticles serve as heterogeneous photocatalysts with high activity and stability, as well as supports for more active phases [52, 53]. Due to their small dimensions, they have large specific areas and numerous exposed active sites, functioning as intermediate homogeneous and heterogeneous catalysts [71]. In the treatment of water with solar photocatalysis, TiO_2 is the most widely used compound, creating nanostructured photocatalysts with higher activity, stability, and selectivity [42, 71, 75].

Photocatalysis promptly removes pollutants and has several advantages, such as: a) without generation of residues or degradation by-products, b) achieves high disinfection rates and low microbial growth rates, c) uses small amounts of the photocatalyst, d) can be recovered and reused, and e) some are activated by sunlight [75]. Depending on the specific energy requirements to achieve electron transfer, they can be activated by UV light alone (*e.g.*, ZnO, TiO_2, ZnO_2, SiO_2, and Al_2O_3), and those activated in the UV-visible light spectrum (*e.g.*, WO_3, CeO_2, Fe_2O_3, Bi_2O_3, and CdS) [71]. Another advantage of using nanometer photocatalysts is that the electrons travel a shorter distance to the semiconductor surface, reducing the probability of material recombination [76].

The materials for photocatalysis must be resistant to photooxidation, chemically inert, and not prone to corrosion processes; therefore, metal oxides and noble metals (Au, Ag, Ir, Hg, Pd, Pt, Rb, Re, Ru, and Os) are suitable for photocatalysis for water purification and disinfection [71, 76]. Also, the shape and dispersion of the metal nanoparticles on the semiconductor surface make it possible to achieve complete mineralization of recalcitrant molecules in a shorter time [44, 71]. Metal nanoparticle-modified semiconductors remove pollutants from water more efficiently.

The immobilization of nanoparticle photocatalysts onto polymers has generated significant attention, like polymer-TiO_2 nanocomposites for water purification [23]. The transition metals oxide (ZnO, SnO, TiO_2) integrated into the organic framework of graphene or nitrides can remove more than 90% of dyes, microbial load, pesticides, and antibiotics, and the graphitic carbon nitride can remove all petrochemicals pollutants [77].

Water disinfection by photocatalysis involves three oxidation processes that compromise the integrity of microbial cells: disruption of cell replication by direct damage to bacterial DNA due to the presence of -OH ions [70, 78], structural damage to the cell wall resulting in the release of cell contents [79], and finally reduction of the enzymatic activity of acetyl coenzyme A, which inhibits cellular respiration and kills the bacterial [70, 79]. A new photocatalysis process called photocatalytic mineralization turns organic substances into inorganic compounds, like CO_2 and H_2O [80].

Nanoparticles with TiO_2 and Ag are the most used in photocatalytic water disinfection. Silver nanoparticles can adhere to the cell wall and membrane, destabilizing the cell function and inactivating pathogens [79]. Other metals, such as Au and Pt, have been employed as catalysts but are less efficient [28, 76, 81]. The use of heterogeneous photocatalysis with TiO_2 and Ag nanoparticles has achieved complete disinfection of water in 2 hours of solar irradiation [71, 74].

When hybrid systems of bimetallic Pb-Au and TiO_2 nanoparticles are used, photocatalysis selectively reduces nitrogenous compounds such as NH_4^+ [66, 70, 72, 75]. These types of photocatalysts produce a few intermediate molecules, such as NO_2^- and NO_3^- ions, and are highly selective; they do not leave a large amount of post-disinfection residues, thus avoiding the production of sludge or waste requiring further treatment [71].

Weathering, rock erosion, and waste from anthropogenic sources such as chemical plants, electroplating, battery industries, pesticides, and fertilizers, among others, pollute water with heavy metals. Adsorption is the simplest and most efficient process for the treatment of industrial effluents and a useful mechanism for the protection of the ecosystem [82]. Activated carbon is the best and most widely used adsorbent for removing heavy metals from water. However, nanoadsorbents offer significant advances in adsorption properties due to their high adsorption kinetics, specific area, and highly active adsorption sites [7, 37, 60, 83]. Nanoadsorbents successfully remove organic compounds and metal ions, and their functionalization can improve their selectivity toward certain contaminants [19, 84]. Photo-, sono-, and electro-catalysis nanomaterials for wastewater treatments inherently provide energy to the nanomaterials to disinfect water [82].

When treating wastewater and drinking water, the hybrid membrane process applies pressure to a membrane with immobilized functionalized materials (*e.g.*, nanocatalysts, chemical oxidants, and phage), eliminating microorganisms *in situ*. The most studied nanocomposition is Ag, followed by graphene and a combined process using phages [82].

Heavy metals are persistent in the environment and have an impact not only on the health and balance of the flora and fauna in the ecosystem but also on the health of humans [83]. Ingestion of a small amount of heavy metals can cause disease and serious metabolic disorders [37].

The mechanical, electrical, optical, physical, and chemical properties of carbon nanotubes (CNTs), along with their large contact surface area, make them ideal for wastewater treatment [57, 58, 85]. These CNTs also serve as a substrate for other adsorbent materials, increasing their efficiency [57, 58, 85, 86]. CNTs have been recognized by scientists as a substitute for activated carbon due to their excellent capacity to adsorb gaseous and liquid phases, such as organic vapors, inorganic contaminants, and various heavy metal ions [7, 57]. This is primarily because their binding sites are more accessible than those of activated carbon [85, 87].

For environmental remediation, CNTs are the best organic adsorbents compared to other carbon-based adsorbents because they function as flexible porous materials, with remarkable adsorption capacity and high removal efficiency of these compounds; however, their efficiency is highly dependent on the total acidity of the medium [19, 57, 71, 83, 88]. The presence of certain organic compounds can increase the adsorption of heavy metals; for example, humic acid (HA), fulvic acid (FA), and hydroxylated and carboxylated fullerenes increase the adsorption of copper (II) and its bioavailability [7, 19, 51]. In addition, -COOH, -OH, and $-NH_2$ functional groups of organic compounds can also form hydrogen bonds with the engraved surface on CNTs, demonstrating a process of chemical sorption of metal ions [89].

One significant drawback in the practical application of CNTs is the difficulty of separating them from the medium. However, magnetic nanotubes can be created by combining them with magnetic nanoparticles. These nanotubes can be easily dispersed in water and can be manipulated due to their weak external magnetic field, allowing them to be separated from water [58, 86, 89]. Therefore, combining the magnetic properties of metal oxides with the adsorption properties of CNTs results in an effective and fast method for separating magnetic adsorbents from aqueous solutions [19]. Their use has been suggested for the adsorption of dioxins by CNTs, being at least three times higher than with activated carbon. Furthermore, nanotubes have a Langmuir constant several orders of magnitude higher than activated carbon [70, 89]. Langmuir adsorption is related to the gas pressure or concentration of a medium on a solid surface at a constant temperature, so it measures the ability of a nanoadsorbent to remove a contaminant from the medium.

In addition, nanoscale metal oxides, such as TIO_2, Fe_2O_3, ZnO_2, and Al_2O_3 have been reported as effective and inexpensive adsorbents for water treatment due to their size and adsorption efficiency, providing a more cost-effective remediation technology [59, 88, 90]. The main way to control adsorption is to form a complex with the surface of nanoscale metal oxides and subject an electron to an oxidation reaction under visible irradiation [6, 91].

Dendrimers are another example of nanostructured materials used as nanoadsorbents [56, 60, 91]. These nanostructured particles are highly branched polymers with controlled architecture and composition. Dendrimers can be employed in ultrafiltration systems in polymeric membranes as chelating agents [56, 59, 60]. Dendrimers can be created to encapsulate zero-valent metals and dissolve them in particular media or adsorb them on surfaces [15, 60]. These organic and inorganic nanocomposites have unique properties to bind specific metal ions and are excellent at adsorbing heavy metal ions, allowing the removal of these contaminants from wastewater [37].

Membrane processes are an effective form of water remediation due to their high separation efficiency, ease of operation as no chemical addition or thermal input is required, no secondary contamination, and no regeneration of spent media [3, 19]. Membrane materials provide a natural balance between selectivity and permeability, which is critical for the performance of membrane filtration systems. The most commonly used polymers for water treatment are polyamide (PA), polyacrylonitrile (PAN), and cellulose acetate (CA) [7].

Membrane processes [30] can be classified into: a) microfiltration to remove suspended solids, protozoa, and bacteria, b) ultrafiltration to remove viruses and colloids, c) nanofiltration to remove hardness, heavy metals, and dissolved organic matter, desalination, water reuse, and production of ultrapure water by reverse osmosis and direct osmosis [19, 75]. Recent advancements in water treatment have led to the development of new membranes incorporating nanomaterials into water filtration membranes, either by mixing or surface grafting [92]. This provides membranes with the required structure and new functionalities, such as high permeability, catalytic reactivity, pollutant degradation, and self-cleaning, in addition to controlling membrane fouling due to the functional groups of the nanoparticles and their hydrophilic properties [19, 35, 47].

Nanocomposites are organic/inorganic compounds whose inorganic phases reach nanometer size [93]. These nanoscale inorganic components include nanotubes, layered silicates, metal nanoparticles (Au and Ag), metal oxides (TiO_2 and Al_2O_3), semiconductors (PbS and CdS), and, most notably, SiO_2. Silica and biopolymer nanocomposites have aroused great interest in both academia and industry [37, 94]. They are among the most frequently cited in the literature and have been used for several purposes. Much effort has also been devoted to the design and fabrication of colloidal silica-biopolymer nanocomposites with tailored morphologies or magnetic activity. Carbon nanocomposites have gained attention for their low preparation cost, high surface area, pores volume, and environmental stability, proving excellent performance in aqueous solution contaminant removal [45, 47, 48, 94 - 97]. The creation of engraved nanocomposites provides additional features, such as metal ion nanoabsorbents, and have been evaluated for the removal of several types of contaminants from water as adsorbents, photocatalysts, magnetic separation, etc [37, 96, 97].

In recent years, a diversity of functional nanomaterials with unique optical, mechanical, chemical, and electronic properties have been synthesized to create nanosensors with fluorescent and electrochemical sensors [98]. These nanosensors may detect single molecules in a range of matrices, including water, food, soil, and even single cells. These include carbon dots, quantum dots, Au-NP, Au-Fe_3O_4-NP, up conversion nanoparticles, fluorescent conjugated polymer nanoparticles, nanosheets, and magnetic nanoparticles [99 - 102]. Additionally, the nanosensors are more efficient in a pH-regulated environment, such as silica nanoparticles, polymers, micellar nanoparticles, proteins, metal-polymer conjugates, and others [103, 104].

Soil Conservation

Soil might contain contaminants in solid, liquid, or gaseous phases. The pH, clay, and organic matter content, as well as the characteristics of the chemical compounds in the soil, are factors that contribute to the transport of contaminants into the soil [105]. It is a sink for several pollutants, particularly heavy metals, which can remain in the soil for decades. They are retained in the organic phase, and both are converted into more stable solid phases [99].

The intricate nature of the soil-water interface renders the decontamination of groundwater and soil a formidable challenge. The lack of suitable environmentally friendly technologies and the high cost of existing ones impede progress in this area [100]. The use of specific nanomaterials (1 to 100 nm) for the degradation of wastes in the environment is known as "nanoremediation" [101].

Since most of the remediation process is carried out *in situ*, the nanoremediation technique is cost-effective because it reduces the need to move contaminated soil. Additionally, these nanomaterials could be recycled and reused to treat environmental pollutants [89]. The main methods used for soil remediation are adsorption, immobilization, Fenton-type oxidation, reduction reaction, and bioremediation [99, 100]. Nanoremediation methods use reactive nanomaterials to transform and inactivate soils of persistent contaminants through chemical and catalytic reduction. Recently, nanotechnology based on nanobiosorbents and nanobiosurfactants has been developed [102]. *In situ* nanoremediation treatment does not require groundwater pumping, and the soil is not moved to other locations for disposal and treatment [20, 62, 103, 104].

The distinctive characteristics of each nanomaterial determine how it may be used for environmental remediation [106]. For example, polymer-based materials are ideal for removing chemical contaminants such as manganese, nitrate, arsenic, and other heavy metals [101]. Graphene-based carbon nitride, Fe_3O_4-NP, TiO_2-NP, and Ag-NP have the photocatalytic ability to remove a spectrum of contaminants [84, 89, 107, 108]. Nanoscale materials such as zeolites, metal oxides, carbon nanotubes and fibers, enzymes, and various noble metals have also been studied for the formation of bimetallic nanoparticles for pollutant removal, with titanium dioxide and nanoscale zerovalent iron (nZVI) being the most widely used [8]. Nanocatalysis made of metal and semiconductor nanoparticles can effectively eliminate bacteria, viruses, and some organic contaminants [7]. When iron-based nanoparticles become valuable nanocomposites with excellent remediation properties when combined with other compounds, there is an exceptional 100% removal effectiveness for hexavalent chromium [7, 89, 102] (Fig. **4**).

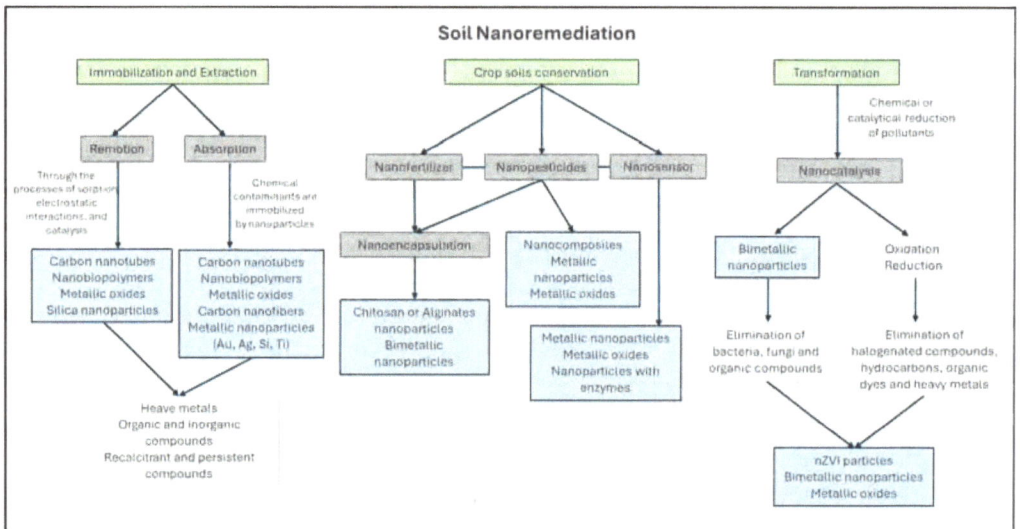

Fig. (4). Soil nanoremediation. Green boxes indicate soil remediation methods, gray boxes indicate nanotechnology methods, and blue boxes indicate the type of nanoparticle used.

The characteristics of the contaminants and the soil condition determine the nanomaterials selected for immobilization remediation. Carbon nanoparticles and metal oxides have been widely used to immobilize and remediate organic and inorganic soil contaminants [89, 92, 99, 104, 109, 110]. However, the immobilization technique for the remediation of contaminated soils presents many challenges. First, soluble phosphates are primarily considered for *in situ* stabilization of heavy metals on a laboratory scale (using very high amounts of phosphoric acid at 3%). This approach is limited not only by the cost of the materials but also by the problems of secondary contamination of groundwater and surface water in the region affected by excessive nutrient inputs (eutrophication) such as phosphate [111].

Reduction reactions are the most widely employed for the removal of pollutants from the environment [99]. nZVI nanoparticles have strong reducing power and react to a variety of organic and inorganic compounds, and heavy metal ions and immobilization [102]. They react rapidly with oxygen and water but tend to agglomerate [112]. To avoid agglomeration, these nZVI nanoparticles are usually coated with a variety of organic materials, such as starch, sodium CMC [43], and polyvinylpyrrolidone (PVP) [113]. Another limitation is the high amount of chemical reagents consumed in the preparation of conventional materials and the engineering applications of iron-based materials [114]. This challenge is currently being replaced by green nanotechnology for the synthesis of nZVI nanoparticles [115].

The nZVI particles are catalysts of 1-100 nm in diameter, usually attached to a noble metal such as palladium, silver, or copper. The catalytic synergy between the second metal and the iron facilitates the distribution and mobility of the nanoparticles when introduced into the soil [18]. These bimetallic particles may contain several metals, with the second metal being less reactive and believed to help oxidize iron or transfer electrons. Palladium and other noble metals might enhance remediation by catalyzing dechlorination and hydrogenation processes [86, 112, 116]. Iron-mediated reactions lead to an increase in pH and a decrease in the redox potential of the solution because of the rapid consumption of oxygen by other potential oxidants and hydrogen production [61, 114]. Modification of iron nanoparticles can increase the efficiency and speed of the remediation process [8].

The use of nZVI nanoparticles in the remediation of sites contaminated with Cr, nitrates, perchlorates, or organochlorine solvents is another example of how nanoparticles may be used to remediate groundwater [115]. Many nations have adopted nanotechnology as a business, but it is not fully accepted [25, 42, 64, 66]. Key factors to consider include concerns about the long-term effects, transformation, and ecotoxicity of nZVI, as well as the lack of comparable studies for varied materials and implementation methods [42, 114, 117].

The nanoparticles remain suspended in the water during the design and manufacturing phases because there are no surface electrostatic forces, which makes them versatile and allows them to be placed directly as a liquid into the contaminated subsoil. Some of the particles flow with the groundwater and remain in suspension for a period of time, while others leach out and adhere to soil particles, creating an *in situ* zone in which to retain contaminants [81, 112]. The nZVI amendment can immobilize heavy metals such as Cd, Cr, and Zn in soil with selective efficiency and stability [89, 103]. In addition, compared to other nanomaterials used for remediation, nZVI is less hazardous to humans and the environment. Different forms of iron are useful for separating and transforming a variety of contaminants metals such as As(III), Pb(II), Cu(II), Ni(II), and Cr(VI), including pesticides, chlorinated organic solvents, organochlorines, organic dyes, and, inorganic compounds [102, 118].

For effective nanoremediation, the use of nZVI and other nanomaterials requires site-specific data and adequate characterization of the contaminated site. This includes information on location, geological conditions, concentration and types of contaminants, geological and hydrogeological conditions, and geochemical properties of the soil [8, 119]. This knowledge makes the application of nanoremediation efficient, as nanoparticles are synthesized with a higher probability of contact with the contaminant; reactions are more efficient, and remediation is cheaper and time-saving. The hydrophilic properties of nZVI allow the removal of aqueous phase contaminants such as DNAPLs (dense nonaqueous phase liquids *i.e.*, petrol, oil), reducing the presence of the contaminant in the soil and increasing the concentration gradient between the aqueous phase and the DNAPLs, improving the efficiency of the process [8, 89].

Other nanomaterials have been used for bioremediation, such as metal oxide nanoparticles, carbon nanotubes, and polymeric materials [89]. These materials are used in nanofiltration, nanoabsorption, catalysis, disinfection, and fertilization processes [112]. The use of nanomaterials for agricultural purposes is essential to increase yields by optimizing nutrient uptake, improving soil quality, enhancing water distribution and storage, and minimizing the need for crop protection products [1].

Traditional sensors analyze a specific soil sample to produce accurate results, but it is costly and time-consuming. Nanosensors may monitor changes caused by a variety of pathogens and pests, consumption of pesticides, fertilizers, and herbicides, and physical soil conditions from real-time videos [22]. They are easy to handle and convenience, highly specific and sensitive, with rapid response, and cheap to run [22, 89]. Certainly, these innovations can help agriculture address complex challenges; further advances in nanotechnology could enable precise farming methods, as well as developments in wireless networks and the miniaturization of sensors to track, analyze, and monitor farming operations in real time [17, 96].

It is possible to implement these nanotechnology applications by using an available device called *"Lab-on-a-chip"*. This device can [1, 17, 97] improve the distribution of pesticides and fertilizers and monitor the quality of plants and agricultural supplies. The software should focus primarily on chemical efficacy and minimize other problems, such as spray drift, soil chelation, pesticide overuse, and water saving, making nanosensors a viable alternative to solve and eradicate these problems. These extremely sensitive devices can be created by combining nanotechnology and biotechnology, allowing faster and more accurate alternatives [17, 98].

Nanofertilizers are less expensive and are used in smaller quantities compared to chemical fertilizers, and sometimes, it is not known exactly how much nutrients the plants absorb [111]. In cropping systems, nanotechnology can provide chemical and biological pesticides, allowing the effective use of prepared agrochemical formulations [120]. Due to their larger contact surface, solubility, and mobility, smaller volumes and quantities of pesticides can be used. The release of pesticides by nanoparticles will improve the quality and efficacy of crops [16, 17, 97].

Nanoscale transporters may deliver fertilizers, herbicides, pesticides, plant growth regulators, and other chemicals to the target site [111]. The most commonly used transporters are nanostructured biopolymers, carbon nanotubes, and dendrimers that bind to pesticides or fertilizers *via* weak ionic bonds and transport them to a specific location in the soil or plant, where the transported compounds are discharged. They adhere to the roots of the plants or in the soil, improving their stability, thus reducing the environmental degradation of compounds and the cost of application [1].

To have a functional delivery system, it must be highly manageable and adjusted to the life cycle of the plant, pathogen, or pest [2, 19, 99, 105]. The nanoencapsulation of insecticides, fungicides, or nematicides in nanotubes or nanoparticules provides a controlled release rate of these products, reducing the risk of their accumulation in the soil [120]. This method also retards the degradation of the agrochemicals, increasing the durability and potency of the formulation and protecting the environment from agricultural pollution.

Clay, carbon, and graphene nanotubes have a low and continuous release rate and use 70% to 80% of the normal pesticide doses [17, 97]. The use of pesticides encapsulated in nanoparticles can solve the droplet size limit, eliminating splashing and spray drift [4, 16, 86, 97]. Nanoparticle-coated fertilizer is released slowly and steadily, reducing the runoff and dissolution during irrigation [22] and maximizing the biological efficacy [19]. Controlled use of nanofertilizers and nanopesticides can improve crop yield and health [17, 63, 97].

The efficiency of nitrogen and phosphorus applied to the soil is 20 to 25% and 10 to 25%, respectively, with the remaining fertilizer accumulating in the soil. Nanofertilizers offer a solution to eliminate the accumulation of nutrients in the soil, preventing contamination and eutrophication of drinking water. In fact, nanotechnology has reduced the cost of environmental protection and increased nutrient efficiency. Recent findings have shown that plant roots and microorganisms can extract nutrient ions directly from the solid mineral phase of nanoparticles. The literature on nanofertilizer technology is very innovative but hardly mentioned. However, there is ample scope for nanofertilizer formulation, as evidenced by several patents and reports [22].

China's National High-Tech R&D Program has been actively working on the development of nanotechnology-based slow or controlled-release fertilizers since the beginning of this century. Some nano-based agrochemicals have been commercialized. Nanometer or nanostructured processing has significantly improved the dispersion and solubility of insoluble mineral micronutrients, such as phosphate fertilizers [3, 92, 97, 103].

The use of native soil microbes, nanoparticles, and plants that hyperaccumulate heavy metals enhances the biodegradation processes, opening new areas for remediation. These processes are named microbially mediated nanoremediation and nanophytoremediation. Research shows that the use of foliar nanoparticles reduces the toxicity of metal contaminants and improves plant growth [121]. As a result, more nitrogen accumulates during plant growth, resulting in a higher accumulation of toxic elements in the plant tissues [4, 100, 101], and thus, heavy metals are removed from the soil and subsequently subtracted from the plant. The foliar application of nanoparticles has significantly improved the release rate of the amount of essential nutrients applied [22, 102].

The effectiveness of nanophytoremediation as a method to remove heavy metals from contaminated soils depends on the level of contamination, bioavailability, and the ability of plants to accumulate metals [122 - 124]. Studies have already been conducted on contaminant removal options with a particular focus on microbially mediated remediation, hyperaccumulator plants, and nanoparticles

[81]. Methods for restoring heavy metal-contaminated soils, as well as the potential benefits and risks associated with nanobioremediation technologies, have also been evaluated [122].

Finally, environmental nanotechnology is fundamental for the establishment of modern environmental engineering and science. The focus on the nanoscale has driven the development and application of innovative and cost-effective technologies for remediation and pollution control [125]. However, the safety of nanoparticles and their potential effects on biota and the environment are the subject of considerable debate, not only among scientists but also among the public. On the other hand, the potential adverse effects of nanoparticles have been the subject of constant discussion in recent years and are among the most important concerns for organisms worldwide [12, 64, 89, 96, 99, 125, 126].

Nevertheless, nanotechnology has shown great benefits in reducing environmental pollution by using smaller amounts of reagents than in bulk, with significant effects on water decontamination, purification, and removal of contaminants from soil, proving to be a useful tool for water and soil conservation. Nanotechnology has the potential to usher in a new era in water treatment and agriculture if it is pursued with caution. Evidence confirms that nanoscale innovations offer revolutionary solutions for feeding the world's growing population sustainably and equitably [120].

Nanoencapsulation

Nanoparticles are defined as colloidal solids formed by macromolecules where the active ingredient is attached by adsorption or chemical bonding on or inside the particle [127]. The procedure to prepare nanoparticles is fundamental to obtain the properties of interest in nano-pesticides and nano-fertilizers [128]. Nanoencapsulation can be defined as the process of trapping a bioactive compound inside another, called wall material. This coating, membrane, or envelope—often referred to as a nanocapsule—is constituted by biopolymers, metallic oxides, or carbon structures. In nanoencapsulation, bioactive compounds are covered and protected from the environment by a physical barrier, preventing their degradation by oxidation or hydrolysis and preserving their biological or chemical function [129, 130].

Polymeric nanoparticles can be synthesized by direct polymerization of monomers and from polymers (dispersion) using methods such as desalination, dialysis, supercritical fluids, or solvent evaporation [128]. The solvent evaporation method involves the formation of an emulsion of biopolymers in volatile solvents, subsequently becoming a suspension of nanoparticles when the solvent evaporates [107, 108]. This method is known as nanoprecipitation and will be described below. Nanoparticles produced by the solvent evaporation method have a *"top-down"* approach, *i.e.*, the bulk polymer is broken down into smaller particles [110, 129] (Fig. **5**).

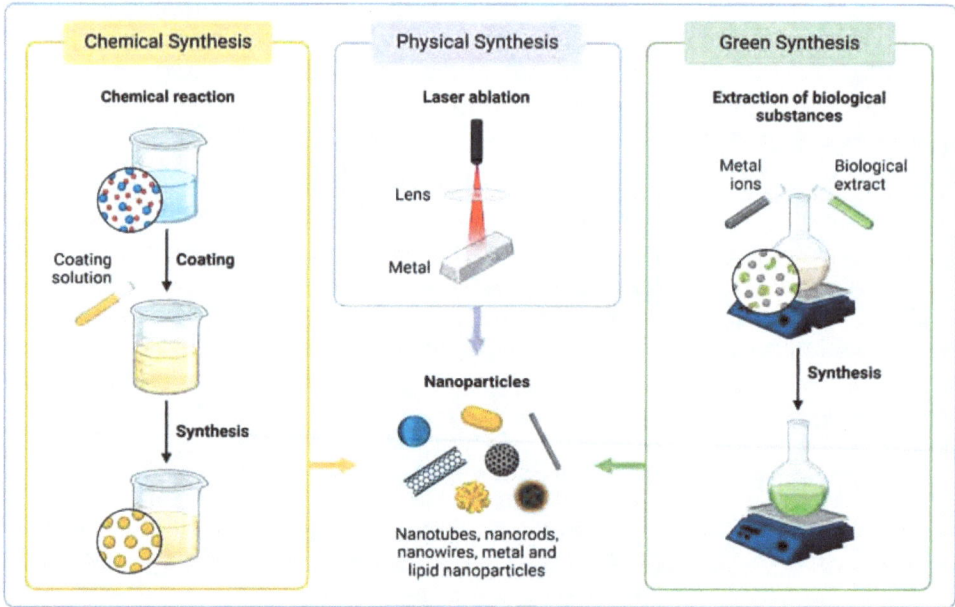

Fig. (5). Synthesis of nanoparticles by "top down" methods Modified to [131].

Nanoprecipitation is one of the most widely used techniques to prepare polymeric nanoparticles or dispersion of preformed polymers. There are two types of emulsions: a) single emulsions (*i.e.*, oil/water) or b) double emulsions (i.e., oil/water/oil). To obtain a suspension/emulsion, homogenization should be carried out at high speed or by sonication at very high frequency. The solvent should be evaporated at room temperature or using a rotary evaporator to reduce the vapor pressure of the solvent. Subsequently, the nanoparticles are obtained by ultracentrifugation to eliminate surfactants and, finally, through lyophilization [128].

In general, the biopolymers are dissolved in an oil phase, while the aqueous phase contains a stabilizer and an active compound. In other cases, nanoparticles are formed by having the biopolymer in the aqueous phase, the active compound encapsulated, and a surfactant in the oil phase [128, 132, 133]. This process involves complex hydrodynamic processes that are explained by the turbulence of the phases of two liquids, in which flow processes, surface interactions, and diffusion are involved. The formation of nanoparticles depends on the experimental conditions (temperature, concentration, volume) and the composition of the phases. This combination determines the properties (Z-potential, particle size, and polydispersity index) of the nanoparticles [127, 128, 134].

Luque-Alcaráz *et al.* [127] proposed a method to prepare chitosan nanoparticles through nanoencapsulation. This involves dissolving the biopolymer (diffusing phase) in a low concentration (<1% *w/v*). To initiate the nanoprecipitation process, alcohols (*e.g.*, methanol, ethanol, etc.) are used for chitosan and $CaCl_2$ for alginate, among others (dispersing phase). The active ingredient (fungicide, insecticide, herbicide, growth promoter) must be dissolved in this dispersing phase solution and a surfactant must be added to allow the formation of an emulsion of synthesized nanoparticles. It is important to mention that the solution with the active ingredient should be kept in constant agitation while small drops of biopolymer (0.01-0.1 ml) are added to the solution. This stirring process allows the particles to become smaller when the polymer comes into contact with the alcohol or $CaCl_2$ solution and the active compound, ultimately forming the capsule with the active compound inside [127, 131 - 133].

The reagents used for nanoparticle synthesis significantly affect the final properties impacting the particle size; these are: a) types of solvents and stabilizers and b) the concentration of the polymer and solvent used to prepare the emulsion [127, 128]. In addition, the intensity and speed of agitation or sonication control the particle size and particle population distribution [128, 135].

Another method of nanoencapsulation is extrusion, which consists of dropping drops of an aqueous solution of a polymer onto a gelling bath using a dropping tool, *i.e.*, any device that allows droplet size control (pipette, peristaltic pump, etc.) [130]. In most nanoencapsulation processes, the components are liquid, so the techniques used are based on spray drying, cold spraying, freeze drying, extrusion and injection in the molten state, ionic gelation, and coacervation, among the most regularly used [129].

Nanotechnology is applied to the synthesis of nano-pesticides to reduce toxicity by making them more efficient in terms of solubility, chemical stability, degradation by light, bioavailability in the soil, and absorption by plants [109]. Different materials have been used for the synthesis of nanoparticles for the formulation of nanopesticides, such as silica nanoparticles, double layers of hydroxides, silver nanoparticles, and polymers [120, 136]. Among these, alginate and chitosan have been the most widely used biopolymers for the synthesis of nanoparticles for disease and pest control, either using chemically active compounds or natural antimicrobials [137, 138].

Chitosan nanoparticles are preferred for their cost-effectiveness, permeability, non-toxicity (GRAS), and biodegradability [139, 140]. Chitosan breaks the cell walls of microorganisms through ionic interactions, causing the cell wall to be ruptured and the cell contents to be expelled. The inhibition of mRNA, protein, and mycotoxin synthesis has been described, and more recently, hydrolysis of cell wall peptidoglycans has been demonstrated in microorganisms [141 - 144]. Other materials like cellulose, chitin, and starch have also been employed to create greener nanocomposites [136].

Nanoencapsulation of pesticides has advantages over conventional pesticides in terms of controlled release of the active compound through the design of the nanoparticle. The approach reduces the amount of active ingredients used, the accumulation of the pesticide in the soil, and the formation of leachates, as well as it offers selective and stable protection over time. The choice of oil or water in preparing the nanoemulsion can further improve its efficacy [111].

The utilization of these nanocomposites and materials facilitates the adsorption and decomposition of an array of pollutants present in wastewater, thereby contributing to the sustainability of water usage. For example, matrix-based hydrogels, nano-encapsulated membranes, and membranes encapsulated by metals have been employed to detect pharmaceuticals and heavy metals in wastewater [136]. Nanoencapsulation in soil treatment represents a sustainable application that may include the protection and growth of plants *via* the encapsulation of bio-active molecules, agrochemicals, and fertilizers, with the aim of delivering them to the intended sites. Polymers that are employed in a variety of forestry, agricultural, industrial, and horticultural applications, as well as in drought management and water conservation, include encapsulated polymers such as absorbent hydrogels and gels [136].

Nanopesticides made from different materials and structures have been developed to control a variety of pests and diseases [145]. The antifungal effect of chitosan nanoparticles synthesized by the ionic gelation method has been successfully evaluated against *Alternaria alternata, Macrophomina phaseolina,* and *Rhizoctonia solani* [139]. Similarly, chitosan nanoparticles synthesized by nanoprecipitation and nanoencapsulation of lime essential oil were evaluated against the pathogenic bacterium *Shigella dysenteriae* [140].

Other materials have been used for the nanoencapsulation of pesticides, such as cypermethrin and natural pyrethrins, using polystyrene nanoparticles with an average diameter of 30 nm and a content of 5% of each active compound. In this formulation, the active ingredient was located between the nanoparticle chains [146]. ZnO-NPs have also been evaluated on different biological parameters of

Spodoptera frugiperda and *Spodoptera litura* [147, 148]. Chitosan and polyacrylic acid nanoparticles synthesized by nanoprecipitation have successfully controlled insect pests and fungal pathogens in soybean [149]. Nanoencapsulation of essential oils of *Cinnamomum zeylanicum* L., *Mentha piperita*, *Origanum vulgare* L., and *Lippia sidoides* in a matrix of chitosan and other polymers [138] has been used for the control of diseases of different crops of economic importance. The active compound, benzyl benzoate nanoencapsulated in polylactide acid synthetized by the nanoprecipitation method, showed pesticidal effects on *Cucumis sativus* and several insects [150 - 154].

CONCLUDING REMARKS

Nanotechnology offers a more promising and sustainable future for water and soil conservation. The implementation of nanoparticles has the potential to reduce the use of agrochemicals on a large scale, leading to the decontamination of soils, groundwater, and drinking water sources. This technology will make water drinkable in less time and decontaminate wastewater for reuse. Nanotechnology facilitates the remediation, treatment, and preservation of water and soil, with photocatalysis being the most widely used decontamination process. This technology creates nanostructured photocatalysts with higher activity, stability, and selectivity, and metallic nanoparticles serve as photocatalysts with high activity and stability.

Soil remediation is one of the main fields where nanotechnology techniques are applied. Adsorption, redox reactions, precipitation, and co-precipitation are very effective nanotechnologies due to their enormous specific surface area. Bimetallic nanoparticles stand out for their interaction with heavy metals governed by sorption-desorption reactions with other soil constituents. In *in situ* soil nano-remediation, the specific characteristics of each nanomaterial determine how they can be used for environmental remediation due to their higher solubility, lower volume and quantity, lower toxicity, and higher mobility.

However, the potential harmful effects of nanoparticles on biota and the environment have been the subject of extensive debate, not only among scientists but also among the general public, making them a topic of constant discussion in recent years and a major concern for organizations around the world due to the complex behavior of nanoparticles in the environment. Nevertheless, nanotechnology has demonstrated its effectiveness in water and soil conservation by reducing pollution. The application of nanotechnology can usher in a new era of water management and technological advances in agriculture, demonstrating that nanoscale innovations offer revolutionary solutions to sustainably and equitably feed the world's growing population.

QUESTIONS

- How can photocatalysis be used for water and soil remediation?
- What is the type of nanotechnology used for water remediation?
- How does nanotechnology improve the bioavailability of nutrients in soil?
- What is nanoencapsulation?
- What are the characteristics of nanoparticles?
- How many nanoencapsulation methods are used to produce nanopesticides, nanoherbicides, and nanofertilizers?
- In general, how can nanotechnology improve the quality of water for future generations?
- Some nanomaterials can be reused; which ones are they, and in which methods are they used?
- What is nanophytoremediation?

ACKNOWLEDGEMENTS

The authors would like to thank Dr. Nadia Salomé Gómez-Domínguez for calling their attention to the topic and Dr. EBS (New Southwest University, Australia) for reviewing, checking, and improving English grammar.

AUTHORS' CONTRIBUTION

The authors confirm their contribution to the chapter as follows:

Conceptualization, bibliographic analysis, writing, original draft, reviewing, and editing: María Alejandra Istúriz-Zapata; Conceptualization, funding, writing, reviewing, and editing: Alfredo Jiménez-Pérez.

LIST OF ABBREVIATIONS

Ag-NP	Silver nanoparticles
Au-NP	Gold nanoparticles
CA	Cellulose acetate
CMC	Carboxymethyl cellulose
CNTs	Carbon nanotubes
DNA	Deoxyribonucleic acid
DNAPLs	Dense Nonaqueous Phase Liquids
FA	Fulvic acid
Fe_3O_4-NP	Magnetite nanoparticles
Fe-NP	Iron nanoparticles

HA	Humic acid
NP	Nanoparticles
nZVI	Nano zerovalent iron
PA	Polyamide
PAN	Polyacrylonitrile
PVP	Polyvinylpyrrolidone
R & D	Research and Development
TiO$_2$-NP	Titanium oxide nanoparticles
UV light	Ultraviolet light
ZnO-NP	Zinc oxide nanoparticles

REFERENCES

[1]   Dhewa T. Nanotechnology Applications in Agriculture: An update. Octa J Environ Res. 2015; 3(2): 204-11. Available from: http://sciencebeingjournal.com/sites/default/files/10-150316_0302_TPD.pdf

[2]   Abobatta WF. Nanotechnology Application in Agriculture. Acta Sci Agric. 2018; 2(6):99-102. Available from: https://actascientific.com/ASAG/ASAG-02-0108.php

[3]   Junejo Y, Safdar M, Ozaslan M. Synthesis of Silver Nanoparticles and Their Applications: Review Eurasia Proc Health Environ Life Sci 2024; 22-48.

[4]   Manjunatha SB, Biradar DP, Aladakatti YR. Nanotechnology and its applications in agriculture: A review. J Farm Sci 2016; 29(1): 1-13.

[5]   Rajput VD, Mïnkïna T, Kumarï A, *et al.* A review on nanobioremediation approaches for restoration of contaminated soil. Eurasian J Soil Sci 2022; 11(1): 43-60.
[http://dx.doi.org/10.18393/ejss.990605] [PMID: 35269257]

[6]   Rathod S, Preetam S, Pandey C, Bera SP. Exploring synthesis and applications of green nanoparticles and the role of nanotechnology in wastewater treatment. Biotechnol Rep (Amst) 2024; 41: e00830.
[http://dx.doi.org/10.1016/j.btre.2024.e00830] [PMID: 38332899]

[7]   Yang L, Yang L, Ding L, Deng F, Luo XB, Luo SL. 1 - Principles for the Application of Nanomaterials in Environmental Pollution Control and Resource Reutilization. En: Luo X, Deng F, editores. Nanomaterials for the Removal of Pollutants and Resource Reutilization. Elsevier; 2019. Pp. 1-23. (Micro and Nano Technologies). Available from: https://www.sciencedirect.com/science/article/pii/B9780128148372000019

[8]   Karn B, Kuiken T, Otto M. Nanotechnology and in situ remediation: a review of the benefits and potential risks. Ciênc Saúde Coletiva. 2011; 16(1): 165-78.
[http://dx.doi.org/10.1590/s1413-81232011000100020] [PMID: 21180825]

[9]   Kumar S, Ahlawat W, Bhanjana G, Heydarifard S, Nazhad MM, Dilbaghi N. Nanotechnology-based water treatment strategies. J Nanosci Nanotechnol 2014; 14(2): 1838-58.
[http://dx.doi.org/10.1166/jnn.2014.9050] [PMID: 24749460]

[10]   Jiang Z, Li L, Huang H, He W, Ming W. Progress in Laser Ablation and Biological Synthesis Processes: "Top-Down" and "Bottom-Up" Approaches for the Green Synthesis of Au/Ag Nanoparticles. Int J Mol Sci 2022; 23(23): 14658.
[http://dx.doi.org/10.3390/ijms232314658] [PMID: 364989860]

[11]   Carnide G, Champouret Y, Valappil D, *et al.* Secured Nanosynthesis–Deposition Aerosol Process for Composite Thin Films Incorporating Highly Dispersed Nanoparticles. Adv Sci (Weinh) 2023; 10(5): 2204929.

[http://dx.doi.org/10.1002/advs.202204929] [PMID: 36529954]

[12]   Hristozov D, Ertel J, Techno Valuation M. Nanotechnology and Sustainability: Benefits and Risks of Nanotechnology for Environmental Sustainability. Forum Forsch 2009; 161-8.

[13]   Borisova ON, Doronkina IG, Feoktistova VM. Resource-saving nanotechnologies in waste water treatment. Nanotechnologies in Construction A Scientific Internet-Journal 2021; 13(2): 124-30.
        [http://dx.doi.org/10.15828/2075-8545-2021-13-2-124-130]

[14]   Chaudhary JP, Jhajharia P. Recent Advances in Wastewater Treatment. Gupta A, Kumar R, Kumar V, editores Integrated Waste Management: A Sustainable Approach from Waste to Wealth. Singapore: Springer Nature 2024; pp. 289-302.
        [http://dx.doi.org/10.1007/978-981-97-0823-9_14]

[15]   Vázquez-Duhalt R. Nanotecnología en procesos ambientales y remediación de la contaminación. Mundo Nano Rev Interdiscip En Nanociencias Nanotecnología. 2015; 8(14):70-80. Available from: http://www.mundonano.unam.mx:80/ojs/index.php/nano/article/view/52514
        [http://dx.doi.org/10.22201/ceiich.24485691e.2015.14.52514]

[16]   Bhattacharya S, Saha I, Mukhopadhyay A, Chattopadhyay D, Chand U, Chatterjee D. Role of nanotechnology in water treatment and purification: Potential applications and implications. Int J Chem Sci Technol 2013; 3(3): 59-64.

[17]   El-Ramady H, Brevik E, Abowaly M, Ali R, Saad Moghanm F, Gharib M, et al. Soil Degradation under a Changing Climate: Management from Traditional to Nano-Approaches. Egypt J Soil Sci. 2024. 64(1): 0-0.
        [http://dx.doi.org/10.21608/ejss.2023.248610.1686]

[18]   Karn B, Kuiken T, Otto M. Nanotechnology and in situ remediation: a review of the benefits and potential risks. Environ Health Perspect 2009; 117(12): 1813-31.
        [http://dx.doi.org/10.1289/ehp.0900793] [PMID: 20049198]

[19]   Ibrahim RK, Hayyan M, AlSaadi MA, Hayyan A, Ibrahim S. Environmental application of nanotechnology: air, soil, and water. Environ Sci Pollut Res Int 2016; 23(14): 13754-88.
        [http://dx.doi.org/10.1007/s11356-016-6457-z] [PMID: 27074929]

[20]   Chinnamuthu CR, Boopathi PM. Nanotechnology and Agroecosystem. Madras Agric J 2009; 96(1-6): 17-31.
        [http://dx.doi.org/10.29321/MAJ.10.100436]

[21]   Jatav GK, De N. Application of nanotechnology in soil-plant system. Asian J Soil Sci 2013; 8(1): 176-84.

[22]   Hamad HT, Al-Sharify ZT, Al-Najjar SZ, Gadooa ZA. A review on nanotechnology and its applications on Fluid Flow in agriculture and water recourses. IOP Conf Ser Mater Sci Eng 2020.
        [http://dx.doi.org/10.1088/1757-899X/870/1/012038]

[23]   Cantarella M, Impellizzeri G, Privitera V. Functional nanomaterials for water purification. Riv Nuovo Cim 2017; 40(12): 595-632.
        [http://dx.doi.org/10.1393/ncr/i2017-10142-8]

[24]   Annan E, Agyei-Tuffour B, Bensah YD, Konadu DS, Yaya A, Onwona-Agyeman B, et al. Application of clay ceramics and nanotechnology in water treatment: A review. Sánchez J, editor. Cogent Eng. 2018; 5(1): 1476017.
        [http://dx.doi.org/10.1080/23311916.2018.1476017]

[25]   Zhang X, Qian J, Pan B. Fabrication of Novel Magnetic Nanoparticles of Multifunctionality for Water Decontamination. Environ Sci Technol 2016; 50(2): 881-9.
        [http://dx.doi.org/10.1021/acs.est.5b04539] [PMID: 26695341]

[26]   Kumpiene J, Lagerkvist A, Maurice C. Stabilization of As, Cr, Cu, Pb and Zn in soil using amendments--a review. Waste Manag 2008; 28(1): 215-25.
        [http://dx.doi.org/10.1016/j.wasman.2006.12.012] [PMID: 17320367]

[27]   Okafor CO, Ude UI, Okoh FN, Eromonsele BO, Okafor CO, Ude UI, et al. Safe Drinking Water: The Need and Challenges in Developing Countries. En: Water Quality - New Perspectives. IntechOpen; 2024. Available at: https://www.intechopen.com/chapters/84994

[28]   Porkodi G, Anand G. Nanotechnology and its Application in Water Treatment. RE:view 2020; 6(5): 59-66.
[http://dx.doi.org/10.5281/zenodo.3868859]

[29]   Ferrovial [Internet]. [cited June 13, 2024]. Conservación del agua: qué es, medidas a tomar e importancia. Available from: https://www.ferrovial.com/es/recursos/conservacion-del-agua/

[30]   Mutegoa E. Efficient techniques and practices for wastewater treatment: an update. Discover Water 2024; 4(1): 69.
[http://dx.doi.org/10.1007/s43832-024-00131-8]

[31]   Nations U. United Nations. United Nations; [cited June 13, 2024]. Objetivo 6—Hacer frente al reto: posibilitar el acceso al agua limpia y potable en todo el mundo | Naciones Unidas. Available at: https://www.un.org/es/chronicle/article/objetivo-6-hacer-frente-al-reto-posibilitar-el-acce-o-al-agua-limpia-y-potable-en-todo-el-mundo

[32]   Ahmed MT, Ali MS, Ahamed T, Suraiya S, Haq M. Exploring the aspects of the application of nanotechnology system in aquaculture: a systematic review. Aquacult Int 2024; 32(4): 4177-206.
[http://dx.doi.org/10.1007/s10499-023-01370-7]

[33]   Shon HK, Vigneswaran S, Kandasamy J, Cho J. Characteristics of Effluent Organic Matter in Wastewater. 2007; 52-100.

[34]   The Ocean Cleanup [Internet]. [cited June 13, 2024]. Projects. Available from: https://theoceancleanup.com/projects/

[35]   Gaya UI, Abdullah AH. Heterogeneous photocatalytic degradation of organic contaminants over titanium dioxide: A review of fundamentals, progress and problems. J Photochem Photobiol C Photochem Rev 2008.
[http://dx.doi.org/10.1016/j.jphotochemrev.2007.12.003]

[36]   Iravani S. Nanomaterials and nanotechnology for water treatment: recent advances. Inorganic and Nano-Metal Chemistry 2021; 51(12): 1615-45.
[http://dx.doi.org/10.1080/24701556.2020.1852253]

[37]   Pandey S, Ramontja J. PTurning to Nanotechnology for Water Pollution Control: Applications of Nanocomposites. Focus on Sciences 2016; 2(2): 1-10.
[http://dx.doi.org/10.20286/focsci-020219]

[38]   Saldivar Tanaka L. Regulando la nanotecnología. Mundo Nano Rev Interdiscip En Nanociencias Nanotecnología 2018.
[http://dx.doi.org/10.22201/ceiich.24485691e.2019.22.63140]

[39]   Solomon NO, Kanchan S, Kesheri M. Nanoparticles as Detoxifiers for Industrial Wastewater. Water Air Soil Pollut 2024; 235(3): 214.
[http://dx.doi.org/10.1007/s11270-024-07016-5]

[40]   Thamarai P, Kamalesh R, Saravanan A, Swaminaathan P, Deivayanai VC. Emerging trends and promising prospects in nanotechnology for improved remediation of wastewater contaminants: Present and future outlooks. Environ Nanotechnol Monit Manag 2024; 21: 100913.
[http://dx.doi.org/10.1016/j.enmm.2024.100913]

[41]   Makhesana MA, Patel KM, Nyabadza A. Applicability of nanomaterials in water and waste-water treatment: A state-of-the-art review and future perspectives. Mater Today Proc 2024.
[http://dx.doi.org/10.1016/j.matpr.2024.01.037]

[42]   Gelover Santiago SL. Nanotecnología, una alternativa para mejorar la calidad del agua. Mundo Nano Revista Interdisciplinaria en Nanociencias y Nanotecnología 2015; 8(14): 40-52.

[http://dx.doi.org/10.22201/ceiich.24485691e.2015.14.52511]

[43]   El Jemli Y, Mansori M, González Diaz O, Barakat A, Solhy A, Abdelouahdi K. Controlling the growth of nanosized titania *via* polymer gelation for photocatalytic applications. RSC Advances 2020; 10(33): 19443-53.
        [http://dx.doi.org/10.1039/D0RA03312J] [PMID: 35515433]

[44]   Doña-Rodríguez JM, Pulido Melián E. Nano-Photocatalytic Materials: Possibilities and Challenges. Nanomaterials (Basel) 2021; 11(3): 688.
        [http://dx.doi.org/10.3390/nano11030688] [PMID: 33803469]

[45]   Olivera S, Muralidhara HB, Venkatesh K, Guna VK, Gopalakrishna K, Kumar K Y. Potential applications of cellulose and chitosan nanoparticles/composites in wastewater treatment: A review. Carbohydr Polym 2016; 153: 600-18.
        [http://dx.doi.org/10.1016/j.carbpol.2016.08.017] [PMID: 27561533]

[46]   Chenab KK, Sohrabi B, Jafari A, Ramakrishna S. Water treatment: functional nanomaterials and applications from adsorption to photodegradation. Mater Today Chem 2020; 16: 100262.
        [http://dx.doi.org/10.1016/j.mtchem.2020.100262]

[47]   Ghaemi N, Khodakarami Z. Nano-biopolymer effect on forward osmosis performance of cellulosic membrane: High water flux and low reverse salt. Carbohydr Polym 2019; 204: 78-88.
        [http://dx.doi.org/10.1016/j.carbpol.2018.10.005] [PMID: 30366545]

[48]   Nemati Giv A, Rastegar S, Özcan M. Influence of nanoclays on water uptake and flexural strength of glass–polyester composites. J Appl Biomater Funct Mater 2020; 18: 2280800020930180.
        [http://dx.doi.org/10.1177/2280800020930180] [PMID: 32946316]

[49]   Cobos Á, Díaz O. Impact of Nanoclays Addition on Chickpea (*Cicer arietinum* L.) Flour Film Properties. Foods 2023; 13(1): 75.
        [http://dx.doi.org/10.3390/foods13010075] [PMID: 38201103]

[50]   Grewe T, Meggouh M, Tüysüz H. Nanocatalysts for Solar Water Splitting and a Perspective on Hydrogen Economy. Chem Asian J 2016; 11(1): 22-42.
        [http://dx.doi.org/10.1002/asia.201500723] [PMID: 26411303]

[51]   Wang W, Nadagouda MN, Mukhopadhyay SM. Advances in Matrix-Supported Palladium Nanocatalysts for Water Treatment. Nanomaterials (Basel) 2022; 12(20): 3593.
        [http://dx.doi.org/10.3390/nano12203593] [PMID: 36296782]

[52]   Xiong Y, Huang X, Lu B, *et al.* Acceleration of floc-water separation and floc reduction with magnetic nanoparticles during demulsification of complex waste cutting emulsions. J Environ Sci (China) 2020; 89: 80-9.
        [http://dx.doi.org/10.1016/j.jes.2019.10.011] [PMID: 31892403]

[53]   Elmobarak WF, Almomani F. Functionalization of silica-coated magnetic nanoparticles as powerful demulsifier to recover oil from oil-in-water emulsion. Chemosphere 2021; 279: 130360.
        [http://dx.doi.org/10.1016/j.chemosphere.2021.130360] [PMID: 33862358]

[54]   Haun JB, Yoon TJ, Lee H, Weissleder R. Magnetic nanoparticle biosensors. Wiley Interdiscip Rev Nanomed Nanobiotechnol 2010; 2(3): 291-304.
        [http://dx.doi.org/10.1002/wnan.84] [PMID: 20336708]

[55]   Jiang Z, Karan S, Livingston AG. Water Transport through Ultrathin Polyamide Nanofilms Used for Reverse Osmosis. Adv Mater 2018; 30(15): 1705973.
        [http://dx.doi.org/10.1002/adma.201705973] [PMID: 29484724]

[56]   Yuan B, Zhao S, Hu P, Cui J, Niu QJ. Asymmetric polyamide nanofilms with highly ordered nanovoids for water purification. Nat Commun 2020; 11(1): 6102.
        [http://dx.doi.org/10.1038/s41467-020-19809-3] [PMID: 33257695]

[57]   Lico D, Vuono D, Siciliano C, B Nagy J, De Luca P. Removal of unleaded gasoline from water by multi-walled carbon nanotubes. J Environ Manage 2019; 237: 636-43.

[http://dx.doi.org/10.1016/j.jenvman.2019.02.062] [PMID: 30851592]

[58]   Sokoloff JB, Lau AWC. Theory of the force of friction acting on water chains flowing through carbon nanotubes. Phys Rev E. 2023; 107(5-2): 055101.
[http://dx.doi.org/10.1103/PhysRevE.107.055101]

[59]   Wang Q, Zhu S, Xi C, Zhang F. A Review: Adsorption and Removal of Heavy Metals Based on Polyamide-amines Composites. Front Chem 2022; 10: 814643.
[http://dx.doi.org/10.3389/fchem.2022.814643] [PMID: 35308790]

[60]   Arkas M, Giannakopoulos K, Favvas EP, *et al.* Comparative Study of the U(VI) Adsorption by Hybrid Silica-Hyperbranched Poly(ethylene imine) Nanoparticles and Xerogels. Nanomaterials (Basel) 2023; 13(11): 1794.
[http://dx.doi.org/10.3390/nano13111794] [PMID: 37299697]

[61]   Gao L, Zhuang J, Nie L, *et al.* Intrinsic peroxidase-like activity of ferromagnetic nanoparticles. Nat Nanotechnol 2007; 2(9): 577-83.
[http://dx.doi.org/10.1038/nnano.2007.260] [PMID: 18654371]

[62]   Ismail M, Akhtar K, Khan MI, *et al.* Pollution, Toxicity and Carcinogenicity of Organic Dyes and their Catalytic Bio-Remediation. Curr Pharm Des 2019; 25(34): 3645-63.
[http://dx.doi.org/10.2174/1381612825666191021142026] [PMID: 31656147]

[63]   Tang SCN, Lo IMC. Magnetic nanoparticles: Essential factors for sustainable environmental applications. Water Res 2013; 47(8): 2613-32.
[http://dx.doi.org/10.1016/j.watres.2013.02.039] [PMID: 23515106]

[64]   Nasrollahzadeh M, Sajjadi M, Iravani S, Varma RS. Green-synthesized nanocatalysts and nanomaterials for water treatment: Current challenges and future perspectives. J Hazard Mater 2021; 401: 123401.
[http://dx.doi.org/10.1016/j.jhazmat.2020.123401] [PMID: 32763697]

[65]   Liu X, Li Y, Chen Z, *et al.* Advanced porous nanomaterials as superior adsorbents for environmental pollutants removal from aqueous solutions. Crit Rev Environ Sci Technol 2023; 53(13): 1289-309.
[http://dx.doi.org/10.1080/10643389.2023.2168473]

[66]   Ngwenya N, Ncube EJ, Parsons J. Recent advances in drinking water disinfection: successes and challenges. Rev Environ Contam Toxicol 2013; 222: 111-70.
[http://dx.doi.org/10.1007/978-1-4614-4717-7_4] [PMID: 22990947]

[67]   Primo JO, Horsth DF, Correa JS, *et al.* Synthesis and Characterization of Ag/ZnO Nanoparticles for Bacteria Disinfection in Water. Nanomaterials (Basel) 2022; 12(10): 1764.
[http://dx.doi.org/10.3390/nano12101764] [PMID: 35630986]

[68]   Wang L, Yu J. Chapter 1 - Principles of photocatalysis. En: Yu J, Zhang L, Wang L, Zhu B. Interface Science and Technology [Internet]. Elsevier; 2023 [citado 1 de octubre de 2024]. p. 1-52. (S-scheme Heterojunction Photocatalysts; vol. 35).Available from: https://www.sciencedirect.com/science/article/pii/B9780443187865000020

[69]   Borbón Jara B, Medel A, Bedolla Valdez Z, Alonso Núñez G, Oropeza Guzmán MT. Evaluación electroquímica de nanoestructuras Fe/MWCNT-Pt y Fe/MWCNT-Pt-Pd como materiales de cátodos multifuncionales con potencial aplicación en el mejoramiento de la calidad de agua tratada. Mundo Nano Revista Interdisciplinaria en Nanociencias y Nanotecnología 2015; 8(14): 6-16.
[http://dx.doi.org/10.22201/ceiich.24485691e.2015.14.52508]

[70]   Coccia M, Bontempi E. New trajectories of technologies for the removal of pollutants and emerging contaminants in the environment. Environ Res 2023; 229: 115938.
[http://dx.doi.org/10.1016/j.envres.2023.115938] [PMID: 37086878]

[71]   Durán-Álvarez JC, Avella E, Zanella R. Descontaminación de agua utilizando nanomateriales y procesos fotocatalíticos. Mundo Nano Revista Interdisciplinaria en Nanociencias y Nanotecnología 2015; 8(14): 17-39.

[http://dx.doi.org/10.22201/ceiich.24485691e.2015.14.52510]

[72] Sosa SM, Huertas R, Pereira VJ. Combination of Zinc Oxide Photocatalysis with Membrane Filtration for Surface Water Disinfection. Membranes (Basel) 2023; 13(1): 56.
[http://dx.doi.org/10.3390/membranes13010056] [PMID: 36676863]

[73] Sharma K, Vaya D, Prasad G, Surolia PK. Photocatalytic process for oily wastewater treatment: a review. Int J Environ Sci Technol 2023; 20(4): 4615-34.
[http://dx.doi.org/10.1007/s13762-021-03874-2]

[74] Noureen L, Xie Z, Gao Y, *et al.* Multifunctional Ag $_3$ PO $_4$ -rGO-Coated Textiles for Clean Water Production by Solar-Driven Evaporation, Photocatalysis, and Disinfection. ACS Appl Mater Interfaces 2020; 12(5): 6343-50.
[http://dx.doi.org/10.1021/acsami.9b16043] [PMID: 31939275]

[75] Khurram R, Javed A, Ke R, Lena C, Wang Z. Visible Light-Driven GO/TiO$_2$-CA Nano-Photocatalytic Membranes: Assessment of Photocatalytic Response, Antifouling Character and Self-Cleaning Ability. Nanomaterials (Basel) 2021; 11(8): 2021.
[http://dx.doi.org/10.3390/nano11082021] [PMID: 34443852]

[76] Rodríguez-Hernández AG, Aguilar Guzmán JC, Vázquez-Duhalt R. Membrana celular y la inespecificidad de las nanopartículas. ¿Hasta dónde puede llegar un nanomaterial dentro de la célula? Mundo Nano Revista Interdisciplinaria en Nanociencia y Nanotecnología 2018; 11(20): 43.
[http://dx.doi.org/10.22201/ceiich.24485691e.2018.20.62711]

[77] Ahtasham Iqbal M, Akram S, khalid S, *et al.* Advanced photocatalysis as a viable and sustainable wastewater treatment process: A comprehensive review. Environ Res 2024; 253: 118947.
[http://dx.doi.org/10.1016/j.envres.2024.118947] [PMID: 38744372]

[78] Hirakawa K, Mori M, Yoshida M, Oikawa S, Kawanishi S. Photo-irradiated titanium dioxide catalyzes site specific DNA damage *via* generation of hydrogen peroxide. Free Radic Res 2004; 38(5): 439-47.
[http://dx.doi.org/10.1080/1071576042000206487] [PMID: 15293551]

[79] Blanchon C, Toulza E, Calvayrac C, Plantard G. Heterogeneous photo-oxidation in microbial inactivation: A promising technology for seawater bio-securing?. Sustain 2023; pp. 1-27.
[http://dx.doi.org/10.1016/j.nxsust.2023.100003]

[80] Rokicka-Konieczna P, Morawski AW. Photocatalytic Bacterial Destruction and Mineralization by TiO$_2$-Based Photocatalysts: A Mini Review. Molecules 2024; 29(10): 2221.
[http://dx.doi.org/10.3390/molecules29102221] [PMID: 38792082]

[81] Arora D, Arora A, Bala R, Panghal V, Kumar S. Enhancement in Phytoremediation Efficiency of Tagetus erecta with the Application of Nano-scale Zero Valent Iron (nZVI) for the Restoration of Lead Contaminated Soil: an Approach Toward Sustainability. Water Air Soil Pollut 2023; 234(8): 535. [Internet].
[http://dx.doi.org/10.1007/s11270-023-06540-0]

[82] Gao R, Gao SH, Li J, *et al.* Emerging Technologies for the Control of Biological Contaminants in Water Treatment: A Critical Review. Engineering (Beijing) 2024 (In press).
[http://dx.doi.org/10.1016/j.eng.2024.08.022]

[83] Tejada-Tovar C, Villabona-Ortiz Á, Garcés-Jaraba L. Adsorción de metales pesados en aguas residuales usando materiales de origen biológico. TecnoLógicas. 1 2015; 18(34): 109.
[http://dx.doi.org/10.22430/22565337.209]

[84] Keshta BE, Gemeay AH, Kumar Sinha D, *et al.* State of the art on the magnetic iron oxide Nanoparticles: Synthesis, Functionalization, and applications in wastewater treatment. Results in Chemistry 2024; 7: 101388.
[http://dx.doi.org/10.1016/j.rechem.2024.101388]

[85] Leon AL, Sacco NA, Zoppas FM, Galindo R, Sandoval EM, Marchesini FA. Dopamine removal from water by advanced oxidative processes with Fe/N-doped carbon nanotubes. Environ Sci Pollut Res Int

2023; 30(19): 55424-36.
[http://dx.doi.org/10.1007/s11356-023-26224-w] [PMID: 36892703]

[86]   Zhou H, Li W, Yu P. Carbon Nanotubes-Based Nanofluidic Devices: Fabrication, Property and Application. ChemistryOpen 2022; 11(11): e202200126.
[http://dx.doi.org/10.1002/open.202200126] [PMID: 36351756]

[87]   Rajput VD, Minkina T, Upadhyay SK, *et al.* Nanotechnology in the Restoration of Polluted Soil. Nanomaterials (Basel) 2022; 12(5): 769.
[http://dx.doi.org/10.3390/nano12050769] [PMID: 35269257]

[88]   Raffa CM, Chiampo F, Shanthakumar S. Remediation of Metal/Metalloid-Polluted Soils: A Short Review. Appl Sci (Basel) 2021; 11(9): 4134.
[http://dx.doi.org/10.3390/app11094134]

[89]   Sathish T, Ahalya N, Thirunavukkarasu M, *et al.* A comprehensive review on the novel approaches using nanomaterials for the remediation of soil and water pollution. Alex Eng J 2024; 86: 373-85.
[http://dx.doi.org/10.1016/j.aej.2023.10.038]

[90]   Liu Y, Vijayakumar P, Liu Q, Sakthivel T, Chen F, Dai Z. Shining Light on Anion-Mixed Nanocatalysts for Efficient Water Electrolysis: Fundamentals, Progress, and Perspectives. Nano-Micro Lett 2022; 14(1): 43.
[http://dx.doi.org/10.1007/s40820-021-00785-2] [PMID: 34981288]

[91]   Kumar A, Devi K. Application of Nanotechnology in Soil Stabilization. Journal of Building Material Science 2023; 5(2): 25-36.
[http://dx.doi.org/10.30564/jbms.v5i2.5913]

[92]   Okmi A, Xiao X, Zhang Y, *et al.* Discovery of Graphene-Water Membrane Structure: Toward High-Quality Graphene Process. Adv Sci (Weinh) 2022; 9(26): 2201336.
[http://dx.doi.org/10.1002/advs.202201336] [PMID: 35856086]

[93]   Joshi S, Kataria N, Garg VK, Kadirvelu K. Pb^{2+} and Cd^{2+} recovery from water using residual tea waste and SiO_2@TW nanocomposites. Chemosphere 2020; 257: 127277.
[http://dx.doi.org/10.1016/j.chemosphere.2020.127277] [PMID: 32702805]

[94]   Hnamte M, Pulikkal AK. Clay-polymer nanocomposites for water and wastewater treatment: A comprehensive review. Chemosphere 2022; 307(Pt 2): 135869.
[http://dx.doi.org/10.1016/j.chemosphere.2022.135869] [PMID: 35948093]

[95]   Sethi J, Wågberg L, Larsson PA. Water-resistant hybrid cellulose nanofibril films prepared by charge reversal on gibbsite nanoclays. Carbohydr Polym. 1 2022; 295: 119867.
[http://dx.doi.org/10.1016/j.carbpol.2022.119867]

[96]   Xia C, Li X, Wu Y, Suharti S, Unpaprom Y, Pugazhendhi A. A review on pollutants remediation competence of nanocomposites on contaminated water. Environ Res 2023; 222: 115318.
[http://dx.doi.org/10.1016/j.envres.2023.115318] [PMID: 36693465]

[97]   Jaspal D, Malviya A. Composites for wastewater purification: A review. Chemosphere 2020; 246: 125788.
[http://dx.doi.org/10.1016/j.chemosphere.2019.125788] [PMID: 31918098]

[98]   Barry S, O'Riordan A. Electrochemical nanosensors: advances and applications. Rep Electrochem. 2 de febrero de 2016; 6:1-14.
[http://dx.doi.org/10.2147/RIE.S80550]

[99]   Kristanti RA, Liong RMY, Hadibarata T. Soil Remediation Applications of Nanotechnology. Tropical Aquatic and Soil Pollution 2021; 1(1): 35-45.
[http://dx.doi.org/10.53623/tasp.v1i1.12]

[100]   Qian Y, Qin C, Chen M, Lin S. Nanotechnology in soil remediation − applications vs. implications. Ecotoxicol Environ Saf 2020; 201: 110815.
[http://dx.doi.org/10.1016/j.ecoenv.2020.110815] [PMID: 32559688]

[101] Zhu L, Ran R, Tadé M, Wang W, Shao Z. Perovskite materials in energy storage and conversion. Asia-Pac J Chem Eng 2016; 11(3): 338-69.
[http://dx.doi.org/10.1002/apj.2000]

[102] Dhanapal AR, Thiruvengadam M, Vairavanathan J, Venkidasamy B, Easwaran M, Ghorbanpour M. Nanotechnology Approaches for the Remediation of Agricultural Polluted Soils. ACS Omega. 26 de marzo de 2024; 9(12):13522-33.
[http://dx.doi.org/10.1021/acsomega.3c09776]

[103] Gil-Díaz M, Pinilla P, Alonso J, Lobo MC. Viability of a nanoremediation process in single or multi-metal(loid) contaminated soils. J Hazard Mater 2017; 321: 812-9.
[http://dx.doi.org/10.1016/j.jhazmat.2016.09.071] [PMID: 27720472]

[104] Olyaie E, Banejad H. Recent nanotechnology applied for soil and water remediation: A critical review. At: Conference: 8th International Soil Science Congress on "Land Degradation and Challenges in Sustainable Soil Management". At: Izmir, Turkey; 2012. Available at: https://www.researchgate.net/publication/259266994_Recent_Nanotechnology_Applied_for_Soil_and_Water_Remediation_A_Critical_Review

[105] Souza LRR, Pomarolli LC, da Veiga MAMS. From classic methodologies to application of nanomaterials for soil remediation: an integrated view of methods for decontamination of toxic metal(oid)s. Environ Sci Pollut Res Int 2020; 27(10): 10205-27.
[http://dx.doi.org/10.1007/s11356-020-08032-8] [PMID: 32064582]

[106] Alex A, Raj S, Sugunan SK, George G. Nanotechnology in Soil Remediation. Nanotechnology for Environmental Remediation. John Wiley & Sons, Ltd 2022; pp. 27-43. Available from: https://onlinelibrary.wiley.com/doi/abs/10.1002/9783527834143.ch3
[http://dx.doi.org/10.1002/9783527834143.ch3]

[107] Meskher H, Ragdi T, Thakur AK, *et al.* A Review on CNTs-Based Electrochemical Sensors and Biosensors: Unique Properties and Potential Applications. Crit Rev Anal Chem 2023; 0(0): 1-24.
[http://dx.doi.org/10.1080/10408347.2023.2171277] [PMID: 36724894]

[108] Meskher H, Belhaouari SB, Deshmukh K, Hussain CM, Sharifianjazi F. A Magnetite Composite of Molecularly Imprinted Polymer and Reduced Graphene Oxide for Sensitive and Selective Electrochemical Detection of Catechol in Water and Milk Samples: An Artificial Neural Network (ANN) Application. J Electrochem Soc 2023; 170(4): 047502.
[http://dx.doi.org/10.1149/1945-7111/acc97c]

[109] Paramo LA, Feregrino-Pérez AA, Guevara R, Mendoza S, Esquivel K. Nanoparticles in Agroindustry: Applications, Toxicity, Challenges, and Trends. Nanomaterials (Basel) 2020; 10(9): 1654.
[http://dx.doi.org/10.3390/nano10091654] [PMID: 32842495]

[110] Rafique M, Sadaf I, Rafique MS, Tahir MB. A review on green synthesis of silver nanoparticles and their applications. Artif Cells Nanomed Biotechnol 2017; 45(7): 1272-91.
[http://dx.doi.org/10.1080/21691401.2016.1241792] [PMID: 27825269]

[111] Chhipa H. Nanofertilizers and nanopesticides for agriculture. Environ Chem Lett 2017; 15(1): 15-22.
[http://dx.doi.org/10.1007/s10311-016-0600-4]

[112] Pasinszki T, Krebsz M. Synthesis and Application of Zero-Valent Iron Nanoparticles in Water Treatment, Environmental Remediation, Catalysis, and Their Biological Effects. Nanomaterials (Basel) 2020; 10(5): 917.
[http://dx.doi.org/10.3390/nano10050917] [PMID: 32397461]

[113] Pastoriza-Santos I, Liz-Marzán LM. Formation of PVP-Protected Metal Nanoparticles in DMF. Langmuir 2002; 18(7): 2888-94.
[http://dx.doi.org/10.1021/la015578g]

[114] Feng J, Lang G, Li T, Zhang J, Li T, Jiang Z. Enhanced removal performance of zero-valent iron towards heavy metal ions by assembling Fe-tannin coating. J Environ Manage 2022; 319: 115619.

[http://dx.doi.org/10.1016/j.jenvman.2022.115619] [PMID: 35810583]

[115]   Kheskwani U, Ahammed MM. Removal of water pollutants using plant-based nanoscale zero-valent iron: A review. Water Sci Technol 2023; 88(5): 1207-31.
[http://dx.doi.org/10.2166/wst.2023.270] [PMID: 37771223]

[116]   Deewan R, Yan DYS, Khamdahsag P, Tanboonchuy V. Remediation of arsenic-contaminated water by green zero-valent iron nanoparticles. Environ Sci Pollut Res Int 2022; 30(39): 90352-61.
[http://dx.doi.org/10.1007/s11356-022-24535-y] [PMID: 36527549]

[117]   Raman CD, Kanmani S. Textile dye degradation using nano zero valent iron: A review. J Environ Manage 2016; 177: 341-55.
[http://dx.doi.org/10.1016/j.jenvman.2016.04.034] [PMID: 27115482]

[118]   Sun H. Grand Challenges in Environmental Nanotechnology. Frontiers in Nanotechnology 2019; 1: 2.
[http://dx.doi.org/10.3389/fnano.2019.00002]

[119]   Latif B. Nanotechnology for Site Remediation: Fate and Transport of Nanoparticles in Soil and Water Systems Beshoy Latif. ECOS Student University of Arizona COS Student University of Arizona 2006.

[120]   Kansotia K, Naresh R, Sharma Y, *et al.* Nanotechnology-driven Solutions: Transforming Agriculture for a Sustainable and Productive Future. J Sci Res Rep 2024; 30(3): 32-51.
[http://dx.doi.org/10.9734/jsrr/2024/v30i31856]

[121]   Al-Shimary AA, Al-Azzawi MA, Abdulateef SM. Systematic Review on Application of Nanoparticles in Agriculture. World J Pharm Sci Res. 2023; 2(6):79-90. Available from: https://wjpsronline.com/abstract/0000000073

[122]   Rajput V, Minkina T, Semenkov I, Klink G, Tarigholizadeh S, Sushkova S. Phylogenetic analysis of hyperaccumulator plant species for heavy metals and polycyclic aromatic hydrocarbons. Environ Geochem Health 2021; 43(4): 1629-54.
[http://dx.doi.org/10.1007/s10653-020-00527-0] [PMID: 32040786]

[123]   Ghazaryan KA, Movsesyan HS, Khachatryan HE, Ghazaryan NP, Minkina T, Mandzhieva S, *et al.* Copper phytoextraction and phytostabilization potential of wild plant species growing in the mine polluted areas of Armenia. Int J Mod Res Eng Technol 2018; 5: 1-20.
[http://dx.doi.org/10.1144/geochem2018-035]

[124]   Madhav S, Mishra R, Kumari A, *et al.* A review on sources identification of heavy metals in soil and remediation measures by phytoremediation-induced methods. Int J Environ Sci Technol 2024; 21(1): 1099-120.
[http://dx.doi.org/10.1007/s13762-023-04950-5]

[125]   Bernd N. Pollution Prevention and Treatment Using Nanotechnology. En: Wiley-VCH Verlag GmbH & Co. KGaA, editor. Nanotechnology. 1st ed. Wiley; 2010. Pp. 1-15. Available at: https://onlinelibrary.wiley.com/doi/10.1002/9783527628155.nanotech010
[http://dx.doi.org/10.1002/9783527628155.nanotech010]

[126]   Maria VL, Barreto A. Ecotoxicity Assessment of Nanomaterials: Latest Advances and Prospects. Nanomaterials (Basel) 2024; 14(4): 326.
[http://dx.doi.org/10.3390/nano14040326] [PMID: 38392699]

[127]   Luque-Alcaraz AG, Lizardi-Mendoza J, Goycoolea FM, Higuera-Ciapara I, Argüelles-Monal W. Preparation of chitosan nanoparticles by nanoprecipitation and their ability as a drug nanocarrier. RSC Advances 2016; 6(64): 59250-6.
[http://dx.doi.org/10.1039/C6RA06563E]

[128]   Rao JP, Geckeler KE. Polymer nanoparticles: Preparation techniques and size-control parameters. Prog Polym Sci 2011; 36(7): 887-913.
[http://dx.doi.org/10.1016/j.progpolymsci.2011.01.001]

[129]   Prasad R, Bhattacharyya A, Nguyen QD. Nanotechnology in Sustainable Agriculture: Recent Developments, Challenges, and Perspectives. Front Microbiol 2017; 8: 1014.

[http://dx.doi.org/10.3389/fmicb.2017.01014] [PMID: 28676790]

[130] Nedovic V, Kalusevic A, Manojlovic V, Levic S, Bugarski B. An overview of encapsulation technologies for food applications. Procedia Food Sci 2011; 1: 1806-15.
[http://dx.doi.org/10.1016/j.profoo.2011.09.265]

[131] Shukla S, Khan R, Daverey A. Synthesis and characterization of magnetic nanoparticles, and their applications in wastewater treatment: A review. Environmental Technology & Innovation 2021; 24: 101924.
[http://dx.doi.org/10.1016/j.eti.2021.101924]

[132] Correa-Pacheco ZN, Bautista-Baños S, Valle-Marquina MÁ, Hernández-López M. The Effect of Nanostructured Chitosan and Chitosan-thyme Essential Oil Coatings on *Colletotrichum gloeosporioides* Growth *in vitro* and on cv Hass Avocado and Fruit Quality. J Phytopathol 2017; 165(5): 297-305.
[http://dx.doi.org/10.1111/jph.12562]

[133] Istúriz-Zapata MA, Correa-Pacheco ZN, Bautista-Baños S, Acosta-Rodríguez JL, Hernández-López M, Barrera-Necha LL. Efficacy of extracts of mango residues loaded in chitosan nanoparticles and their nanocoatings on in vitro and in vivo postharvest fungal. J Phytopathol 2022; 170(10): 661-74.
[http://dx.doi.org/10.1111/jph.13130]

[134] Anton N, Benoit JP, Saulnier P. Design and production of nanoparticles formulated from nano-emulsion templates—A review. J Control Release 2008; 128(3): 185-99.
[http://dx.doi.org/10.1016/j.jconrel.2008.02.007] [PMID: 18374443]

[135] Bilati U, Allémann E, Doelker E. Sonication Parameters for the Preparation of Biodegradable Nanocapsulesof Controlled Size by the Double Emulsion Method. Pharm Dev Technol 2003; 8(1): 1-9.
[http://dx.doi.org/10.1081/PDT-120017517] [PMID: 12665192]

[136] Gulati S, Amar A, Chhabra L, Katiyar R. Meenakshi, Sahu T, et al Greener nanobiopolymers and nanoencapsulation: environmental implications and future prospects. RSC Sustain 2024; 2: pp. 2805-32.

[137] Divya K, Jisha MS. Chitosan nanoparticles preparation and applications. Environ Chem Lett 2018; 16(1): 101-12.
[http://dx.doi.org/10.1007/s10311-017-0670-y]

[138] Natrajan D, Srinivasan S, Sundar K, Ravindran A. Formulation of essential oil-loaded chitosan–alginate nanocapsules. J Food Drug Anal 2015; 23(3): 560-8.
[http://dx.doi.org/10.1016/j.jfda.2015.01.001] [PMID: 28911716]

[139] Saharan V, Mehrotra A, Khatik R, Rawal P, Sharma SS, Pal A. Synthesis of chitosan based nanoparticles and their in vitro evaluation against phytopathogenic fungi. Int J Biol Macromol 2013; 62: 677-83.
[http://dx.doi.org/10.1016/j.ijbiomac.2013.10.012] [PMID: 24141067]

[140] Sotelo-Boyás ME, Correa-Pacheco ZN, Bautista-Baños S, Corona-Rangel ML. Physicochemical characterization of chitosan nanoparticles and nanocapsules incorporated with lime essential oil and their antibacterial activity against food-borne pathogens. Lebensm Wiss Technol 2017; 77: 15-20.
[http://dx.doi.org/10.1016/j.lwt.2016.11.022]

[141] Sharif R, Mujtaba M, Ur Rahman M, *et al.* The Multifunctional Role of Chitosan in Horticultural Crops; A Review. Molecules 2018; 23(4): 872.
[http://dx.doi.org/10.3390/molecules23040872] [PMID: 29642651]

[142] Lopez-Moya F, Suarez-Fernandez M, Lopez-Llorca LV. Molecular Mechanisms of Chitosan Interactions with Fungi and Plants. Int J Mol Sci 2019; 20(2): 332.
[http://dx.doi.org/10.3390/ijms20020332] [PMID: 30650540]

[143] Maluin FN, Hussein MZ. Chitosan-Based Agronanochemicals as a Sustainable Alternative in Crop

Protection. Molecules 2020; 25(7): 1611.
[http://dx.doi.org/10.3390/molecules25071611] [PMID: 32244664]

[144] Goy RC, Morais STB, Assis OBG. Evaluation of the antimicrobial activity of chitosan and its quaternized derivative on *E. coli* and *S. aureus* growth. Rev Bras Farmacogn 2016; 26(1): 122-7.
[http://dx.doi.org/10.1016/j.bjp.2015.09.010]

[145] Castro-Restrepo D. Nanotecnología en la agricultura. Bionatura 2017; 2(3): 384-9.
[http://dx.doi.org/10.21931/RB/2017.03.03.9]

[146] Hened Saade, Janett Valdez, Fabiola Castellanos, Javier Enríquez. Plastics Technology Mexico. 2022. Nanotecnología revoluciona el uso de pesticidas. Available at: https://www.pt-mexico.com/columnas/nanoencapsulacion-de-pesticidas-una-agropcion-con-futuro

[147] Pittarate S, Rajula J, Rahman A, *et al.* Insecticidal Effect of Zinc Oxide Nanoparticles against *Spodoptera frugiperda* under Laboratory Conditions. Insects 2021; 12(11): 1017.
[http://dx.doi.org/10.3390/insects12111017] [PMID: 34821816]

[148] Jameel M, Shoeb M, Khan MT, *et al.* Enhanced Insecticidal Activity of Thiamethoxam by Zinc Oxide Nanoparticles: A Novel Nanotechnology Approach for Pest Control. ACS Omega 2020; 5(3): 1607-15.
[http://dx.doi.org/10.1021/acsomega.9b03680] [PMID: 32010835]

[149] Sabbour M, Sahab, A. F.1; Waly, Sabbour, M. M.3 And Lubna S. Nawar. Synthesis, antifungal and insecticidal potential of Chitosan (CS)-g-poly (acrylic acid) (PAA) nanoparticles against some seed borne fungi and insects of soybean [Internet]. Unpublished; 2015.
[http://dx.doi.org/10.13140/RG.2.1.1198.8325]

[150] Istúriz-Zapata MA, Hernández-López M, Correa-Pacheco ZN, Barrera-Necha LL. Quality of cold-stored cucumber as affected by nanostructured coatings of chitosan with cinnamon essential oil and cinnamaldehyde. Lebensm Wiss Technol 2020; 123: 109089.
[http://dx.doi.org/10.1016/j.lwt.2020.109089]

[151] Mohammadi A, Hashemi M, Hosseini SM. Chitosan nanoparticles loaded with Cinnamomum zeylanicum essential oil enhance the shelf life of cucumber during cold storage. Postharvest Biol Technol 2015; 110: 203-13.
[http://dx.doi.org/10.1016/j.postharvbio.2015.08.019]

[152] Asbahani AE, Miladi K, Badri W, *et al.* Essential oils: From extraction to encapsulation. Int J Pharm 2015; 483(1-2): 220-43.
[http://dx.doi.org/10.1016/j.ijpharm.2014.12.069] [PMID: 25683145]

[153] Ladj-Minost A. Répulsifs d'arthropodes à durée d'action prolongée : étude pharmacotechnique, devenir in situ et efficacité [phdthesis]. Université Claude Bernard - Lyon I ; 2012. Available at: https://theses.hal.science/tel-00943382

[154] Abreu FOMS, Oliveira EF, Paula HCB, de Paula RCM. Chitosan/cashew gum nanogels for essential oil encapsulation. Carbohydr Polym 2012; 89(4): 1277-82.
[http://dx.doi.org/10.1016/j.carbpol.2012.04.048] [PMID: 24750942]

CHAPTER 13

Conservation and Reuse of Water in Agriculture: Biotechnological Techniques for Efficient Use

Israel Valencia Quiroz[1,*], Diana Violeta Sánchez Oropeza[1], María Fernanda Trujillo Lira[1], Miriam Arlette López Pérez[1], Casandra Rosales García[1] and **Susana Rafael Maya[1]**

[1] *Phytochemistry Laboratory, UBIPRO, Superior Studies Faculty (FES)-Iztacala, National Autonomous University of Mexico (UNAM), Tlalnepantla de Baz, Mexico State, Mexico*

Abstract: Agricultural water conservation involves implementing sustainable practices that reduce water loss and optimize water use efficiency. Techniques such as drip irrigation, precision agriculture, and mulching play a crucial role in minimizing water wastage. Additionally, treating water for reuse and utilizing low-quality water for irrigation are essential strategies to ensure safe and effective water use. Proper soil management enhances the absorption of wastewater, preventing salt accumulation that could harm crops. Traditional methods like plant breeding and optimizing planting times complement modern biotechnological approaches, such as genetic modifications, to improve water efficiency. Agriculture consumes the majority of the world's water resources, highlighting the need for efficient water management to prevent contamination and ensure sustainability. Water scarcity poses significant challenges to food security, particularly in regions reliant on rain-fed agriculture. To address these challenges, strategies such as developing drought-tolerant crop varieties, integrated water resource management, and the reuse of treated wastewater are being employed. Emerging biotechnological techniques, including the use of transgenic plants and innovative water treatment technologies, offer promising solutions for water conservation in agriculture. Case studies demonstrate successful applications of hydroponics, low-pressure irrigation systems, and the integration of biotechnological solutions in real-world settings. Future directions emphasize the importance of continued research and innovation in biotechnology to enhance water use efficiency, promote sustainable agricultural practices, and address the global water scarcity crisis. By adopting these advanced techniques, the agricultural sector can significantly contribute to the conservation and efficient use of water resources, ensuring long-term food security and environmental sustainability.

* **Corresponding author Israel Valencia Quiroz:** Phytochemistry Laboratory, UBIPRO, Superior Studies Faculty (FES)-Iztacala, National Autonomous University of Mexico (UNAM), Tlalnepantla de Baz, Mexico State, Mexico; Tel: +525556231137; E-mail: israelv@unam.mx

Keywords: Agricultural water conservation, Biotechnological techniques, Drought-tolerant crops, Drip irrigation, Genetic modifications, Precision agriculture, Soil management, Sustainable agriculture, Water reuse, Water efficiency.

INTRODUCTION

The term "agricultural water conservation" refers to the beneficial use of water to promote sustainable agriculture by reducing water loss through evaporation from the soil and providing water to crops at optimal times for growth while considering the surrounding environment. Water conservation measures can be implemented at all stages of crop production, including soil preparation and irrigation practices. Implementing water-saving techniques such as drip irrigation, precision agriculture, and mulching can greatly reduce water wastage. Additionally, the use of water treatment techniques is crucial to enable water reuse or the utilization of low-quality water for irrigation purposes. Treating water ensures that it is safe for plant uptake while minimizing any potential harm to the environment. Proper soil management is also essential for agricultural water conservation. Adequate preparation of the soil allows for efficient absorption of wastewater or other low-quality water, preventing excessive accumulation of salts that could potentially damage crops and hinder productivity. By implementing comprehensive water conservation strategies and adopting sustainable farming practices, farmers can mitigate the impact of water scarcity and enhance the long-term viability of agricultural systems [1, 2] (Fig. **1**).

The growing scarcity of water for agriculture has driven the search for techniques to reduce water use by increasing efficiency in crop production. This chapter discusses several traditional and modern biotechnological techniques, especially those involving genetic modifications. Traditional methods include plant breeding and selection, optimizing planting times, using organic amendments, and integrated water management. Modern techniques involve collaborative action with companies producing seeds and agrochemicals [3, 4].

Importance of Water in Agriculture

Globally, approximately 70% of water resources are used for agriculture, 16% for industry, and 14% for domestic purposes. However, these proportions vary across different regions of the world. Agriculture consumes most of the water available for human use, underscoring its great social importance in water resource management. Efforts to increase water use efficiency and reduce percolation are crucial in minimizing contamination of aquifers and surface water. It is estimated that about 40% of the world's irrigation water is used on crops that are highly sensitive to water shortages, resulting in decreased yields (Fig. **2**) [5, 6].

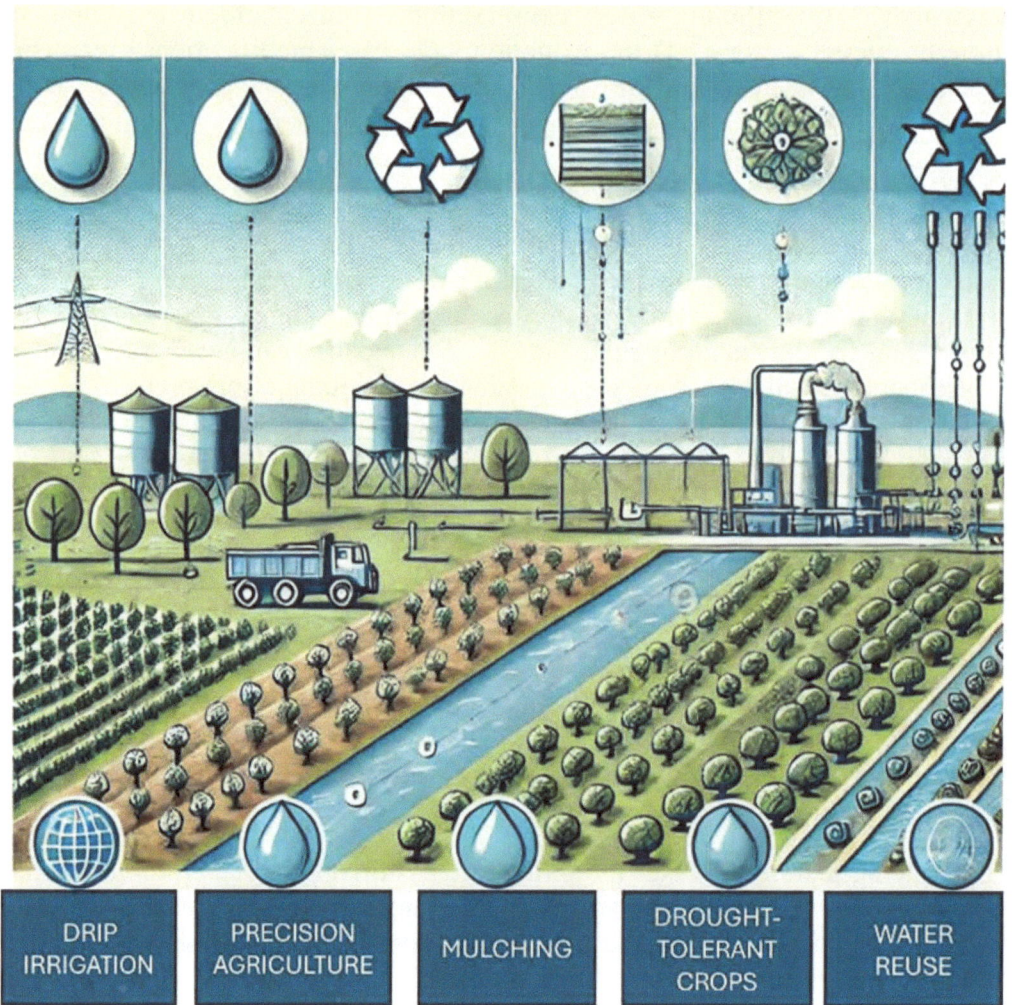

Fig. (1). Various water conservation techniques employed in agriculture. The figure illustrates methods such as drip irrigation, precision agriculture, mulching, water reuse, and drought-tolerant crops. These strategies help reduce water waste and enhance sustainable agricultural practices. Figure made with DALL-E.

Fulfilling the right to an adequate standard of living includes access to a sufficient amount of good-quality water. Nearly 70% of the world's poor people live in rural areas and depend on agriculture for their livelihood. However, most poor farmers cannot ensure stable food production due to inadequate, unreliable, or unaffordable water supplies in their rain-fed or irrigated agricultural systems. Indeed, high risks of food insecurity due to water shortages are significant for both rain-fed and irrigated agriculture. It is known that higher levels of investment are needed to reduce these risks linked to water shortages in rain-fed areas compared to investment needed in areas with supplementary irrigation. In countries with dry climates, the introduction of supplementary irrigation can increase the proportion of the year's food production, decreasing the risk of food shortages and famine [7, 8].

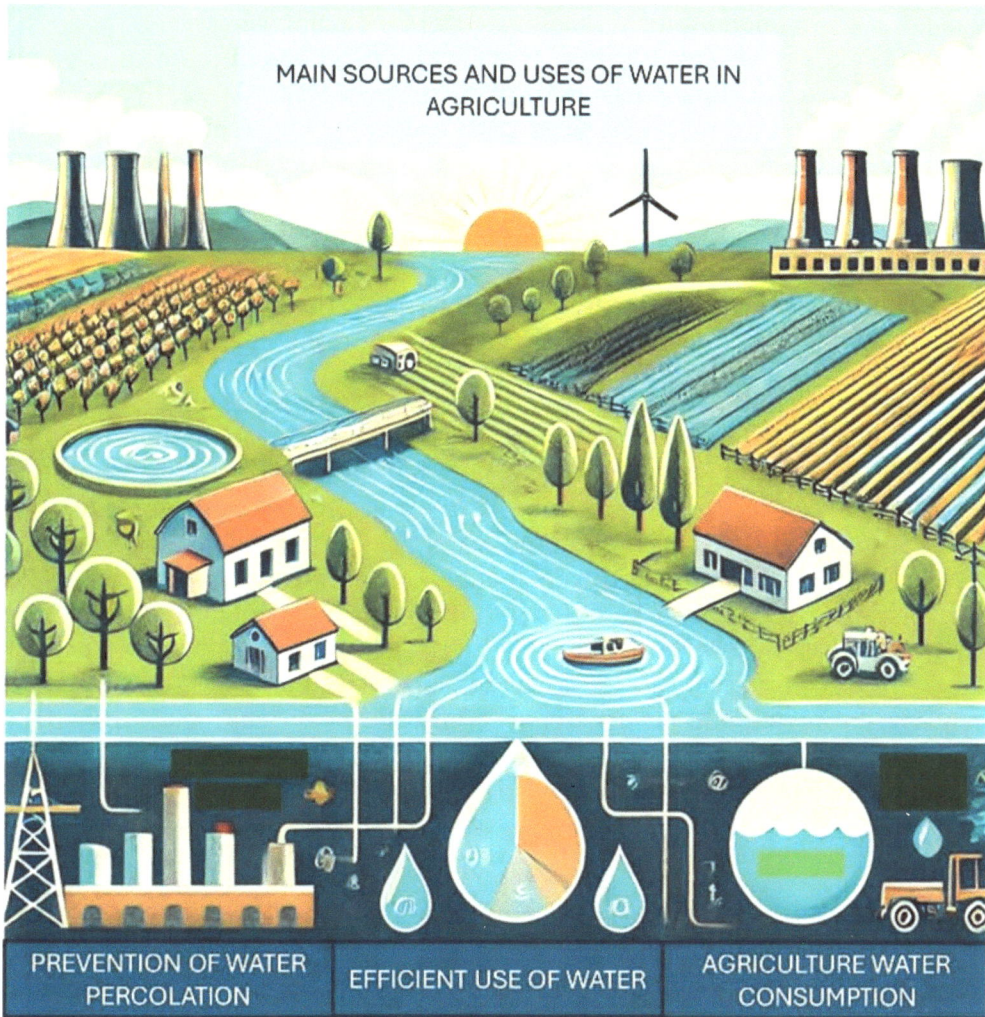

Fig. (2). Main sources and uses of water in agriculture. The figure shows the proportion of water resources used for agriculture, industry, and domestic purposes. It emphasizes the significant water consumption by agriculture and the need for efficient water management to prevent percolation and contamination of aquifers and surface water. Figure made with DALL-E.

Challenges in Water Management in Agriculture

Efforts are underway to overcome these constraints by developing plant genotypes suitable for different growing conditions and improving irrigation management strategies, including the use of plant growth regulators. Additionally, advances in understanding the regulatory systems of plant water status, combined with biotechnological tools, will provide rapid indications of the best strategies tailored to specific locations and climate conditions. Water preservation techniques, such as rainwater harvesting and sustainable use of degraded water, require the integration of several systems within the concept of Integrated Water Resource Management (IWRM). This is the proposal for Agro-ecological Hydro-spatial Planning (AHSP), aimed at integrating water resources with the potential of agricultural production. The use of several biotechnological tools, integrated with adequate water management, will enable the agricultural sector to produce the expected results in a scenario of increasing water resource limitations [9, 10].

The growing demand for water across the major productive sectors is generating increasing social conflicts over water use, resulting in economic policies that tend to penalize agricultural production, especially high-water-consuming crops. Over the past few decades, agriculture's escalating water use has made it one of the most unsustainable sectors. Agricultural water use has become inefficient due to the concentration of soluble fertilizers that have accumulated in the water bodies. These eutrophic waters reduce the supply of water for agricultural production. Climate change, with its various negative effects on environmental agricultural production, mainly related to the availability and quality of water, has further incentivized research into the efficient use of water in agriculture. The challenge is not only to increase water use efficiency but also to adapt to new climate conditions to ensure sustainable crop production [11, 12].

Scarcity of Freshwater Resources

Agriculture has the potential to move to closed systems in a stepwise manner, making water recycling a feasible option. A significant reduction, or even the total exclusion, of the pollution of drainage water from greenhouses and cold stores can be relatively easy to achieve and bring these agricultural activities to a closed water system. Animal production units can make use of part of the water discharged by hydroponically cultivated forage crops to reach at least a semiclosed water balance. The treated water from animal manure slurry or digestate could be applied to forage production in a closed system, and the forage produced may be fed to the animals, thereby completing the cycle [13, 14].

Global freshwater resources are increasingly unable to meet the growing demand for water. Nearly two-thirds of the world's population is likely to suffer from moderate to high water stress conditions by the early 21st century. Over 70% of the global freshwater resources are currently being used in agriculture, the activity that also contributes most to the pollution of underground and superficial water. About half of the Earth's arable lands rely on some form of irrigation for crop production, and the area under irrigation is increasing. It is estimated that 40% of the world's food is produced on 18% of the world's cultivated land that is irrigated. In Europe, the demand for water is continually increasing, driven largely by the increased use of water in industry, services, and tourism. This growing demand is depleting ecosystems in certain river basins at an alarming rate, intensifying competition for water between agriculture, other economic activities, and natural ecosystems in the coming years [15, 16].

Biotechnological Approaches for Water Conservation

Reusing greywater can be a viable water conservation strategy, particularly when used with appropriate biological and/or physical treatment systems, such as constructed wetlands, sand filters, or membrane bioreactors. However, greywater reuse poses health risks (microbial contamination) that have to be managed, particularly if there is high public exposure, such as in urban landscaping. The need for less concentrated (physically or biologically diluted) greywater to be used in an unrestricted manner has driven research and development of new treatment technologies, such as membrane distillation and electrocoagulation. The less concentrated effluents produced by these technologies can be further diluted with fresh water, and the resulting fractions may be used to maintain soil moisture or for other less critical purposes [17, 18].

The demand for water by agriculture competes directly with the needs of municipalities and industries. Reducing agriculture's reliance on freshwater by increasing the use of marginal-quality water and implementing techniques that decrease water loss at all steps in the agricultural production process can help alleviate this competitive pressure. Water recycling is a key strategy for conservation. Closed hydroponic or aeroponic systems in greenhouses enable very efficient use of water and chemicals by recirculating the nutrient solutions in which plants are grown. The market for greenhouse-grown products using these advanced systems—such as tomatoes, cucumbers, peppers, lettuce, other salad greens, and herbs—is expanding. In some cases, commercial products require desalination of the recirculating nutrient solutions, but in other cases, special salt-tolerant nutrient solutions are used— known as "half-strength" greenhouse nutrient solutions [19, 20].

Drought-Tolerant Crop Varieties

Drought-tolerant crop varieties are essential components of the effective farm-water management. In fact, they are the best single, long-term defense against the impact of intermittent drought on agricultural water use. When a drought-stressed farm is permitted to reduce the loss of water and improve its conservation, the stored soil moisture can often provide enough benefit to allow the planting of a subsequent (sequentially less valuable) crop, support tree or vine growth closer to the harvest period, or serve as a crucial buffer when surface water allocations are reduced. All the genetic, physiological, and root biotechnological approaches described in this chapter may, in the future, contribute not only to the creation of drought-tolerant crop varieties but also to the dynamic management of agricultural water ranging from gene selection to on-farm variety use [21, 22].

Plant breeding is an ancient practice that started with the transition from hunting and gathering to the first agricultural societies. The significance of selecting the best plant individuals to form the foundation of modern agriculture cannot be overstated. Today, society faces a huge challenge, which is to double food production in the next 40 years to feed a population of more than 9 billion people. However, this increase in food production has to be achieved with less water. To cope with this scenario, plant breeding for drought-tolerant crop varieties is a viable and sustainable solution [23, 24].

Biotechnological Techniques for Water Reuse

Due to the increased urbanization of the global population, there is a growing need to treat, reuse, or discharge municipal wastewater, which has a significant environmental impact. Based on the characteristics of effluents and the problems they present (*e.g.*, the presence of xenobiotic chemicals and the potential for the selection of pathogenic microorganisms), several biotechnological techniques can be applied to minimize these issues and enable the safe reuse of effluents in agriculture. The specific biotechnological techniques that address these challenges are discussed and presented. The text is easily understandable by non-specialists in the field [25, 26].

The reuse of water in agriculture is crucial for the conservation of water at the global level. The main sources of water for reuse in agriculture are agricultural drainage water, treated municipal wastewater, and industrial wastewater. However, each of these sources presents unique characteristics and challenges, requiring the application of specific biotechnological techniques to address them. The biotechnological techniques applied to solve or mitigate the specific problems associated with the reuse of each source in agriculture are presented and discussed, including biofiltration, constructed wetlands, and aquatic plants used for phyto-purification are described in detail. In addition, the potential use of certain halophyte crops for the reuse of saline drainage water from agricultural areas is explored [27, 28] (Fig. **3**). An example of different wetlands systems is shown in Fig. (**4**).

Fig. (3). Halophytes are salt-tolerant plants that can grow in saline environments, making them suitable for the reuse of saline drainage water in agriculture. These plants have specialized adaptations that allow them to thrive in high-salinity conditions by efficiently managing salt uptake and water retention. Halophyte crops are valuable for reusing saline water because they help absorb excess salts from the soil and water, making these resources more usable for irrigation in regions with saline water challenges. Figure made with DALL-E.

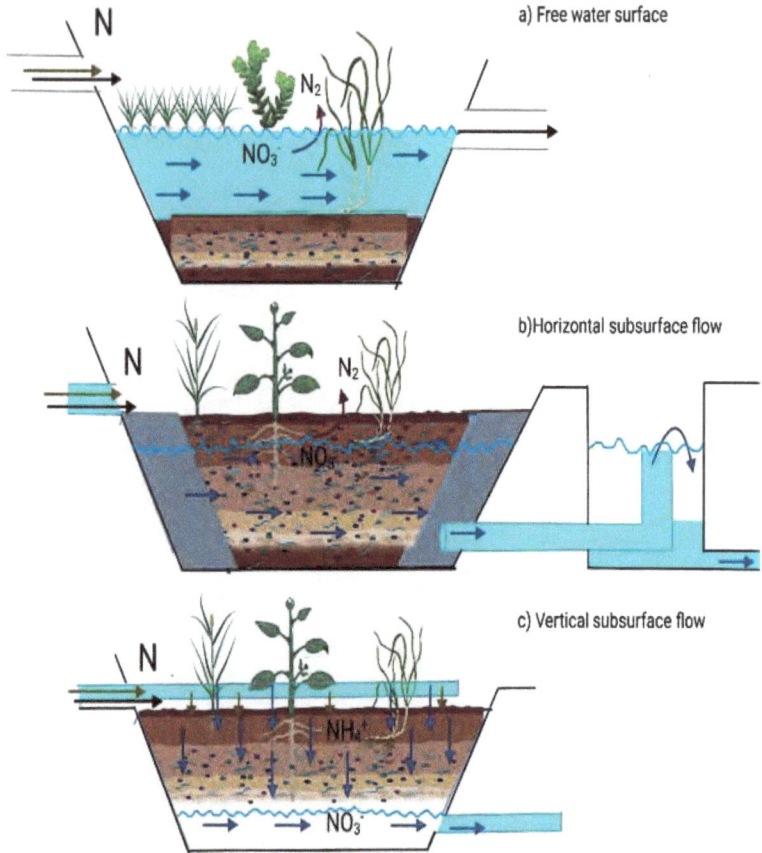

Fig. (4). Constructed Wetlands: These are engineered ecosystems designed to mimic natural wetlands and are used for treating and purifying wastewater. They consist of layers of vegetation, soil, and microorganisms that help filter out pollutants and contaminants from the water. Wetlands remove organic matter, nutrients, heavy metals, and pathogens through biological processes, making the water safe for reuse in agriculture. The plants in the wetland absorb nutrients and facilitate the breakdown of contaminants, promoting water purification. Three different wetlands models are shown in the figure: a) Free Water Surface Wetlands: In this system, water flows above the soil surface through vegetation. The exposed water allows for natural sunlight and wind to assist in the treatment process. It mimics a natural marsh, where plants and microorganisms break down pollutants in the water. b) Horizontal Subsurface Flow Wetlands: In this type, water flows horizontally below the surface of a gravel or soil layer, where plant roots and microorganisms treat the water. The absence of surface water reduces odors and mosquito breeding while maintaining high treatment efficiency. c)Vertical Subsurface Flow Wetlands: In this system, water moves vertically through several layers of substrate, usually from the top down, before being collected at the bottom. This design ensures better aeration and is effective for nutrient and pathogen removal. Adapted from [28]. Figure made with Biorender.

As explained in the introduction, the reuse of water in agriculture is of paramount importance for sustainable agriculture and, consequently, for the conservation of water at the global level. In general, agricultural drainage water, treated municipal wastewater, and industrial wastewater are the most relevant sources of water for reuse in agriculture. Each of these sources presents specific characteristics and challenges, requiring the application of appropriate biotechnological techniques. The biotechnological techniques that can be used to solve or mitigate the specific problems associated with the reuse of each water source in agriculture are presented in this chapter [29, 30].

Wastewater Treatment and Recycling

Agriculture can significantly contribute to the recycling of water used in other economic activities. The use of domestic sewage effluent and industrial wastewater for crop irrigation is widespread. However, as with other forms of treated water, certain conditions have to be met to ensure that recycling municipal and industrial wastewater does not cause environmental harm. The use of wastewater in agriculture has to be carefully planned and monitored to prevent the introduction of alien species into aquatic systems, chemicals that interfere with the processing of food, harmful substances, and the growth of different pathogenic microorganisms that may harm aquatic biota. Animal production, especially aquaculture, generates a large amount of nutrient-rich water that is particularly well-suited to be used in intersystem water recycling [28, 31].

Modern biotechnological research has developed plants that can be used in phytoremediation. Phytoremediation is a process that uses plants to clean up contaminants in the environment or render them harmless. Wastewater from aquaculture contains high levels of nutrients, especially in the form of fish manure. Treatment systems that use wetland plants to remove or inactivate nutrients and other contaminants are an attractive option for treatment. Suspended algae (phytoplankton) are very efficient in effecting the initial conversion of nutrients in wastewater, particularly nitrogen and phosphorus. Several plant species can accumulate high levels of heavy metals, becoming hyperaccumulators. Transgenic plants can be improved through biotechnological measures to achieve better results and increased stability in the uptake of substances from the soil medium [32, 33].

Case Studies and Success Stories

The Use of Hydroponics Techniques to Reduce the Amount of Water Used in the Growth of Forage Crops in Qatar

The project comprised components: the development of circulating hydroponic techniques and non-circulating hydroponic techniques. The development of circulating hydroponic techniques significantly improved the system and reduced the cost and difficulties inherent in standing up the non-circulating technique. The inorganic matter supply was based on the use of diluted seawater, which ensured the supply of necessary components for plant growth. This development will enable Qatar to reduce its substantial water footprint in the production of forage grass. With this technique, the stakeholders could reduce the abstracted water for the production of forage grass, thereby benefiting from reduced reliance on depleting freshwater resources. The non-circulating technique that uses disposable trays has allowed quick seeding and has been used for the rapid development of green areas in Doha [34, 35].

Low-pressure Center Pivot: A Case Study on Water Savings in Sugarcane Commercial Plantations

Sugarcane is one of the most water-demanding crops and is grown in regions with very limited water availability. The Low-Pressure System (LPS) was developed through the reduction of the operating pressure of the original system. When installed, the center pivot had an operating pressure of between 45 and 65 psi. This pressure allowed the selection of only sprinklers designed to operate at a high pressure. The LPS involved the substitution of high-pressure operation nozzles with medium- and low-pressure nozzles. An operating pressure of 10-28 psi was achieved. The innovation started on one farm and then spread to others through the sharing of experiences. As a result of this innovation, the water savings were estimated to be between 20 and 40% [36, 37]. An example diagram of two different LPS is shown in Fig. (5).

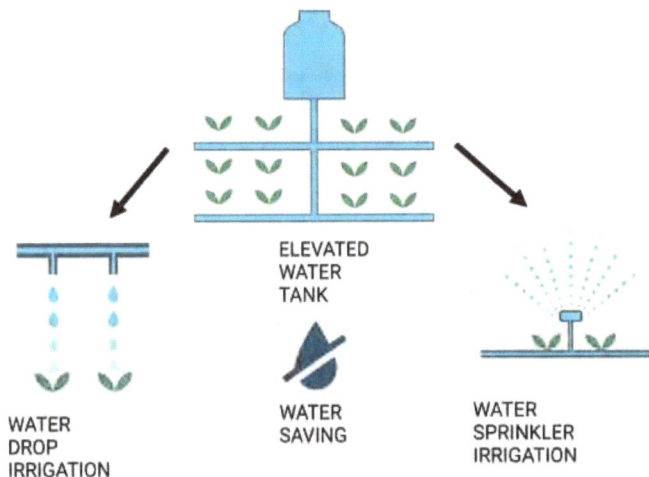

Fig. (5). Two irrigation methods: water drop irrigation and sprinkler irrigation. Both systems are powered by low-pressure water sources, such as elevated water tanks or energy-efficient water pumps, demonstrating sustainable water use and energy-saving techniques in agricultural irrigation. Figure made with Biorender.

Implementation of Biotechnological Solutions in Real-world Settings

The key to success in the coming era of modern biotechnology lies in its harmonious combination with existing conventional and simple methodologies through participative innovation models involving the stakeholders at various levels. An ecological understanding of the relevance of biotechnological solutions for improved agricultural water management situated within a knowledge-based adaptive framework specific to geographical regions, local ecosystems, and socio-economic systems is the need of the hour to prevent any evil consequences of rapid advances in modern biotechnology. Only then can the inherent potentials of advanced biotechnological applications cascade to accelerate knowledge-based innovative solutions for efficient conservation and reuse of water in agriculture for global welfare [38, 39].

The practical utilization of biotechnological breakthroughs in agricultural water management for full realization remains mostly theoretical. However, in the days to come, the implementation of advanced biotechnological approaches could enhance the application of some existing conventional methods for better results. Transgenic crops can play a vital role in overcoming the problem of water shortages by helping to grow more food with fewer inputs, reducing loss and waste, and contributing to the overall sustainability of agricultural systems. Even in the complex limited water flow scenario in agriculture, some low-cost, non-commercially formulated biostimulants could perform a salvage operation to a certain degree by efficiently utilizing available water. The development of low-cost, labor-effective field phenotyping techniques could expedite the progress of open-field research on crops related to stress management and the real effect of developed biotechnological solutions under prevalent stress conditions [40].

Future Directions and Innovations

Water deficits are the most important factor limiting crop growth and yield in large areas of the world. Among the available technologies to increase water use efficiency, easier and faster gains can be obtained from the application of molecular techniques, with or without biotechnological components. In recent times, the environmental movement has shifted the focus of agricultural research, emphasizing the development of technologies that are not only efficient but also environmentally friendly. In this context, biotechnology has demonstrated its capacity to support agricultural practices that minimize negative environmental impacts [40, 41].

The pace of innovation in the field of plant biotechnology has outpaced that of nearly any other sector related to agriculture. Unforeseen advances have validated predictions that plant biotechnology would revolutionize not only the seed industry but also all agriculture as a whole. Looking ahead, success will come to companies that continue to invest in new technologies. Innovation in product development, whether through biotechnology or the chemistry platform, is the key to growth in the agricultural sector. The development of input-efficient crop varieties, which maximize resource utilization, represents a new and rapidly expanding area within biotechnology [42].

Emerging Technologies in Water Conservation

Likewise, the growth substances that regulate the growth and development of plants can be utilized to control the loss of water and trigger responses to drought, thereby promoting water conservation. The growth inhibitors might also be employed to manage the growth of the roots and shoots of the plant, diminishing the demand for water and minimizing the loss of abscisic acid. Water conservation in horticulture or agronomic crops can also be achieved by designing irrigation systems that deliver water directly to the roots of the plants using bioengineering and biotechnological techniques. Molecular genetics and genetic engineering techniques also help in the conservation of water by engineering crop species [43, 44].

Water conservation for sustainable agriculture may be achieved by designing crop plants, whether existing or newly developed, in such a way that they grow, mature, and produce a high yield with minimal water input. Emerging technologies, biotechnologies, and molecular techniques, in collaboration with other sciences, can help crop plants utilize less water and reduce water loss and soil evaporation through more efficient control of the transpiration process. The development of new varieties of crops by using biotechnological techniques, for example, plant breeding, can result in a lower requirement of water for growth and development. Genetically modified crops can exhibit resistance to drought and the absorption of water from great depths in the soil [45, 46].

New Technologies for Agriculture: Optimization of Water Resources

The agricultural sector faces numerous challenges in improving decision-making due to the need for timely information and appropriate decisions. Technologies such as GPS satellites, drones, and sensors allow data collection to create predictive models for selecting suitable crops for higher yields [47]. This data-driven decision-making process enhances accuracy and results. Integration of drone data with satellite imagery, ground sensors, and autonomous vehicles enables comprehensive field analysis and management [48]. Drones equipped with various sensors, including multispectral cameras and NDVI sensors, can assess crop health, optimize resource use, and improve overall agricultural productivity [49]. Additionally, understanding weather conditions, soil nature, and crop features helps provide the exact needs of the crops, leading to the development of computerized, automated systems for accurate prediction and improved decision-making in agriculture [47].

Innovations like India's AgriBot, which performs essential farming functions by using image processing and motor control, and New Zealand's Vitirover, a solar-powered robot for weed removal, showcase advancements in automation and technology. Agribots are autonomous robots designed for various agricultural tasks, aiming to increase efficiency and reduce manual labor in farming [50]. These robots can perform functions such as ploughing, seeding, fertilizing, and harvesting. Advanced agribots like Vitirover use solar power for grass control in vineyards and can monitor crop health and environmental conditions [51]. Many agribots utilize microcontrollers like Arduino or Raspberry Pi for operation and control [52]. They often incorporate sensors for obstacle avoidance and precision farming. These agricultural robots offer promising solutions to labor-intensive tasks, improving productivity and reducing costs in various farming practices worldwide [47].

Researchers have utilized artificial neural networks and image processing techniques for weed classification in agriculture, significantly reducing manual labor and herbicide usage through autonomous field scouting and selective weed detection. Neural networks have shown promising results in weed classification for precision agriculture. Various approaches have been explored, including backpropagation networks using color co-occurrence method texture statistics [53], artificial neural networks with shape features, and convolutional neural networks (CNNs) [54]. Deep learning models have achieved high accuracy, with some studies reporting over 97% accuracy [54]. Drones equipped with GPS, cameras, and sensors are also used to monitor plant health, water supply, and the presence of cattle, with an expected adoption rate of 80% among farmers in the coming years [47].

Rule-Based Agriculture System (RBAS)

The RBAS methodology integrates smart and precision agriculture to optimize agricultural practices in response to changing environmental conditions. This approach involves a comprehensive examination of environmental factors such as climate, soil quality, and weather patterns to inform crop selection. Smart technology and data-driven decision-making optimize all stages of the agricultural lifecycle, from planting to harvest and sale, with the goal of increasing productivity, sustainability, and profitability [47]. RBAS has emerged as a valuable tool for enhancing agricultural practices. These expert systems utilize IF-THEN rules and inference engines to provide decision support for various agricultural challenges [55]. RBAS applications include pest and disease management for crops like rice, wheat, and coconut [56], soil nutrient management [57], and automated fertilization and irrigation. These systems can diagnose problems, recommend solutions, and even generate alerts for farmers [58]. RBAS combines expertise from multiple disciplines, such as plant pathology and entomology, to address the specific on-site needs of farmers.

The RBAS methodology is divided into three phases: survey, growth, and warehouse [47]:

In the survey phase, before the commencement of planting activities, a series of tests are carried out to assess the quality of water, soil composition, and the suitability of fertilizers to be used. The process of data collection involves the utilization of GPS-based yield monitoring systems, soil testing procedures, and the gathering of weather-related data to effectively monitor the prevailing field conditions and the moisture content of the soil.

The growth phase involves the meticulous selection of high-grade seeds, the implementation of appropriate spacing techniques, and the application of supplementary fertilizers to facilitate the initial growth of crops. Furthermore, protective measures such as fencing are employed to safeguard the crops against potential threats posed by birds, insects, and other animals. Strategic decisions based on predictive models serve as a guiding force for farmers regarding the optimal timing for sowing, watering, fertilizing, and harvesting activities.

Lastly, in the warehouse phase, activities related to harvesting, storage, procurement, and trading are executed in a timely manner to ensure efficient management. A substantial volume of agricultural data is meticulously stored in databases, a resource that proves instrumental in aiding farmers in making well-informed decisions in a timely manner, thereby contributing to the enhancement of crop yields.

Precision Agriculture and Data-Driven Farming

Precision agriculture utilizes technology and data to enhance farming methods at a field-specific or individual plant level. Information is gathered from various sources, such as weather stations, GPS, drones, and satellites, for the assessment of soil quality, moisture content, temperature, and crop well-being. By employing data analytics and machine learning, tailored insights are derived for precise agricultural techniques. Automation and robotic systems are integrated into precision agriculture for tasks like sowing, reaping, and other agricultural activities, thereby promoting efficiency in resource utilization and upholding environmental sustainability [47]. These systems utilize sensors to monitor environmental conditions, automate irrigation, and detect crop diseases. The Internet of Things (IoT) enables remote monitoring and data-driven decision-making, optimizing resource utilization and minimizing manual tasks [59, 60]. Advanced technologies such as aerial imagery, multispectral cameras, and agricultural robots contribute to precise crop management and increased yields [61]. Smart agriculture systems can address challenges like climate change, pest attacks, and water management while improving soil fertility [62].

Data Collection in Smart Farming

Smart agriculture depends on the utilization of sensors for the acquisition of up-to-date information regarding the well-being of crops, livestock, meteorological patterns, and soil quality. The consistent gathering of data plays a crucial role in supplying necessary insights for prompt decision-making processes. To illustrate, the utilization of soil moisture sensors aids in determining the appropriate timing for crop irrigation [47]. Cloud-based systems and big data processing architectures are being developed to handle the challenges of data acquisition, storage, and visualization [63]. The integration of these technologies aims to maximize productivity and sustainability in agriculture while addressing challenges such as data governance and confidentiality [64].

Machine Learning in Agriculture

Machine learning algorithms aid agricultural practitioners in the selection of economically viable crops and the anticipation of agricultural outputs. Models such as random forest, support vector machine, multivariate linear regression, artificial neural network, and K-nearest neighbor are utilized to approximate crop yield. The integration of GPS-enabled mobile applications supports farmers in the identification of appropriate crops and the enhancement of fertilizer applications to maximize agricultural productivity [47]. The technology is particularly useful in pre-harvesting, harvesting, and post-harvesting activities, allowing for more precise and efficient farming with less human labor [65]. However, challenges remain, such as the need for large amounts of high-quality data, especially in remote areas [66]. Despite these obstacles, ML has the potential to revolutionize agriculture and contribute to food security.

Advantages and Challenges of Precision Agriculture

Advantages

Precision agriculture offers numerous benefits, including improved crop yields, better soil quality management, precise pest management, and real-time data collection. The technology enhances resource efficiency, reduces environmental impact, and increases profitability by optimizing fertilizer and pesticide application, water usage, and crop management strategies. By tailoring inputs such as fertilizers, water, and pesticides to specific field conditions, precision farming minimizes waste and increases efficiency [67, 68]. The technology has shown promising results in various crops, including sugar beet, sugarcane, tea, and coffee [69].

Challenges

Despite its benefits, precision agriculture faces challenges such as high initial capital costs, the need for extensive data collection and analysis, and the requirement for technical expertise. In developing countries, its adoption is often hindered by cultural perceptions, small farm sizes, a lack of success stories, and limited access to finance and credit facilities. While precision agriculture offers significant benefits, challenges related to cost, accessibility, and data management need to be addressed [70]. Nevertheless, the potential environmental and economic advantages of precision farming justify greater public and private sector incentives to encourage adoption, particularly in small-scale farming systems in developing countries [71].

Artificial Intelligence (AI) and Smart Agriculture for Sustainable Development

AI technologies are transforming farming by increasing efficiency and overcoming traditional challenges through enhanced crop and soil monitoring, disease diagnosis, yield prediction, and decision-making. They also aid in weather forecasting, livestock management, and precision agriculture, providing real-time data for better resource management and cost savings. AI is emerging as a powerful tool for sustainable agriculture, offering solutions to meet growing food demands while addressing environmental challenges. Its applications include crop monitoring, disease detection, resource management, and decision support systems [72 - 74]. These technologies can enhance productivity, reduce environmental impact, and improve farmers' livelihoods [73, 75].

CONCLUDING REMARKS

The agricultural sector is undergoing a transformation through smart farming and precision agriculture, integrating advanced technologies and data-driven insights. These innovations are significantly improving decision-making, boosting productivity, and promoting sustainability. However, meeting the increasing demand for food and improving the quality of life in the face of decreasing freshwater availability remains a critical challenge.

The development of crops that use water more efficiently is essential for sustainable food production, especially in regions with low and unpredictable precipitation. Biotechnological interventions, including genetic modification and less complex alternatives, have been pivotal in enhancing the water use efficiency of crops. Field trials of genetically modified (GM) crops and innovative combined approaches involving both biotechnology and membrane technology show promise in propelling water reuse in agriculture. However, these biotechnological tools and processes must be carefully managed to ensure a favorable benefit-to-risk ratio, particularly when the genes used are derived from sources closely related to the recipient plant.

Water recycling in anthropic processes, such as in urban centers, can increase the availability of water for agriculture. However, wastewater from urban activities often contains high levels of organic matter, salts, heavy metals, and other toxic compounds that can adversely affect soil, plants, and human health. Technological convergence, utilizing both biotechnological tools for plant development and physicochemical techniques like membranes to improve treated wastewater quality, offers a promising pathway for implementing water recycling in agriculture.

In regions experiencing water scarcity or during specific periods of the year, it is crucial to allocate the limited water available for high-value crops. The reuse of wastewater in agriculture, after appropriate treatment, can mitigate chemical and physical challenges associated with low-quality irrigation water. Biotechnological research plays a vital role in this process by developing highly efficient plant purification systems. This chapter reviews various biotechnological interventions to conserve water in agriculture, including reducing water loss in soil-plant systems, selecting crops based on their water use efficiency, reusing wastewater, and employing phytoremediation of pollutants.

Water scarcity poses a significant threat to humankind. Although three-quarters of the earth's surface is covered by water, 97% is saline, and much of the remaining 3% of freshwater is locked in ice caps and glaciers. Less than 1% of water resources are available to meet the needs of the growing population for drinking, sanitation, and food production. Agriculture accounts for about 70% of global freshwater consumption. Promoting research to enhance knowledge in plant water use, developing water-saving biotechnologies, and implementing these innovations are crucial steps in addressing water scarcity. Biotechnologies hold substantial potential in developing crop plants with increased resistance to various environmental stresses. Molecular studies on transgenic plants have shown that manipulating single genes controlling characteristics related to water scarcity response can improve crops' water use efficiency.

In summary, smart farming, precision agriculture, and biotechnological advancements are key to overcoming the challenges posed by water scarcity and ensuring sustainable agricultural practices. These innovations not only enhance productivity and sustainability but also ensure food security and environmental conservation for the future.

AUTHORS' CONTRIBUTION

The authors confirm their contribution to the chapter as follows:

Conceptualization, methodology, supervision, writing-original draft preparation, writing review, and editing: Israel Valencia Quiroz; Methodology, investigation, data curation, validation, writing-original draft preparation, writing review, and editing: Diana Violeta Sánchez Oropeza; Formal analysis, investigation, data curation, visualization, writing-original draft preparation, writing review, and editing: María Fernanda Trujillo Lira; Investigation, resources, data curation, writing-original draft preparation, writing review, and editing: Miriam Arlette López Pérez; Methodology, validation, data curation, writing-original draft preparation, writing review, and editing: Casandra Rosales García; Conceptualization, supervision, validation, writing review, and editing: Susana Rafael Maya.

QUESTIONS RELATED TO THE TEXT

- What are the primary techniques mentioned in the chapter for reducing water waste in agriculture, and how do they contribute to sustainable agricultural practices?
- Explain how biotechnological approaches, such as genetic modifications, can enhance water use efficiency in crops. Provide examples from the chapter to illustrate your answer.
- Discuss the role of treated wastewater in agricultural irrigation. What biotechnological methods are used to ensure the safe and effective reuse of low-quality water for crops?
- What are the challenges associated with water scarcity in agriculture, and how do the strategies presented in the chapter address these challenges to improve food security?
- Describe the concept of Integrated Water Resource Management (IWRM) as presented in the chapter. How does it integrate water conservation techniques into agricultural practices?
- Review the case studies provided in the chapter. What successful applications of biotechnological solutions were highlighted, and what future directions are suggested for enhancing water use efficiency in agriculture?

LIST OF ABBREVIATIONS

AHSP Agro-ecological Hydro-spatial Planning

AI Artificial Intelligence

GM Genetically Modified

IoT Internet of Things

IWRM Integrated Water Resource Management

LPS Low-Pressure System

NDVI Normalized Difference Vegetation Index

RBAS Rule-Based Agriculture System

REFERENCES

[1]   Fang L, Zhang L. Does the trading of water rights encourage technology improvement and agricultural water conservation? Agric Water Manage 2020; 233: 106097.
[http://dx.doi.org/10.1016/j.agwat.2020.106097]

[2]   Xu H, Yang R. Does agricultural water conservation policy necessarily reduce agricultural water extraction? Evidence from China. Agric Water Manage 2022; 274: 107987.
[http://dx.doi.org/10.1016/j.agwat.2022.107987]

[3]   He C, Liu Z, Wu J, *et al.* Future global urban water scarcity and potential solutions. Nat Commun 2021; 12(1): 4667.
[http://dx.doi.org/10.1038/s41467-021-25026-3] [PMID: 34344898]

[4]   Huang Z, Yuan X, Liu X. The key drivers for the changes in global water scarcity: Water withdrawal *versus* water availability. J Hydrol (Amst) 2021; 601: 126658.
[http://dx.doi.org/10.1016/j.jhydrol.2021.126658]

[5]   Rosa L, Chiarelli DD, Rulli MC, Dell'Angelo J, D'Odorico P. Global agricultural economic water scarcity. Sci Adv 2020; 6(18): eaaz6031.
[http://dx.doi.org/10.1126/sciadv.aaz6031] [PMID: 32494678]

[6]   Ulian T, Diazgranados M, Pironon S, *et al.* Unlocking plant resources to support food security and promote sustainable agriculture. Plants People Planet 2020; 2(5): 421-45.
[http://dx.doi.org/10.1002/ppp3.10145]

[7]   Choithani C, van Duijne RJ, Nijman J. Changing livelihoods at India's rural–urban transition. World Dev 2021; 146: 105617.
[http://dx.doi.org/10.1016/j.worlddev.2021.105617]

[8]   Deng Q, Li E, Zhang P. Livelihood sustainability and dynamic mechanisms of rural households out of poverty: An empirical analysis of Hua County, Henan Province, China. Habitat Int 2020; 99: 102160.
[http://dx.doi.org/10.1016/j.habitatint.2020.102160]

[9]   Eyni-Nargeseh H, AghaAlikhani M, Shirani Rad AH, Mokhtassi-Bidgoli A, Modarres Sanavy SAM. Late season deficit irrigation for water-saving: selection of rapeseed (*Brassica napus*) genotypes based on quantitative and qualitative features. Arch Agron Soil Sci 2020; 66(1): 126-37.
[http://dx.doi.org/10.1080/03650340.2019.1602866]

[10]   Zhou H, Chen J, Wang F, Li X, Génard M, Kang S. An integrated irrigation strategy for water-saving and quality-improving of cash crops: Theory and practice in China. Agric Water Manage 2020; 241: 106331.
[http://dx.doi.org/10.1016/j.agwat.2020.106331]

[11]   Guo Y, Zhang M, Liu Z, *et al.* Applying and Optimizing Water-Soluble, Slow-Release Nitrogen Fertilizers for Water-Saving Agriculture. ACS Omega 2020; 5(20): 11342-51.
[http://dx.doi.org/10.1021/acsomega.0c00303] [PMID: 32478222]

[12]   Martínez-Dalmau J, Berbel J, Ordóñez-Fernández R. Nitrogen Fertilization. A Review of the Risks Associated with the Inefficiency of Its Use and Policy Responses. Sustainability (Basel) 2021; 13(10): 5625.
[http://dx.doi.org/10.3390/su13105625]

[13]   Altamira-Algarra B, Puigagut J, Day JW, *et al.* A review of technologies for closing the P loop in agriculture runoff: Contributing to the transition towards a circular economy. Ecol Eng 2022; 177: 106571.
[http://dx.doi.org/10.1016/j.ecoleng.2022.106571]

[14]   Li S, Chan CY, Sharbatmaleki M, Trejo H, Delagah S. Engineered Biochar Production and Its Potential Benefits in a Closed-Loop Water-Reuse Agriculture System. Water 2020; 12(10): 2847.
[http://dx.doi.org/10.3390/w12102847]

[15]   Adak S, Mandal N, Mukhopadhyay A, Maity PP, Sen S. Current State and Prediction of Future Global Climate Change and Variability in Terms of CO2 Levels and Temperature. In: Naorem A, Machiwal D, editores. Enhancing Resilience of Dryland Agriculture Under Changing Climate: Interdisciplinary and Convergence Approaches [Internet]. Singapore: Springer Nature Singapore; 2023. p. 15-43.
[http://dx.doi.org/10.1007/978-981-19-9159-2_2]

[16]   Poonia V, Kumar Goyal M, Jha S, Dubey S. Terrestrial ecosystem response to flash droughts over India. J Hydrol (Amst) 2022; 605: 127402.
[http://dx.doi.org/10.1016/j.jhydrol.2021.127402]

[17]   Filali H, Barsan N, Souguir D, Nedeff V, Tomozei C, Hachicha M. Greywater as an Alternative Solution for a Sustainable Management of Water Resources—A Review. Sustainability (Basel) 2022; 14(2): 665.
[http://dx.doi.org/10.3390/su14020665]

[18]   Khor CS, Akinbola G, Shah N. A model-based optimization study on greywater reuse as an alternative urban water resource. Sustainable Production and Consumption 2020; 22: 186-94.
[http://dx.doi.org/10.1016/j.spc.2020.03.008]

[19]   Carotti L, Pistillo A, Zauli I, *et al.* Improving water use efficiency in vertical farming: Effects of growing systems, far-red radiation and planting density on lettuce cultivation. Agric Water Manage 2023; 285: 108365.
[http://dx.doi.org/10.1016/j.agwat.2023.108365]

[20]   Massa D, Magán JJ, Montesano FF, Tzortzakis N. Minimizing water and nutrient losses from soilless cropping in southern Europe. Agric Water Manage 2020; 241: 106395.
[http://dx.doi.org/10.1016/j.agwat.2020.106395]

[21]   Ray RL, Ampim PAY, Gao M. Crop protection under drought stress. In: Jabran K, Florentine S, Chauhan BS, editors. Crop Protection Under Changing Climate [Internet]. Cham: Springer International Publishing; 2020. 145-70.
[http://dx.doi.org/10.1007/978-3-030-46111-9_6]

[22]   Reddy YAN, Reddy YNP, Ramya V, Suma LS, Reddy ABN, Krishna SS. Chapter 8 - Drought adaptation: Approaches for crop improvement. In: Singh M, Sood S, editores. Millets and Pseudo Cereals [Internet]. Woodhead Publishing; 2021. p. 143-58. Availble from: https://www.sciencedirect.com/science/article/pii/B9780128200896000082

[23]   Simtowe F, Makumbi D, Worku M, Mawia H, Rahut DB. Scalability of Adaptation strategies to drought stress: the case of drought tolerant maize varieties in Kenya. Int J Agric Sustain 2021; 19(1): 91-105.
[http://dx.doi.org/10.1080/14735903.2020.1823699]

[24]   Yahaya MA, Shimelis H. Drought stress in sorghum: Mitigation strategies, breeding methods and technologies—A review. J Agron Crop Sci 2022; 208(2): 127-42.
[http://dx.doi.org/10.1111/jac.12573]

[25]   Kesari KK, Soni R, Jamal QMS, *et al.* Wastewater Treatment and Reuse: a Review of its Applications and Health Implications. Water Air Soil Pollut 2021; 232(5): 208.
[http://dx.doi.org/10.1007/s11270-021-05154-8]

[26]   Khan MM, Siddiqi SA, Farooque AA, *et al.* Towards Sustainable Application of Wastewater in Agriculture: A Review on Reusability and Risk Assessment. Agronomy (Basel) 2022; 12(6): 1397.
[http://dx.doi.org/10.3390/agronomy12061397]

[27]   Shoushtarian F, Negahban-Azar M. Worldwide Regulations and Guidelines for Agricultural Water Reuse: A Critical Review. Water 2020; 12(4): 971.
[http://dx.doi.org/10.3390/w12040971]

[28]   Passeport E, Vidon P, Forshay KJ, *et al.* Ecological engineering practices for the reduction of excess nitrogen in human-influenced landscapes: a guide for watershed managers. Environ Manage 2013; 51(2): 392-413.
[http://dx.doi.org/10.1007/s00267-012-9970-y] [PMID: 23180248]

[29]   Castellini M, Diacono M, Gattullo CE, Stellacci AM. Sustainable Agriculture and Soil Conservation. Appl Sci (Basel) 2021; 11(9): 4146.
[http://dx.doi.org/10.3390/app11094146]

[30]   Srivastav AL, Dhyani R, Ranjan M, Madhav S, Sillanpää M. Climate-resilient strategies for sustainable management of water resources and agriculture. Environ Sci Pollut Res Int 2021; 28(31): 41576-95.
[http://dx.doi.org/10.1007/s11356-021-14332-4] [PMID: 34097218]

[31]   Singh A. A review of wastewater irrigation: Environmental implications. Resour Conserv Recycling 2021; 168: 105454.
[http://dx.doi.org/10.1016/j.resconrec.2021.105454]

[32]   Anwar A, Kim JK. Transgenic Breeding Approaches for Improving Abiotic Stress Tolerance: Recent Progress and Future Perspectives. Int J Mol Sci 2020; 21(8): 2695.
[http://dx.doi.org/10.3390/ijms21082695] [PMID: 32295026]

[33]   Raza A, Tabassum J, Fakhar AZ, *et al.* Smart reprograming of plants against salinity stress using modern biotechnological tools. Crit Rev Biotechnol 2023; 43(7): 1035-62.
[http://dx.doi.org/10.1080/07388551.2022.2093695] [PMID: 35968922]

[34]   Ahamed MS, Sultan M, Shamshiri RR, Rahman MM, Aleem M, Balasundram SK. Present status and challenges of fodder production in controlled environments: A review. Smart Agricultural Technology 2023; 3: 100080.
[http://dx.doi.org/10.1016/j.atech.2022.100080]

[35]   Bouadila S, Baddadi S, Skouri S, Ayed R. Assessing heating and cooling needs of hydroponic sheltered system in mediterranean climate: A case study sustainable fodder production. Energy 2022; 261: 125274.
[http://dx.doi.org/10.1016/j.energy.2022.125274]

[36]   Barnard JH, Matthews N, du Preez CC. Formulating and assessing best water and salt management practices: Lessons from non-saline and water-logged irrigated fields. Agric Water Manage 2021; 247: 106706.
[http://dx.doi.org/10.1016/j.agwat.2020.106706]

[37]   Mao L, Yang T, Zhang H, *et al.* Fully Textured, Production-Line Compatible Monolithic Perovskite/Silicon Tandem Solar Cells Approaching 29% Efficiency. Adv Mater 2022; 34(40): 2206193.
[http://dx.doi.org/10.1002/adma.202206193] [PMID: 35985840]

[38]   Holzinger A, Keiblinger K, Holub P, Zatloukal K, Müller H. AI for life: Trends in artificial intelligence for biotechnology. N Biotechnol 2023; 74: 16-24.
[http://dx.doi.org/10.1016/j.nbt.2023.02.001] [PMID: 36754147]

[39]   Yong JJJY, Chew KW, Khoo KS, Show PL, Chang JS. Prospects and development of algal-bacterial biotechnology in environmental management and protection. Biotechnol Adv 2021; 47: 107684.
[http://dx.doi.org/10.1016/j.biotechadv.2020.107684] [PMID: 33387639]

[40]   Mbava N, Mutema M, Zengeni R, Shimelis H, Chaplot V. Factors affecting crop water use efficiency: A worldwide meta-analysis. Agric Water Manage 2020; 228: 105878.
[http://dx.doi.org/10.1016/j.agwat.2019.105878]

[41]   Sah RP, Chakraborty M, Prasad K, *et al.* Impact of water deficit stress in maize: Phenology and yield components. Sci Rep 2020; 10(1): 2944.
[http://dx.doi.org/10.1038/s41598-020-59689-7] [PMID: 32076012]

[42]   Sharma A, Banyal A, Sirjohn N, Kulshreshtha S, Kumar P. Nano-biotechnology and Its Applications in Maintaining Soil Health. In: Bhatia RK, Walia A, editores. Advancements in Microbial Biotechnology for Soil Health [Internet]. Singapore: Springer Nature Singapore; 2024. p. 323-42.
[http://dx.doi.org/10.1007/978-981-99-9482-3_14]

[43]   Farooqi ZUR, Ayub MA, Zia ur Rehman M, Sohail MI, Usman M, Khalid H, et al. Chapter 4 - Regulation of drought stress in plants. In: Tripathi DK, Pratap Singh V, Chauhan DK, Sharma S, Prasad SM, Dubey NK, et al., editores. Plant Life Under Changing Environment [Internet]. Academic Press;   2020.   p.   77-104.   Available   at: https://www.sciencedirect.com/science/article/pii/B9780128182048000047

[44]   Saberi Riseh R, Ebrahimi-Zarandi M, Gholizadeh Vazvani M, Skorik YA. Reducing Drought Stress in Plants by Encapsulating Plant Growth-Promoting Bacteria with Polysaccharides. Int J Mol Sci 2021; 22(23): 12979.
[http://dx.doi.org/10.3390/ijms222312979] [PMID: 34884785]

[45]   Munaweera TIK, Jayawardana NU, Rajaratnam R, Dissanayake N. Modern plant biotechnology as a strategy in addressing climate change and attaining food security. Agric Food Secur 2022; 11(1): 26.
[http://dx.doi.org/10.1186/s40066-022-00369-2]

[46]   Steinwand MA, Ronald PC. Crop biotechnology and the future of food. Nat Food 2020; 1(5): 273-83.
[http://dx.doi.org/10.1038/s43016-020-0072-3]

[47]   Balasubramanian S, Natarajan G, Raj P. Intelligent robots and drones for precision agriculture [Internet]. Cham, Switzerland: Springer; 2024. (Signals and communication technology).
[http://dx.doi.org/10.1007/978-3-031-51195-0]

[48]   Barrile V, Simonetti S, Citroni R, Fotia A, Bilotta G. Experimenting Agriculture 4.0 with Sensors: A Data Fusion Approach between Remote Sensing, UAVs and Self-Driving Tractors. Italian National Conference on Sensors 2022.
[http://dx.doi.org/10.3390/s22207910]

[49]   Puri VP, Raja L. Agriculture drones: a modern breakthrough in precision agriculture. J Stat Manag Syst. 2017; 20(4): 507-18.
[http://dx.doi.org/10.1080/09720510.2017.1395171]

[50]   Singh A, Gupta A, Bhosale A, Poddar S. Agribot: an agriculture robot. Int J Adv Res Comput Commun Eng. 2015; 4(1): 317.
[http://dx.doi.org/10.17148/IJARCCE.2015.4173]

[51]   Keresztes B, Germain C, Da Costa JP, Grenier G, Beaulieu XD. Vineyard Vigilant & INNovative Ecological Rover (VVINNER): an autonomous robot for automated scoring of vineyards. Proc Int Conf Agric Eng. Zurich; 2014 Jul 6-10. Available at: www.eurageng.eu

[52]   Bhorge PS. Wi-Fi Based Multipurpose Agri-bot. Int J Res Appl Sci Eng Technol 2024; 12(5): 4023-7.
[http://dx.doi.org/10.22214/ijraset.2024.62383]

[53]   Burks TF, Shearer SA, Gates RS, Donohue KD. Backpropagation Neural Network Design and Evaluation for Classifying Weed Species Using Color Image Texture. Trans ASAE 2000; 43(4): 1029-37.
[http://dx.doi.org/10.13031/2013.2971]

[54]   Hoang Trong V, Gwang-hyun Y, Thanh Vu D, Jin-young K. Late fusion of multimodal deep neural networks for weeds classification. Comput Electron Agric 2020; 175: 105506.
[http://dx.doi.org/10.1016/j.compag.2020.105506]

[55]   Grosan C, Abraham A. Rule-based expert systems. In: Intelligent Systems Reference Library [Internet]. Springer Berlin Heidelberg; 2011. 149-85.
[http://dx.doi.org/10.1007/978-3-642-21004-4_7]

[56]   Masri N, Abu Sultan Y, Akkila AN, Almasri A, Ahmed A, Mahmoud AY, Zaqout I, Abu-Naser SS. Survey of rule-based systems. Int J Acad Inf Syst Res. 2019; 3(7): 1-22.

[57]   Jadhav SK, Yelapure SJ. Rule-based expert system in the use of inorganic fertilizers for sugarcane crop. Int J Comput Appl. 2011; 36(4): 34-42.

[58]   Gogos C, Alefragis P, Housos E. Sensor enabled rule based alarm system for the agricultural industry. 2007 IEEE Conference on Emerging Technologies & Factory Automation (EFTA 2007) 2007.
[http://dx.doi.org/10.1109/EFTA.2007.4416880]

[59]   Prakash C, Singh LP, Gupta A, Lohan SK. Advancements in smart farming: A comprehensive review of IoT, wireless communication, sensors, and hardware for agricultural automation. Sens Actuators A Phys 2023; 362: 114605.
[http://dx.doi.org/10.1016/j.sna.2023.114605]

[60]   Misra T, Anwarul S, Srivastava D, Cheng X. Automation techniques for smart and sustainable agriculture and its challenges. Appl Comput Eng. 2023; 2: 749-758.
[http://dx.doi.org/10.54254/2755-2721/2/20220673]

[61]   Hassan SI, Alam MM, Illahi U, Al Ghamdi MA, Almotiri SH, Su'ud MM. A Systematic Review on Monitoring and Advanced Control Strategies in Smart Agriculture. IEEE Access 2021; 9: 32517-48.
[http://dx.doi.org/10.1109/ACCESS.2021.3057865]

[62]   Katiyar S, Farhana A. Smart Agriculture: The Future of Agriculture using AI and IoT. J Comput Sci 2021; 17(10): 984-99.
[http://dx.doi.org/10.3844/jcssp.2021.984.999]

[63]   Roukha A, Nolack Fote F, Mahmoudi SA, Mahmoudi S. Big data processing architecture for smart farming. Procedia Comput Sci. 2020; 177: 78-85.

[64]   Saiz-Rubio V, Rovira-Más F. From smart farming towards Agriculture 5.0: a review on crop data management. Agronomy [Internet]. 2020; 10(2).
[http://dx.doi.org/10.3390/agronomy10020207]

[65]   Meshram V, Patil K, Meshram V, Hanchate D, Ramkteke SD. Machine learning in agriculture domain: A state-of-art survey. Artificial Intelligence in the Life Sciences. 2021; 1:100010.
[http://dx.doi.org/10.1016/j.ailsci.2021.100010]

[66]   Bhardwaj B, Tiwari S. Exploring the Potential of Machine Learning in Agriculture: A Review of its Applications and Results. Research & Review: Machine Learning and Cloud Computing 2022; 2(1): 7-11.
[http://dx.doi.org/10.46610/RRMLCC.2023.v02i01.002]

[67]   McBratney A, Whelan B. Precision Ag. - Oz style. 2001.

[68]   Chen Z. Precision agriculture: technology, advantages and limitations. J Agric Sci Bot 2021; 5(10).

[69]   Rimpika, Anushi, Manasa S, Anusha KN, Sharma S, Thakur A, Shilpa, Sood A. An overview of precision farming. Int J Environ Clim Change [Internet]. 2023; 13(12): 441-56. Available at: https://journalijecc.com/index.php/IJECC/article/view/3701

[70] Sharma S. Precision agriculture: reviewing the advancements, technologies, and applications in precision agriculture for improved crop productivity and resource management. Rev Food Agric. 2023;4(2):45-49.
[http://dx.doi.org/10.26480/rfna.02.2023.45.49]

[71] Finger R, Swinton SM, El Benni N, Walter A. Precision Farming at the Nexus of Agricultural Production and the Environment. Annu Rev Resour Econ 2019; 11(1): 313-35.
[http://dx.doi.org/10.1146/annurev-resource-100518-093929]

[72] Sood A, Bhardwaj AK, Sharma RK. Towards sustainable agriculture: key determinants of adopting artificial intelligence in agriculture. J Decis Syst 2022; 1-45.
[http://dx.doi.org/10.1080/12460125.2022.2154419]

[73] Ibrahim U. Artificial intelligence in agricultural extension for sustainable development. Int J Appl Sci Res 2023; 1(3): 259-68.
[http://dx.doi.org/10.59890/ijasr.v1i3.740]

[74] Sachithra V, Subhashini LDCS. How artificial intelligence uses to achieve agriculture sustainability: systematic review. Artif Intell Agric. 2023; 8:46-59.
[http://dx.doi.org/10.1016/j.aiia.2023.04.002]

[75] Mathur R. Artificial Intelligence in Sustainable Agriculture. Int J Res Appl Sci Eng Technol 2023; 11(6): 4047-52.
[http://dx.doi.org/10.22214/ijraset.2023.54360]

CHAPTER 14

Ethical and Social Challenges of Environmental Biotechnology

Israel Valencia Quiroz[1,*]

[1] *Phytochemistry Laboratory, UBIPRO, Superior Studies Faculty (FES)-Iztacala, National Autonomous University of Mexico (UNAM), Tlalnepantla de Baz, Mexico State, Mexico*

Abstract: Environmental biotechnology offers the potential to develop and commercialize clean biological processes to restore, preserve, and improve the environment. However, these advancements also bring ethical and social challenges that require careful consideration. A primary ethical issue is the equitable distribution of the benefits of biotechnological processes among affected communities. It is essential to respect the autonomy of individuals and communities who oppose these technologies. Dialogue and democratic procedures are essential for addressing concerns and ensuring public participation in decision-making. Environmental biotechnology is subject to public scrutiny and criticism, particularly from influential interest groups concerned about the effects of these technologies on the environment and human health. Ethical considerations include the equitable distribution of benefits, public participation, and informed consent. The implementation of these principles depends on the legal frameworks and governmental support in different countries. The social implications of environmental biotechnology are significant. Public debate is often heated, with concerns arising from fears and uncertainties about new technologies. The bioprocessing industry must address these concerns through transparency and by involving communities in project decision-making. Traditional practices may also be affected, raising social concerns and potentially eroding cultural values. Regulatory frameworks and policies play a crucial role in managing the risks associated with environmental biotechnology. International agreements and conventions aim to establish common principles and rules to ensure safe and ethical applications of biotechnology. Case studies and examples, such as the use of biochar in soil amendment, highlight both the potential benefits and challenges of these technologies.

Keywords: Bioremediation, Bioethics, Community concern, Environmental biotechnology, Ethical considerations, Public participation, Regulatory frameworks, Sustainable practices, Social challenges, Technology implementation.

* **Corresponding author Israel Valencia Quiroz:** Phytochemistry Laboratory, UBIPRO, Superior Studies Faculty (FES)-Iztacala, National Autonomous University of Mexico (UNAM), Tlalnepantla de Baz, Mexico State, Mexico; Tel: +525556231137; E-mail: israelv@unam.mx

INTRODUCTION

Environmental biotechnology encompasses the application of biological processes and organisms to address and remediate environmental challenges. This multidisciplinary field combines principles from biology, chemistry, and engineering to develop sustainable solutions for pollution control, waste management, and resource conservation. By leveraging natural processes, environmental biotechnology aims to restore, preserve, and enhance the environment. The scope of this field is broad, including bioremediation of contaminated sites, the development of eco-friendly industrial processes, and the enhancement of agricultural practices through bio-based technologies. Ethical considerations play a crucial role in this domain, as the implementation of biotechnological solutions must be guided by principles of fairness, public participation, and respect for community autonomy. Addressing the social challenges and ensuring equitable distribution of benefits is essential for gaining public trust and fostering acceptance of these technologies. This chapter delves into the ethical and social dimensions of environmental biotechnology, highlighting the importance of ethical principles, regulatory frameworks, and community engagement in the successful implementation of biotechnological innovations.

Ethical Considerations in Environmental Biotechnology

If the installation of a clean process to ensure the decontamination of a nearby degraded ecosystem generates concern among the local or nearby inhabitants, the company should consider the reasons for this opposition and begin a dialogue with the affected individuals to address their concerns. In this case, the ethical principle recommended is to respect the autonomy of individuals who oppose the technology. If the reasons for this opposition are well-founded and the company that is implementing the technology has not taken the appropriate measures to ensure a just distribution of the benefits of the clean process or if the potential risks of the technology for the environment or human health have not been adequately evaluated, the technology enterprise must reconsider its project. Even when bioremediation is considered a safe technology and no risks to the population or the environment are anticipated, public concern against the implementation of this technology can arise. This concern often stems from fears among the affected population, primarily because they lack control over the technology and its effects. To address these concerns, it is necessary to establish democratic procedures that allow the interested parties to express their opinions and participate in the decisions [1] (Fig. **1**).

Fig. (1). Various ethical principles exist in environmental biotechnology, including respect for autonomy, just distribution of benefits, and public participation. The illustration depicts community meetings, discussions about bioremediation projects, and diverse groups of people expressing their opinions in an environmental setting. Figure made with DALL-E.

One of the main goals in the field of environmental biotechnology is the development and commercialization of clean biological processes to restore, preserve, and improve the environment. However, this process may raise some ethical issues. The first is a consequence of the pollution or alteration of an environment that affects two or more communities. In this case, determining which community is entitled to use a certain environment or has the authority to make decisions about the management of that environment can generate a social conflict. Ethical frameworks are required to resolve this potential conflict in the distribution of the benefits of biotechnological processes applied to environmental clean-up. Practical consequences stemming from these ethical considerations include the implementation of environmental biotechnology following the prior informed agreement (PIA) and ensuring a just distribution of the economic benefits derived from clean processes between the companies implementing these technologies and the communities affected by the pollution. It should be noted that a considerable part of these formulations has an idealistic character and that the implementation of these ethical considerations depends on the legislation and the governments of the countries where the technologies are applied [2, 3].

Principles of Bioethics

The term "environmental biotechnology" encompasses a complex international enterprise devoted to the application of novel biological systems and organisms to address environmental problems. Scientific and engineering efforts have developed field-tested technologies for the remediation of contaminated land, air, and water, showing promise for sustainable energy use and production. Commercial ventures have begun marketing the environmental benefits of these technologies—for example, by offering "green" cleanup of hazardous waste sites, reducing agricultural reliance on chemical fertilizer by developing beneficial plant-microbe nutrient cycles, and minimizing industrial pollution by harnessing naturally occurring microorganisms that metabolize toxic compounds. However, with success in commercial applications, biotechnology has attracted public scrutiny and criticism. Influential interest groups raise ethical, social, and legal concerns about the effects of environmental applications and their regulatory oversight. Their concerns are part of larger debates at both national and international levels, encompassing the commercial uses of biotechnology, the conduct of research involving recombinant DNA, and trade relationships among nations with varying levels of support for biotechnology innovation [4, 5].

Resolving the ethical and social challenges of environmental biotechnology, which comprises a wide array of field-tested techniques and commercial applications, may rely on a simple set of shared moral principles. Consensus on core values may unite diverse groups involved in deciphering and applying life-science knowledge. Such stakeholders include companies developing transgenic crops that require less fertilizer and fewer chemical pesticides, research consortia studying biofilms that decontaminate subsurface environments, and nations debating trade agreements that affect access to pharmaceuticals developed from genetically modified organisms. Respecting professionally derived and internationally recognized principles of bioethics can help mitigate public controversies and regulatory hurdles that currently impede biotechnology innovation [6, 7].

Social Implications of Environmental Biotechnology

The public debate on this issue is considerable and heated. A second area of social concern involves the bioprocessing industry itself, as communities may be less than excited about hosting a bioremediation project. Concerns raised by community groups have prompted the industry to engage in negotiated rulemaking for developing guidelines for field-testing genetically engineered organisms used to clean up Superfund sites. This has also led the industry to work more closely with social scientists to better understand the dynamics of community concerns. The industry's own challenge lies in its slow response to the issue of public participation in project decision-making and its focus on technical, rather than social, solutions to community concerns. In the case of bioprocessing, the industry has tended to be secretive about specific microorganisms and processes, which has increased social concern. In conclusion, the social challenges of environmental biotechnology arise from both internal and external sources [8].

Environmental biotechnologies, much like previous advances, are likely to face some social concern, even though environmental biotechnology is anticipated to provide society with benefits such as the cleanup of environmental pollutants and more efficient and cost-effective methods for producing chemicals. One area of social concern stems from groups having concerns about the release of genetically engineered organisms into the environment [9, 10].

Impact on Traditional Practices

Developing societies are not homogenous. New technologies have varying and often reciprocal effects on various sections of society. Mechanization may free some members of farming communities from arduous labor in the fields while at the same time destroying the livelihoods of other members from whom the community derives its social cohesion. The resolution of one problem through progress may imply the emergence of new challenges. Groups with specific cultural interests may feel threatened by the homogenizing impact of modern technologies. Urban development and industrial growth are already eroding many of the value systems that protected natural resources in traditional societies. The weakening or loss of these unique bases for environmental ethics should be a matter of global concern. However, the promotion of biotechnologies may reinforce rather than break down these social and cultural barriers. The use of such technologies needs to be managed carefully to avoid detriment to progress in other more urgent areas [11, 12].

Biotechnologies have the potential to improve the lives of people in developing countries by offering environmentally sound solutions to pressing health, agricultural, and industrial problems. However, realizing this potential will require addressing a range of ethical as well as social, economic, and political challenges. In addition, pioneering technologies in fields such as environmental biotechnology can have adverse short-term or long-term effects. A grim difficulty is that these effects may be unpredictable. For example, one major concern about environmental applications of recombinant DNA technology is the possible accumulation of heavy metals in the environment as a side effect of the mass introduction of genetically engineered plants designed to clean up pollutant chemicals [13, 14].

Regulatory Frameworks and Policies

A common theme across these different regulatory approaches is the emphasis on risk assessment. However, not all countries or international bodies have established uniform procedures for assessing the risks associated with biotechnological products. The burden of showing that a product is safe is typically placed on the developer or proponent of the technology. This stringent approach has been criticized on several grounds. First, it stymies innovation by imposing high regulatory costs on developers, especially small firms and public research institutions. Secondly, it relies on an empirical-inductive methodology of risk assessment that is considered to be of limited value in forecasting or estimating complex, long-term, or highly uncertain environmental and human health impacts. A postulate of the Theory of Life, for instance, is that the biosphere is dynamic, non-linear, hierarchical, and complex, making it impossible to establish the safety of a technology through empirical induction. Moreover, the inductive model of risk assessment is predicated on the assumption that nature is the ultimate authority concerning the safety of biotechnological products. This is a highly contested view as nature is, in a sense, "unnatural" because it is replete with examples of "mistakes" made by the process of evolution (*e.g.*, neurotoxins in certain plants, and deadly pathogens) [9, 15].

Several governmental and international bodies have produced guidelines on the release and use of GMOs and their products. In the USA, the Coordinated Framework for Regulation of Biotechnology outlines the regulatory policies of the EPA, USDA, and FDA. The European Union has established a comprehensive set of regulations that cover the contained use, deliberate release, and marketing of GMOs. The Cartagena Protocol on Biosafety is an international agreement that seeks to regulate the transboundary movement of LMOs. These regulatory frameworks primarily focus on GMOs and LMOs. Non-GMO environmental biotechnology, such as the use of enzymes or indigenous microorganisms, does not usually fall under these regulations unless there is a deliberate attempt to introduce a non-indigenous microorganism into the environment. It could be argued that more policy attention needs to be directed towards non-GMO technologies, especially considering the increasing relevance of their products in the field of environmental biotechnology [16, 17].

International Agreements and Conventions

The ethical and social challenges of environmental biotechnology reflect the multidimensional nature of the objectives that the field pursues: preserving territory (solid waste management), means of subsistence (clean water, clean air, fertile soil), and biodiversity (emerging technologies like phytoremediation) (Fig. **2**). The implementation of all procedures involved in the above-mentioned remediation technologies at field scale may encounter social concerns that can delay, stop, or inflate the cost of remediation projects. Public attitudes toward risk and trust in the involved stakeholders (technology developer, promoter, and implementer) are significant and are influenced by the level of knowledge and information about the remediation technology. Since reliable stakeholders will have to deal with uncertainty, new knowledge should be developed through continuous monitoring and research during technology implementation at the field level [18, 19].

To address the challenges arising from the growing human population and its impact on the environment, science must respond to both the advancement of knowledge and the development of new technologies. With biotechnology, and in particular with environmental biotechnology, a series of international agreements have been established to implement international laws among the different countries, aiming to create a formal structure guiding these efforts. International conventions have been agreed upon to develop standard rules to ensure that the use of biotechnology is performed with respect and benefits for present and future generations. The main international agreements that have addressed biotechnology and biosafety are outlined, and some particularities of the signatory countries are briefly presented. The aim is to establish a common set of principles and rules so that the applications of biotechnology are managed in a transparent and orderly manner, allowing cooperation and sharing of knowledge and benefits among the countries that are developing these technologies [20, 21].

Bioremediation of contaminated sites

Despite the commercial and practical pressures to deploy these technologies quickly, the perceived risks associated with intentional releases of genetically engineered organisms developed for environmental applications have slowed their implementation. In fact, to date, no field demonstrations of genetically engineered plants have been specifically conducted to evaluate the risks of unintentional spreading of the modification to native plant populations; releases of genetically engineered hyperaccumulating plants have been restricted to greenhouse trials. The public debate and controversy surrounding the use of genetically engineered organisms for environmental applications have resulted in a new form of adaptive management known as "deliberate release". In these trials, the genetically engineered organisms are released in the field but are "contained" and destroyed at the end of the experiment [22, 23].

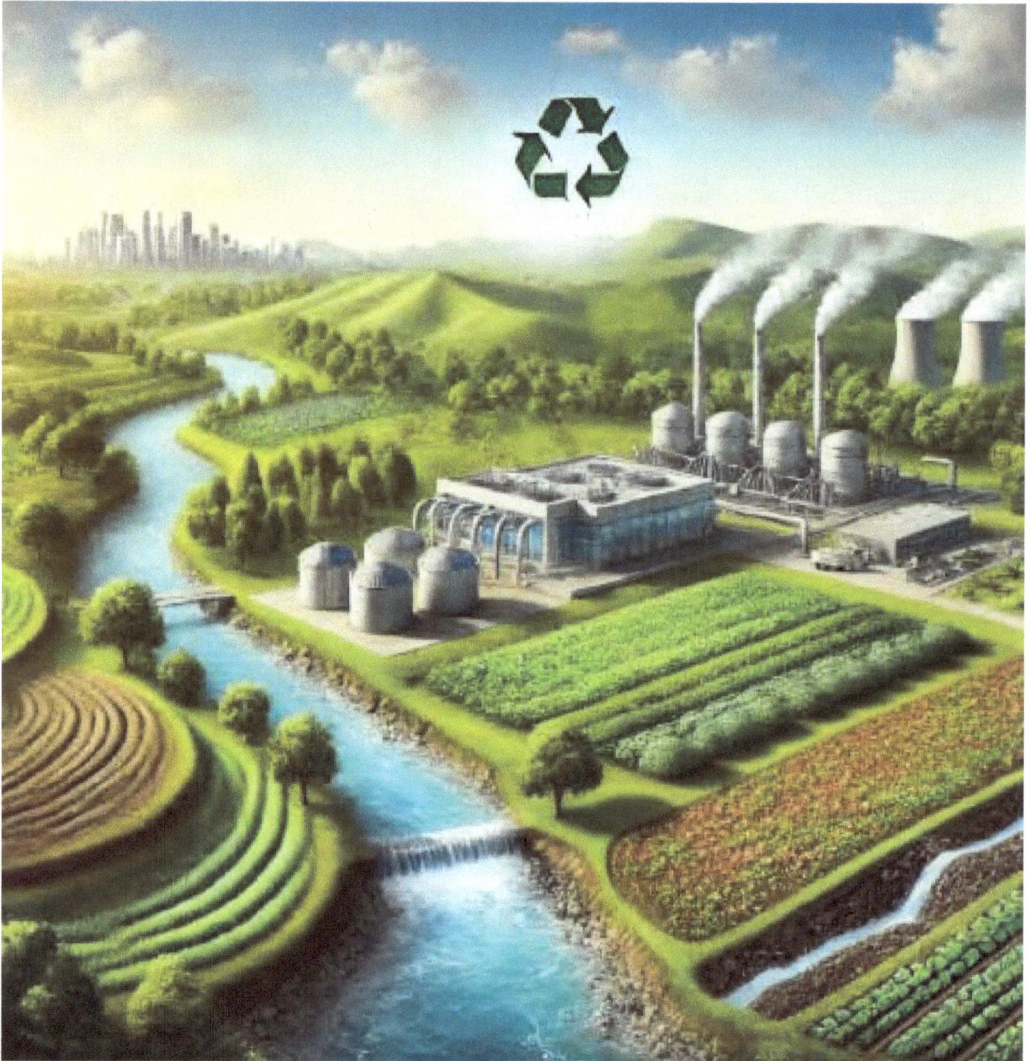

Fig. (2). Depiction of environmental biotechnology goals: preserving territory through solid waste management, sustaining life with clean water, clean air, and fertile soil, and protecting biodiversity with technologies like phytoremediation. These elements are integrated into a balanced and thriving ecosystem, emphasizing the harmony between technological advancement and environmental stewardship. Figure made with DALL-E.

Despite our best intentions and the use of a range of sophisticated technologies, undesirable outcomes (side effects, mistakes, and disasters) still occasionally occur in the stewardship of the earth's biological systems. These misfortunes can be particularly disheartening when they occur during attempts to repair or alleviate earlier damages to the environment. The examples provided in this chapter are drawn from those applications of biotechnology intended to address clearly defined environmental problems: the treatment of hazardous wastes, oil bioremediation, and the cleanup of organic chemical spills. Although these bioremediation applications are relatively new, the use of microorganisms in green plants to detoxify and accumulate heavy metals in the process of phytoremediation has a longer history. Indeed, the commercial deployment of plants for the cleanup of metals from contaminated soil is currently the most widespread example of the application of environmental biotechnology [24, 25].

Estimates of the number of contaminated sites in the United States alone range from about 300,000 to more than 7 million. Many of these sites are abandoned landfills or have been used for waste disposal by industry, particularly the chemical and petrochemical industries. As many as 35,000 of these sites are considered to be heavily contaminated. The World Health Organization has estimated that more than 3 million people are killed each year by the products of incomplete combustion of fossil fuels, with many more suffering from respiratory illnesses. Clearly, the removal and/or destruction of environmental pollutants at these and other sites is of paramount importance to ensure the quality of both human and ecosystem environments. The use of biological processes for pollutant removal can often be more cost-effective and environmentally benign than other physicochemical processes. For many pollutants, biological processes are the only viable solution. Even in cases where pretreatment of the pollutant may be necessary, the use of bioreactors containing immobilized microorganisms can often lower the costs of subsequent, more conventional treatment operations such as activated sludge systems [26].

Case Study: The Application of Biochar in Soil Amendment

A significant case study in the field of edaphology and biotechnology is the use of biochar as a soil amendment to improve soil health at the molecular level [27]. Biochar, a carbon-rich product obtained from the thermal decomposition of organic material under limited oxygen conditions (pyrolysis) [28], is being studied for its potential to enhance soil properties, sequester carbon, reduce greenhouse gas emissions, and treat contaminated water [29, 30].

Production and Properties of Biochar

Biochar is generated through the process of pyrolysis, which may vary in speed depending on factors such as temperature, duration of biomass residence, and rates of heating. Biochar produced through high-temperature pyrolysis and subsequent activation by water vapor typically exhibits elevated pH levels and a decreased capacity for cation exchange. Such biochar varieties are characterized by a substantial surface area that governs the process of chemical adsorption. The alkalinity of biochar is influenced by the presence of inorganic carbonates and organic anions, contributing to a reduction in the levels of accessible heavy metals and thereby facilitating biochar-assisted phytoremediation [31] (Fig. **3**).

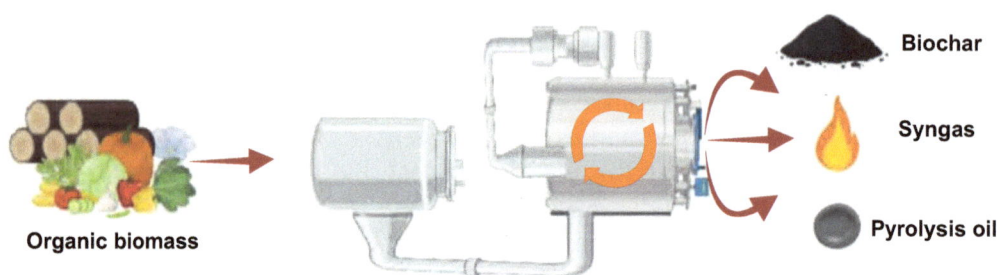

Execution of pyrolysis in a plant

Fig. (3). Schematic representation of the pyrolysis process, where organic mass undergoes pyrolysis in a processing plant, resulting in the production of pyrolysis oil, syngas, and biochar. This process illustrates the conversion of organic materials into valuable byproducts through thermal decomposition in an oxygen-limited environment. Figure made with Biorender.

The characteristics of biochar, including specific surface area, elemental composition, surface chemical properties, and pH, exhibit notable variations depending on the conditions of pyrolysis. Notably, biochar derived from chicken manure has shown efficacy in the immobilization of chromium within contaminated soils, thereby mitigating its leaching [31].

Molecular Interactions and Benefits of Biochar in Soil

Biochar, a carbon-rich product derived from the pyrolysis of biomass, has gained attention as a soil amendment inspired by the fertile Amazonian Terra Preta soils [32, 33]. These ancient soils, enriched with charcoal over 800 years ago, demonstrate biochar's potential to enhance soil fertility and sequester carbon [34]. Research suggests that biochar can improve soil quality, increase crop production, and mitigate climate change by storing carbon in a stable form [35, 36]. Research conducted on degraded soils in the Amazon Basin, where a form of biochar known as "terra preta" was traditionally utilized, revealed that biochar notably elevated the soil's cation exchange capacity, enhanced water retention, and promoted microbial activity. These alterations contribute to an enhanced availability of nutrients for vegetation and increased agricultural productivity. Biochar exhibits high environmental stability, persisting in the soil for thousands of years, which renders it well-suited for prolonged application [29].

Biochar in Water Treatment

Biochar has shown promise in water treatment, particularly in the adsorption of heavy metals and organic pollutants from wastewater. Its effectiveness is attributed to its high surface area, porous structure, and functional groups that facilitate the adsorption of contaminants [37, 38]. The adsorption mechanisms involve surface area and pore distribution, the surface chemistry of the adsorbent, the pH of the solution, and the presence of other electrolytes. Acidic oxygen surface groups on biochar ionize in aqueous solutions to produce negative charge sites that adsorb positively charged metal ions through cation exchange. Biochar's structure makes it effective for removing a wide range of contaminants, including pharmaceuticals, personal care products, surfactants, and pesticides, from municipal wastewater [39, 40]. Modified biochar, produced through chemical or physical processes, exhibits enhanced adsorption capacity and a broader application range [37]. Low-temperature torrefaction offers an energy-efficient method for biochar production [41]. Biochar has been successfully applied in treating industrial, municipal, and agricultural wastewater [40, 42, 43] and shows potential in energy applications such as batteries and fuel cells [44]. Overall, biochar technology represents a cost-effective and environmentally friendly solution for wastewater treatment and environmental remediation (Fig. **4**).

Procedure for enhancing soil properties using biochar approach

To enhance soil properties at the molecular level using the biochar approach, the following steps are recommended:

Soil Sampling and Analysis

Collect soil samples and perform a comprehensive analysis to determine the current physical, chemical, and biological status of the soil.

Biochar Production

Produce biochar from suitable biomass using pyrolysis. Tailor the properties of the biochar to the specific needs of the soil based on the initial analysis.

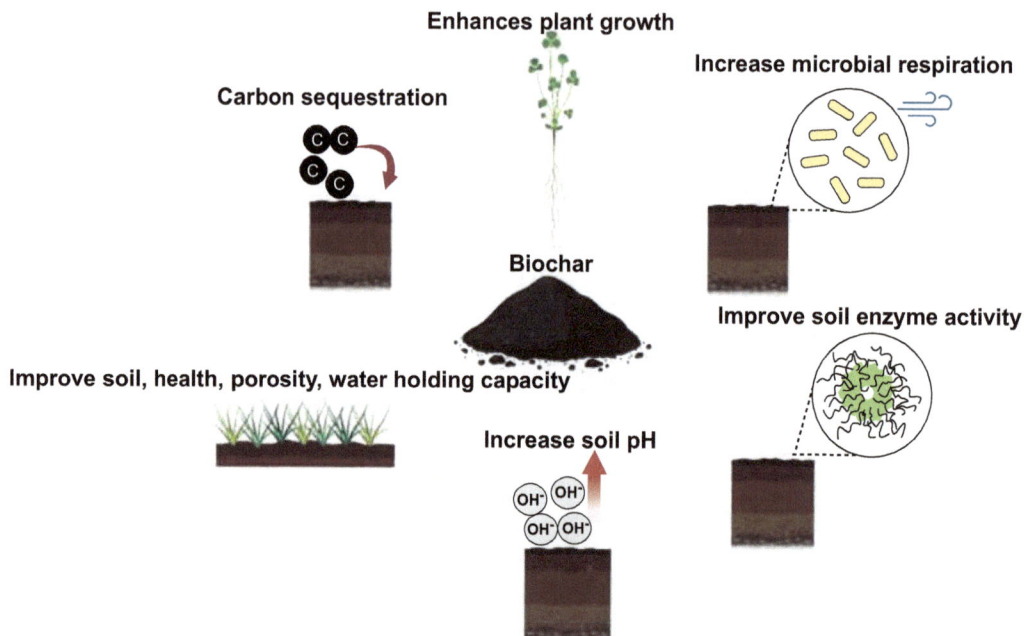

Fig. (4). Overview of biochar benefits, highlighting its roles in raising soil pH, enhancing soil health, increasing porosity and water retention, boosting soil enzyme activity, sequestering carbon, promoting plant growth, and stimulating microbial respiration. These attributes make biochar a valuable tool for sustainable soil management. Figure made with Biorender.

Application of Biochar

Apply the biochar to the soil according to the results of the soil analysis. Determine the application rate based on the desired improvement in soil properties.

Molecular Monitoring

Use molecular techniques such as spectroscopy and chromatography to monitor changes in the soil's organic matter and microbial community structure after biochar application.

Assessment of Soil Health

Evaluate the soil's health through bioassays and plant growth trials to determine the impact of biochar on soil fertility and plant productivity.

Adjustment and Optimization

Adjust the biochar application strategy based on feedback from soil health assessments to optimize the benefits of soil improvement.

Long-term Monitoring and Management

Implement a long-term soil monitoring program to assess the sustainability of biochar use and its long-term effects on soil properties.

Ethical and Social Considerations of Biochar

The application of biochar and other biotechnological interventions in agriculture raises important ethical and social considerations, such as the impact on smallholder farmers, the potential for inequitable access to technology, and the broader implications for food security and environmental sustainability. Biochar offers potential benefits for sustainable agriculture, climate change mitigation, and environmental remediation [45, 46]. It can improve soil fertility, increase crop productivity, and sequester carbon [47, 48]. However, implementation faces challenges, including adaptation barriers, socio-economic feasibility, and environmental risks [46, 49].

To address these concerns, researchers propose sustainability certification frameworks and holistic methodologies for evaluating biochar systems [50, 51]. Small-scale biochar projects in developing countries may be particularly relevant, leveraging current knowledge to explore future potential for climate-smart agriculture. Despite promising applications in soil improvement, waste management, and energy production, biochar adoption requires overcoming financial, social, and environmental constraints, as well as addressing knowledge gaps through further research [47, 48].

Impact on Smallholder Farmers

Smallholder farmers often lack the resources and technical knowledge to implement advanced biotechnological solutions like biochar. Ensuring equitable access and providing support for these farmers is crucial to avoid exacerbating existing inequalities in agriculture. Successful adoption requires transdisciplinary research, consideration of local contexts, and integration of farmers' perspectives [46, 52]. Small-scale, low-tech biochar production methods may be more suitable for resource-poor farmers [53]. While biochar shows promise for improving agricultural sustainability and mitigating climate change, further research is needed to address adaptation barriers and develop context-specific, economically viable biochar systems for smallholder farmers [46, 54]. The benefits of biotechnological innovations must be accessible to all farmers, regardless of their socioeconomic status. Policies and programs should aim to bridge the gap between large-scale commercial farms and smallholder operations. This requires creating frameworks that facilitate equitable access to biochar technology and providing the necessary support to smallholder farmers to implement and benefit from these advancements.

Food Security and Environmental Sustainability

The use of biochar can contribute to sustainable agricultural practices by improving soil health and reducing reliance on chemical fertilizers. It holds significant potential for sustainable agriculture and climate change mitigation. It also can improve soil fertility, increase crop productivity, and enhance food security [48, 55, 56]. Biochar applications can sequester carbon in soil, reduce greenhouse gas emissions, and provide renewable energy [57, 58]. However, the implementation of biochar systems faces challenges, including environmental risks and adaptation barriers in agricultural communities [46, 50]. To address these concerns, researchers propose sustainability certification, effective regulation, and holistic approaches that consider social-ecological systems [46, 50]. However, it is crucial to carefully monitor the long-term environmental impacts to prevent potential negative consequences.

Balancing Benefits and Risks

While biochar offers numerous benefits, it is essential to balance these with potential risks, such as the introduction of contaminants through the use of waste-derived biochar. Comprehensive risk assessments and regulatory frameworks are necessary to ensure its safe and effective application.

In pioneering efforts to address environmental problems, researchers have explored the potential of biochar to mitigate the adverse effects of pollutants. Oil spills, like the Exxon Valdez disaster, have highlighted the environmental devastation caused by petrochemicals. However, numerous other organic and inorganic chemicals, metals, and radionuclides also pose significant threats to human and ecosystem health. These pollutants often exist at concentrations that natural processes struggle to degrade.

Environmental biotechnology, particularly through the application of biochar, offers solutions to challenges related to soil degradation, pollution, and water quality. Biochar can enhance soil properties, immobilize heavy metals, and improve water quality, acting as a key tool in bioremediation (biological clean-up of pollutants) and phytoremediation (use of plants to clean up the environment). It improves soil fertility, increases crop yield, and boosts microbial activity while immobilizing contaminants [59]. In sub-Saharan Africa, biochar could address both soil degradation and energy access issues [47].

However, biochar application also presents risks, including potential negative impacts on soil microorganisms, nutrient depletion, and the presence of toxic compounds [60, 61]. The effectiveness of biochar varies with soil type and feedstock properties [60]. To maximize benefits and minimize risks, careful consideration of trade-offs is necessary [62]. Future research should focus on long-term consequences, standardized reporting, and risk evaluation methods to ensure sustainable and safe biochar use [63, 64].

Future Directions and Emerging Issues

Ensuring public trust and societal acceptance of biochar-enabled environmental biotechnologies requires the proactive integration of social science research into biotechnological agendas. Recent studies emphasize the importance of enhanced education and outreach, including early dialogue between the public and technology developers. Independent oversight mechanisms can further ensure that societal values are respected in the deployment of these technologies. As biochar and related biotechnologies advance rapidly, it is crucial to adopt responsible innovation practices. Future research should focus on optimizing biochar properties for specific applications, assessing long-term impacts, and addressing trade-offs between potential benefits [62]. Additionally, life cycle assessments and techno-economic analyses are crucial for evaluating the overall sustainability and feasibility of biochar systems [65, 66]. Policymakers and researchers must consider these factors to effectively implement biochar technologies.

Technological Innovations

Biotechnological research has led to the emergence of new technologies that significantly impact various economic sectors. The first wave involved relatively simple technologies using microorganisms and genetically engineered plant and animal cells to break down compounds, produce energy, and develop medical products. The second wave introduced more complex technologies, including genetic engineering, cell technologies, nerve technologies, and nanobiotechnology.

The rapid advancements in fields such as biology, chemistry, and information technology have given rise to a third wave of technologies known as environmental biotechnology. This field applies biotechnology to monitor, prevent, and remediate pollution in air, land, and water, produce environmentally safe energy and chemicals and protect the environment. Biochar, as a component of environmental biotechnology, plays a crucial role by enhancing soil health, sequestering carbon, and treating contaminants, contributing to a sustainable and resilient environmental future.

Small-scale biochar production technologies are being developed to address technical and socio-economic challenges in both developed and developing countries [54, 67]. Implementing biochar systems requires consideration of local social-ecological contexts and potential adaptation barriers [46]. Biochar's multifaceted applications extend to water treatment, organic farming, and contaminant removal [65]. Despite its promise, challenges remain in biomass collection, transportation, and conversion costs [65]. Ongoing research aims to enhance the sustainability benefits of biochar across food-energy-water systems and improve its economic viability [54, 65].

Social and ethical challenges in science often arise from its application. Many scientists engage in applied research and collaborate with industry to develop products that address social, environmental, and health problems. The process of transferring scientific discoveries into new products, processes, or services involves innovation. While not all scientific discoveries lead to innovation, many countries strive to be at the cutting edge of science, leveraging new knowledge to drive economic growth through innovation. Establishing close links between science, education, industry, and government is critical for facilitating technology transfer and fostering an environment conducive to developing new technologies. However, the rapid development of new technologies can have social, ethical, and legal implications that must be addressed alongside efforts to create a supportive climate for innovation [68, 69].

CONCLUDING REMARKS

The integration of biochar as a soil amendment highlights the potential of biotechnological interventions to improve agricultural practices sustainably. Biochar enhances soil properties, sequesters carbon, and treats contaminated water, contributing significantly to environmental sustainability and food security. However, addressing the ethical and social challenges associated with these technologies is crucial to ensure their equitable and responsible use.

As the agricultural sector continues to evolve with smart farming and precision agriculture, these innovations must be implemented thoughtfully to maximize benefits and minimize risks. Ensuring equitable access to technology, supporting smallholder farmers, and balancing the benefits and risks are essential steps. Comprehensive risk assessments and regulatory frameworks are necessary to safeguard against potential negative consequences.

By fostering sustainable agricultural practices, these biotechnological innovations can promote a more sustainable future for all stakeholders, ensuring that the benefits of advanced agricultural technologies are accessible and beneficial to everyone.

AUTHORS' CONTRIBUTION

Israel Valencia Quiroz: Conceptualization, methodology, investigation, supervision, writing-original draft preparation, writing review and editing, figure editing.

QUESTIONS RELATED TO THE TEXT

- What are the main ethical challenges associated with environmental biotechnology, and how do they relate to the fair distribution of benefits among affected communities?
- Why is it important to respect the autonomy of individuals and communities who oppose the implementation of environmental biotechnological technologies?
- How can biotechnology companies address social concerns with their projects, and what role do democratic processes play in this dynamic?
- What impact might traditional practices have on the acceptance of new biotechnological technologies in developing societies, and how should these challenges be addressed?
- What is the role of regulatory frameworks and international agreements in managing the risks associated with environmental biotechnology?
- Discuss the ethical and social dilemmas that may arise from the application of environmental biotechnology, particularly in the context of bioremediation and the release of genetically modified organisms into the environment.

ACKNOWLEDGEMENTS

The authors would like to acknowledge Ismael Manuel Urrutia Ortega for reviewing, methodology, data curation, formal analysis, investigation, draft preparation, and figure editing.

LIST OF ABBREVIATIONS

EPA   Environmental Protection Agency

FDA   Food and Drug Administration

GM   Genetically Modified

GMO   Genetically Modified Organism

LMO   Living Modified Organism

PIA   Prior Informed Agreement

USDA United States Department of Agriculture

REFERENCES

[1]   Lal R, Bouma J, Brevik E, *et al.* Soils and sustainable development goals of the United Nations: An International Union of Soil Sciences perspective. Geoderma Reg 2021; 25: e00398.
[http://dx.doi.org/10.1016/j.geodrs.2021.e00398]

[2]   Dolezal C, Novelli M. Power in community-based tourism: empowerment and partnership in Bali. J Sustain Tour 2022; 30(10): 2352-70.
[http://dx.doi.org/10.1080/09669582.2020.1838527]

[3]   Seddon N, Smith A, Smith P, *et al.* Getting the message right on nature-based solutions to climate change. Glob Change Biol 2021; 27(8): 1518-46.
[http://dx.doi.org/10.1111/gcb.15513] [PMID: 33522071]

[4]   Ali S, Paul Peter A, Chew KW, Munawaroh HSH, Show PL. Resource recovery from industrial effluents through the cultivation of microalgae: A review. Bioresour Technol 2021; 337: 125461.
[http://dx.doi.org/10.1016/j.biortech.2021.125461] [PMID: 34198241]

[5]   Hamad HN, Idrus S. Recent Developments in the Application of Bio-Waste-Derived Adsorbents for the Removal of Methylene Blue from Wastewater: A Review. Polymers (Basel) 2022; 14(4): 783.
[http://dx.doi.org/10.3390/polym14040783] [PMID: 35215695]

[6]   Fischer K, Stenius T, Holmgren S. Swedish Forests in the Bioeconomy: Stories from the National Forest Program. Soc Nat Resour 2020; 33(7): 896-913.
[http://dx.doi.org/10.1080/08941920.2020.1725202]

[7]   Harfouche AL, Petousi V, Meilan R, Sweet J, Twardowski T, Altman A. Promoting Ethically Responsible Use of Agricultural Biotechnology. Trends Plant Sci 2021; 26(6): 546-59.
[http://dx.doi.org/10.1016/j.tplants.2020.12.015] [PMID: 33483266]

[8]   Mu'azzam K, Santos da Silva FV, Murtagh J, Sousa Gallagher MJ. A roadmap for model-based bioprocess development. Biotechnol Adv 2024; 73: 108378.
[http://dx.doi.org/10.1016/j.biotechadv.2024.108378] [PMID: 38754797]

[9]   Akinbo O, Obukosia S, Ouedraogo J, *et al.* Commercial Release of Genetically Modified Crops in Africa: Interface Between Biosafety Regulatory Systems and Varietal Release Systems. Front Plant Sci 2021; 12: 605937. Available from: https://www.frontiersin.org/journals/plant-science/articles/10.3389/fpls.2021.605937
[http://dx.doi.org/10.3389/fpls.2021.605937] [PMID: 33828569]

[10] Muzhinji N, Ntuli V. Genetically modified organisms and food security in Southern Africa: conundrum and discourse. GM Crops Food 2021; 12(1): 25-35.
[http://dx.doi.org/10.1080/21645698.2020.1794489] [PMID: 32687427]

[11] Wassie SB. Natural resource degradation tendencies in Ethiopia: a review. Environ Syst Res 2020; 9(1): 33.
[http://dx.doi.org/10.1186/s40068-020-00194-1]

[12] Xu L, Tan J. Financial development, industrial structure and natural resource utilization efficiency in China. Resour Policy 2020; 66(C): 101642. Available from: https://ideas.repec.org/a/eee/jrpoli/v66y2020ics0301420719310189.html
[http://dx.doi.org/10.1016/j.resourpol.2020.101642]

[13] Iravani S, Varma RS. Genetically Engineered Organisms: Possibilities and Challenges of Heavy Metal Removal and Nanoparticle Synthesis. Cleanroom Technol 2022; 4(2): 502-11.
[http://dx.doi.org/10.3390/cleantechnol4020030]

[14]   Nedjimi B. Phytoremediation: a sustainable environmental technology for heavy metals decontamination. SN Applied Sciences 2021; 3(3): 286.
[http://dx.doi.org/10.1007/s42452-021-04301-4]

[15]   Sundh I, Eilenberg J. Why has the authorization of microbial biological control agents been slower in the EU than in comparable jurisdictions? Pest Manag Sci 2021; 77(5): 2170-8.
[http://dx.doi.org/10.1002/ps.6177] [PMID: 33201551]

[16]   Bakhsh A, Zainab R, Ali MA, Chung G, Golokhvast KS, Nawaz MA. Chapter 10 - Genetically modified organisms in Europe: state of affairs, birth, research, and the regulatory process(es). En: Nawaz MA, Chung G, Golokhvast KS, Tsatsakis AM, editores. GMOs and Political Stance [Internet]. Academic Press; 2023; 165-72.
[http://dx.doi.org/10.1016/B978-0-12-823903-2.00012-3]

[17]   Ichim MC. The more favorable attitude of the citizens toward GMOs supports a new regulatory framework in the European Union. GM Crops Food 2021; 12(1): 18-24.
[http://dx.doi.org/10.1080/21645698.2020.1795525] [PMID: 32787504]

[18]   Bouchaut B, Asveld L. Safe-by-Design: Stakeholders' Perceptions and Expectations of How to Deal with Uncertain Risks of Emerging Biotechnologies in the Netherlands. Risk Anal 2020; 40(8): 1632-44.
[http://dx.doi.org/10.1111/risa.13501] [PMID: 32421209]

[19]   Butkowski OK, Baum CM, Pakseresht A, Bröring S, Lagerkvist CJ. Examining the social acceptance of genetically modified bioenergy in Germany: Labels, information valence, corporate actors, and consumer decisions. Energy Res Soc Sci 2020; 60: 101308.
[http://dx.doi.org/10.1016/j.erss.2019.101308]

[20]   Wohlgemuth R, Twardowski T, Aguilar A. Bioeconomy moving forward step by step – A global journey. N Biotechnol 2021; 61: 22-8.
[http://dx.doi.org/10.1016/j.nbt.2020.11.006] [PMID: 33197617]

[21]   Zhang D, Hussain A, Manghwar H, *et al.* Genome editing with the CRISPR-Cas system: an art, ethics and global regulatory perspective. Plant Biotechnol J 2020; 18(8): 1651-69.
[http://dx.doi.org/10.1111/pbi.13383] [PMID: 32271968]

[22]   Jeon DW, Park JR, Jang YH, Kim EG, Ryu T, Kim KM. Safety verification of genetically modified rice morphology, hereditary nature, and quality. Environ Sci Eur 2021; 33(1): 73.
[http://dx.doi.org/10.1186/s12302-021-00516-9]

[23]   Legros M, Marshall JM, Macfadyen S, Hayes KR, Sheppard A, Barrett LG. Gene drive strategies of pest control in agricultural systems: Challenges and opportunities. Evol Appl 2021; 14(9): 2162-78.
[http://dx.doi.org/10.1111/eva.13285] [PMID: 34603490]

[24]   Padhye LP, Srivastava P, Jasemizad T, *et al.* Contaminant containment for sustainable remediation of persistent contaminants in soil and groundwater. J Hazard Mater 2023; 455: 131575.
[http://dx.doi.org/10.1016/j.jhazmat.2023.131575] [PMID: 37172380]

[25]   Phang LY, Mohammadi M, Mingyuan L. Underutilised Plants as Potential Phytoremediators for Inorganic Pollutants Decontamination. Water Air Soil Pollut 2023; 234(5): 306.
[http://dx.doi.org/10.1007/s11270-023-06322-8]

[26]   Sánchez-Castro I, Molina L, Prieto-Fernández MÁ, Segura A. Past, present and future trends in the remediation of heavy-metal contaminated soil - Remediation techniques applied in real soil-contamination events. Heliyon 2023; 9(6): e16692.
[http://dx.doi.org/10.1016/j.heliyon.2023.e16692] [PMID: 37484356]

[27]   Fu G, Qiu X, Xu X, Zhang W, Zang F, Zhao C. The role of biochar particle size and application rate in promoting the hydraulic and physical properties of sandy desert soil. Catena 2021; 207: 105607.
[http://dx.doi.org/10.1016/j.catena.2021.105607]

[28]   Shaw DR, Tobon Gonzalez J, Bibiano Guadarrama C, Saikaly PE. Emerging biotechnological

applications of anaerobic ammonium oxidation. Trends Biotechnol 2024; 42(9): 1128-43.
[http://dx.doi.org/10.1016/j.tibtech.2024.02.013] [PMID: 38519307]

[29]   Lebrun M, Nandillon R, Miard F, Bourgerie S, Morabito D. Chapter 4 - Biochar assisted phytoremediation for metal(loid) contaminated soils. In: Pandey V, editor. Assisted Phytoremediation [Internet]. Elsevier; 2022. 101-30.
[http://dx.doi.org/10.1016/B978-0-12-822893-7.00010-0]

[30]   Wang T, Stewart CE, Ma J, Zheng J, Zhang X. Applicability of five models to simulate water infiltration into soil with added biochar. J Arid Land 2017; 9(5): 701-11.
[http://dx.doi.org/10.1007/s40333-017-0025-3]

[31]   Malik G, Hooda S, Majeed S, Pandey VC. Understanding assisted phytoremediation: Potential tools to enhance plant performance. In: Pandey V, Ed. Assisted Phytoremediation. Elsevier 2022; pp. 1-24. Available from: https://www.sciencedirect.com/science/article/pii/B978012822893700015X
[http://dx.doi.org/10.1016/B978-0-12-822893-7.00015-X]

[32]   Bezerra J, Turnhout E, Vasquez IM, Rittl TF, Arts B, Kuyper TW. The promises of the Amazonian soil: shifts in discourses of Terra Preta and biochar. J Environ Policy Plann 2019; 21(5): 623-35.
[http://dx.doi.org/10.1080/1523908X.2016.1269644]

[33]   Tenenbaum DJ. Biochar: carbon mitigation from the ground up. Environ Health Perspect 2009; 117(2): A70-3.
[http://dx.doi.org/10.1289/ehp.117-a70] [PMID: 19270777]

[34]   Mao JD, Johnson RL, Lehmann J, *et al.* Abundant and stable char residues in soils: implications for soil fertility and carbon sequestration. Environ Sci Technol 2012; 46(17): 9571-6.
[http://dx.doi.org/10.1021/es301107c] [PMID: 22834642]

[35]   Novotny EH, Maia CMBF, Carvalho MTM, Madari BE. Biochar: Pyrogenic carbon for agricultural use - a critical review. Rev Bras Ciênc Solo 2015; 39(2): 321-44.
[http://dx.doi.org/10.1590/01000683rbcs20140818]

[36]   Rhodes CJ. Biochar, and its potential contribution to improving soil quality and carbon capture. Sci Prog 2012; 95(3): 330-40.
[http://dx.doi.org/10.3184/003685012X13445960110308] [PMID: 23094328]

[37]   Mohit A, Remya N, Priyadarshini U. Multifaceted application of modified biochar for water and wastewater treatment. Environ Qual Manage 2024; 33(4): 945-53.
[http://dx.doi.org/10.1002/tqem.22170]

[38]   Enaime G, Baçaoui A, Yaacoubi A, Lübken M. Biochar for Wastewater Treatment—Conversion Technologies and Applications. Appl Sci (Basel) 2020; 10(10): 3492.
[http://dx.doi.org/10.3390/app10103492]

[39]   González Delgado JC, González Delgado AX, González Delgado FJ, Cruz Cerro GJF. Systematization of the impact of wastewater on society and the application of biochar as an alternative to the current crisis. Centro Sur. 2021; 5(4). Available at: http://portal.amelica.org/ameli/jatsRepo/384/3842826008/index.html

[40]   Xiang W, Zhang X, Chen J, *et al.* Biochar technology in wastewater treatment: A critical review. Chemosphere 2020; 252: 126539.
[http://dx.doi.org/10.1016/j.chemosphere.2020.126539] [PMID: 32220719]

[41]   Lin SL, Zhang H, Chen WH, Song M, Kwon EE. Low-temperature biochar production from torrefaction for wastewater treatment: A review. Bioresour Technol 2023; 387: 129588.
[http://dx.doi.org/10.1016/j.biortech.2023.129588] [PMID: 37558107]

[42]   De Caprariis B, Di Filippis P, Hernandez AD, Petrucci E, Petrullo A, Scarsella M, et al. Pyrolysis wastewater treatment by adsorption on biochars produced by poplar biomass. J Environ Manage. 2017; 197: 231-8.
[http://dx.doi.org/10.1016/j.jenvman.2017.04.007] [PMID: 28391096]

[43] Deng, Y., Zhang, T., & Wang, Q. (2017). Biochar Adsorption Treatment for Typical Pollutants Removal in Livestock Wastewater: A Review. IntechOpen. Available at: https://www.intechopen.com/chapters/55175

[44] Gupta M, Savla N, Pandit C, *et al.* Use of biomass-derived biochar in wastewater treatment and power production: A promising solution for a sustainable environment. Sci Total Environ 2022; 825: 153892. [http://dx.doi.org/10.1016/j.scitotenv.2022.153892] [PMID: 35181360]

[45] Joseph S, Anh ML, Clare A, Shackley S. Socio-economic feasibility, implementation and evaluation of small-scale biochar projects 2015.

[46] Müller S, Backhaus N, Nagabovanalli P, Abiven S. A social-ecological system evaluation to implement sustainably a biochar system in South India. Agron Sustain Dev 2019; 39(4): 43. [http://dx.doi.org/10.1007/s13593-019-0586-y]

[47] Gwenzi W, Chaukura N, Mukome FND, Machado S, Nyamasoka B. Biochar production and applications in sub-Saharan Africa: Opportunities, constraints, risks and uncertainties. J Environ Manage 2015; 150: 250-61. [http://dx.doi.org/10.1016/j.jenvman.2014.11.027] [PMID: 25521347]

[48] Semida WM, Beheiry HR, Sétamou M, Simpson CR, El-Mageed TAA, Rady MM, et al. Biochar implications for sustainable agriculture and environment: A review. South African Journal of Botany. 2019;127:333–47. [http://dx.doi.org/10.1016/j.sajb.2019.11.015]

[49] Bjerregaard PP. The social shaping of technology - a case study of biochar in Denmark: a case study of the innovation journey of biochar in Denmark: insights for businesses, researchers, investors and policy-makers [Master's thesis]. Copenhagen: Copenhagen Business School; [cited 2024 Nov 6]. Available from: https://research.cbs.dk/en/studentProjects/3d892ad3-b2fe-4e55-9edc-566c303a9d6f

[50] Cowie A, Downie A, George B, Singh B, Zwieten L, O'Connell D. Is sustainability certification for biochar the answer to environmental risks. Crit Rev Environ Sci Technol 2012; 42(3): 225-50.

[51] Veres J, Kolonicny J, Ochodek T. Biochar status under international law and regulatory issues for the practical application. Chem Eng Trans. 2014; 37: 799-804. [http://dx.doi.org/10.3303/CET1437134]

[52] Hagemann N. Biochar for smallholder farmers in East Africa: arguing for transdisciplinary research. Climate change and sustainable development. Wageningen Academic Publishers 2012; pp. 400-4. Internet [http://dx.doi.org/10.3920/9789086867530_0062]

[53] Shafer M. Whose Carbon Capture? A Bit of Good News. Environmental Science & Sustainable Development 2023; 8(1): 25-30. [http://dx.doi.org/10.21625/essd.v8i1.933]

[54] Scholz SM, Sembres T, Roberts K, Whitman T, Wilson K, Lehmann J. Biochar systems for smallholders in developing countries: leveraging current knowledge and exploring future potential for climate-smart agriculture. World Bank Study. Washington, DC: World Bank; 2014. Available at: http://hdl.handle.net/10986/18781

[55] Chen WF, Zhang WF, Meng J. Biochar and agro-ecological environment: review and prospect. J Agro-Environ Sci. 2014; 33: 821-8.

[56] Chukwuka SK, Akanmu AO, Umukoro BO, Asemoloye DA, Odebode CA. Biochar: A Vital Source for Sustainable Agriculture. En: Biostimulants in Plant Science [Internet]. IntechOpen; 2020. [http://dx.doi.org/10.5772/intechopen.86568]

[57] Waters D, Van Zwieten L, Singh BP, Downie A, Cowie AL, Lehmann J. Biochar in Soil for Climate Change Mitigation and Adaptation. Soil Biology. Springer Berlin Heidelberg 2011; pp. 345-68. Internet [http://dx.doi.org/10.1007/978-3-642-20256-8_15]

[58] Woolf D, Amonette JE, Street-Perrott FA, Lehmann J, Joseph S. Sustainable biochar to mitigate global climate change. Nat Commun 2010; 1(1): 56.
[http://dx.doi.org/10.1038/ncomms1053] [PMID: 20975722]

[59] Jatav HS, Rajput VD, Minkina T, *et al.* Sustainable Approach and Safe Use of Biochar and Its Possible Consequences. Sustainability (Basel) 2021; 13(18): 10362.
[http://dx.doi.org/10.3390/su131810362]

[60] Subedi R, Bertora C, Zavattaro L, Grignani C. Crop response to soils amended with biochar: expected benefits and unintended risks. Ital J Agronomy [Internet]. 2017; 12(2). Available at: https://www.agronomy.it/agro/article/view/794
[http://dx.doi.org/10.4081/ija.2017.794]

[61] Kuppusamy S, Thavamani P, Megharaj M, Venkateswarlu K, Naidu R. Agronomic and remedial benefits and risks of applying biochar to soil: Current knowledge and future research directions. Environ Int 2016; 87: 1-12.
[http://dx.doi.org/10.1016/j.envint.2015.10.018] [PMID: 26638014]

[62] Jeffery S, Bezemer TM, Cornelissen G, *et al.* The way forward in biochar research: targeting trade-offs between the potential wins. GCB Bioenergy. 2013; 7(1): 1-13.
[http://dx.doi.org/10.1111/gcbb.12132]

[63] Downie A, Munroe P, Cowie A, Van Zwieten L, Lau DMS. Biochar as a Geoengineering Climate Solution: Hazard Identification and Risk Management. Crit Rev Environ Sci Technol 2012; 42(3): 225-50.
[http://dx.doi.org/10.1080/10643389.2010.507980]

[64] Ndirangu SM, Liu Y, Xu K, Song S. Risk evaluation of pyrolyzed biochar from multiple wastes. J Chem. 2019; 2019(1): 4506314.
[http://dx.doi.org/10.1155/2019/4506314]

[65] Hersh B, Mirkouei A, Sessions J, Rezaie B, You Y. A review and future directions on enhancing sustainability benefits across food-energy-water systems: the potential role of biochar-derived products. AIMS Environ Sci 2019; 6(5): 379-416.
[http://dx.doi.org/10.3934/environsci.2019.5.379]

[66] Zhu H, An Q, Syafika Mohd Nasir A, *et al.* Emerging applications of biochar: A review on techno-environmental-economic aspects. Bioresour Technol 2023; 388: 129745.
[http://dx.doi.org/10.1016/j.biortech.2023.129745] [PMID: 37690489]

[67] Nsamba H, Hale S, Cornelissen G, Bachmann R. Sustainable technologies for small-scale biochar production—a review. J Sustain Bioenergy Syst. 2015; 5: 10-31.
[http://dx.doi.org/10.4236/jsbs.2015.51002]

[68] Barrett CB. Overcoming Global Food Security Challenges through Science and Solidarity. Am J Agric Econ 2021; 103(2): 422-47.
[http://dx.doi.org/10.1111/ajae.12160]

[69] Freddi F, Galasso C, Cremen G, *et al.* Innovations in earthquake risk reduction for resilience: Recent advances and challenges. Int J Disaster Risk Reduct 2021; 60: 102267.
[http://dx.doi.org/10.1016/j.ijdrr.2021.102267]

SUBJECT INDEX

A

Acids 77, 80, 148, 149, 150, 151, 211, 247, 249, 272, 276, 284
  carbonic 149
  carboxylic 151
  fulvic (FA) 272
  oleic 150
  phosphoric 276
  polylactide 284
  pyruvic 77, 80
  salicylic 148
  terephthalic 247
  trimesic 249
Action, metabolic 144
Activated sludge 186, 187, 330
  method 186
  systems 187, 330
Activation energy 269
Activity 15, 60, 77, 78, 80, 82, 88, 120, 121, 123, 124, 125, 126, 127, 137, 166, 171, 197, 205, 210, 218, 231, 250, 266, 271, 274, 311, 312
  anthropogenic 77, 88
  antigerminative 126
  enzymatic 80, 121, 218, 250, 271
  industrial 60, 137, 197, 231
  magnetic 274
  pesticidal 127
  planting 311
  post-harvesting 312
  socioeconomic 15
Advanced oxidation processes (AOPs) 196, 269
Aerobic treatment 186, 187
  processes 186
  systems 187
Aggregate-associated microbial communities 7
Agricultural 17, 81, 301, 311
  activities 81, 301, 311
  policies 17

Agriculture soils 77
Agro-ecological hydro-spatial planning (AHSP) 301
Agrochemical(s) 12, 13, 125, 139, 230, 231, 232, 252, 278, 283, 284, 298
  applications 13
  toxic 230
Airborne attraction 166
Algae 164, 167, 168, 169, 170, 189, 190, 191, 192, 208, 209, 211, 218, 220
  fast-growing 209
Allelopathic bioactivity 126
Anaerobic treatment 187, 188
  processes 187
  technologies 187
Analysis 79, 81, 83, 85, 86
  metabolomic 79, 83, 85, 86
  metagenomic 81
  protein 86
  proteomic 85
  soil metabolome 86
  transcriptomic 83

B

Bacteria, heterotrophic 146
Biochar 332, 335
  in water treatment 332
  waste-derived 335
Biodegradation 137, 141, 143, 144, 151, 183, 187, 209, 212, 213, 215
  anaerobic 151
  microalgae-mediated 144
  of wastewater 151
Bioremediation techniques 212, 214, 222
Biosensors 251, 253
  enzyme-based 253
  fluorescence-based 251
Biotechnology, water desalination 159
Branched electron transport systems 192

C

Carbon quantum dots (CQDs) 252
Chemical(s) 3, 4, 13, 126, 139, 141, 145, 146,
    147, 148, 164, 187, 189, 207, 211, 221,
    231, 263, 267, 271, 302, 306, 335
  components of soil 3
  compounds, inorganic 267
  harmful 231
  inorganic 335
  organic 211, 221
  oxidants 271
  oxygen demand (COD) 139, 145, 146, 147,
    187
  stresses 13
  weathering processes 4
Chromatographic 86, 231
  conditions 86
  -mass spectrometry 231
Climate change 1, 2, 6, 8, 9, 16, 18, 49, 51,
    57, 58, 59, 61, 62, 158, 159, 334, 335
  mitigating 334
  mitigation 6, 9, 334, 335
Communities, agricultural 335
Components 40, 172
  organic 172
  photosynthetic 40
Composition, metabolic 83
Contaminants 212, 269, 276
  hydrocarbon 212
  inorganic soil 276
  microbiological 269
Convolutional neural networks (CNNs) 310

D

Degradation 129, 144, 266
  enzymatic 129
  metabolic 144
  microbial 266
Dehydratase 144
Dehydrogenase 144, 218
Desalination facilities 159, 162, 163, 169
  thermal 159
  traditional 169
Desert 34, 38, 40, 41, 42, 44, 50
  environments 34, 41, 42
  soils 38, 40, 41, 42, 44, 50
Devices, sensitive 278

Diseases 6, 17, 115, 116, 120, 122, 123, 140,
    150, 272, 282, 283, 284
  bone 140
  chronic-degenerative 17
DNA 191, 249
  aptamers 249
  degradation 191
  repair protein activity 191
Droughts 2, 49, 57, 58, 59, 61, 158, 267, 303,
    309
  intermittent 303
  prolonged 57, 267
Drug(s) 147
  anti-inflammatory 147
  manufacturing process 147

E

Economy 118
Ecosystem(s) 3, 6, 9, 12, 58, 64, 78, 266
  agricultural 3, 12
  management 6
  restoration 58
  restore freshwater 266
  sustainability 9, 64
  terrestrial 78
Effluents, industrial 146, 183, 184, 196, 211,
    271
Electrochemiluminescent 245
Electron transfer 172, 246, 247, 248, 270
Electrophoresis 86
Electrothermal atomic absorption
    spectrometry 140
Energy   65, 80, 138, 159, 163, 164, 169, 171
  consumption 138, 163, 164, 169
  efficiency 159
  metabolism 80
  solar 65, 171
Energy production 194, 334
  nuclear 194
Environmental 12, 81, 116, 274, 314, 326,
    327, 330
  pollutants 274, 326, 330
  protection agency (EPA) 81, 327
  stresses 12, 116, 314
Enzyme(s) 185, 214
  metal-degrading 214
  microbial 214
  protein system 185
Essential oils (EOs) 125, 126, 283, 284

F

Farming 118, 310, 311, 326, 337
  communities 326
  organic 118, 337
Fertilizer(s) 82, 113, 114, 115, 116, 117, 138,
        207, 212, 265, 271, 277, 278, 279, 283,
        311, 312, 313, 325, 335
  applications 312
  bioorganic 82
  chemical 113, 115, 116, 117, 138, 278, 325,
        335
Fluorescence spectroscopy 243
Fluorescent sensors 232, 248, 251
Food 94, 114, 230, 231
  industry 94, 230, 231
  security information network (FSIN) 114
Food production 9, 16, 114, 125, 167, 216,
        303, 314
  global 114
  sustainable 314
Forest ecosystems 6
Freshwater 60, 61, 157, 160, 161, 166, 168,
        171, 218, 234, 266, 302, 314
  bodies 168
  ecosystem 218
  sources 60, 266
Freshwater resources 157, 301, 302, 307
  depleting 307
  global 301, 302

G

Generation, hydride 140
Gene(s) 78, 79, 81, 83, 86, 87, 127, 213, 214,
        251, 314
  antibiotic-resistance 81
  expression 86
  metal transporter 214
  reductive dehalogenase 213
  toxin-producing 127
Genetic modifications 297, 298, 314
Genetically 115, 123, 127, 208, 309, 314, 325
  modified (GM) 115, 123, 127, 208, 309,
        314, 325
  modified crops (GMC) 123, 127, 309
GI effectiveness 50
Glucosyltransferase 144
Glutamyl-tRNA reductase 144
Glyphosate pesticide 243, 245

H

Halophytes plants 164
Halophytic algae surpasses 168
Hydroponics techniques 306

I

Indicator-displacement assay (IDA) 239, 241,
        242, 243
Industrial 16, 146, 169, 170, 184, 185, 186,
        196, 211, 267, 303, 306, 326
  agriculture system 16
  growth 326
  wastewater 146, 169, 170, 184, 185, 186,
        196, 211, 303, 306
  waters 146, 267
Industry 59, 62, 117, 138, 139, 158, 160, 185,
        205, 207, 298, 300, 302, 326, 337
  biofertilizer 117
Innate immunity 217
Inorganic 121, 138, 146, 190, 191, 192
  nanofertilizers 121
  nitrogen 146
  pollutants 138, 190, 191, 192
Insect pests 122
Integrated 63, 64, 114, 123, 125, 301
  pest management (IPM) 114, 123, 125
  water resources management (IWRM) 63,
        64, 301
Ion transport mechanisms 172
Ionic gelation method 283
Iron-mediated reactions 276
Irrigation 47, 231, 246, 253, 307
  methods 307
  water 47, 231, 246, 253

L

Lipase-producing microorganisms 214
Low-energy consumption 188
Luminescent 241, 244, 253
  chemodetection 241
  metal-organic frameworks (LMOFS) 244
  sensors 253

M

Machine learning 15, 311, 312

algorithms aid 312
  in Agriculture 312
Macrocycle 234
  -based sensors for pesticides 234
Macronutrients 40, 141
Marine, hydrocarbon-contaminated 217
Matter, dry 148, 188
  degradation 188
Mechanisms 6, 7, 119, 121, 122, 141, 142,
      143, 144, 145, 164, 165, 166, 168, 187,
      191, 242, 244, 245, 246
  biochemical 168
  chemodosimetric 242
  microbiological 187
  sensing 244, 245
Membranes 273
  polymeric 273
  water filtration 273
Metabolic 145, 169, 272
  activities 145, 169
  disorders 272
Metabolites 77, 79, 80, 82, 83, 84, 87, 144,
      146, 149
  biomarker 77, 82, 83
  profile 79
  pathways 84
Metal 230, 231, 232, 243, 244, 246, 247, 249,
      253, 268, 274
  complexes, functional transition 243
  ion nanoabsorbents 274
  -organic frameworks (MOFs) 230, 231,
      232, 244, 246, 247, 249, 253, 268
Microalgae 137, 138, 141, 147, 148, 149, 150
  ability 138
  -based wastewater treatment systems 137
  biomass 138, 147, 150
  biomass cultivation process 149
  biotechnology 141
  cultivation 148
  development 141
  growth 147
  heterotrophic 149
Microbes 40, 81, 83, 115, 119, 121, 279
  native soil 279
Microbial 4, 5, 6, 7, 40, 79, 80, 81, 83, 159,
      172, 173, 174, 183, 195, 209, 218, 333,
      335
  activity 4, 5, 6, 7, 79, 183, 195, 209, 218,
      335
  bioremediation techniques 209

communities 40, 80, 81, 83, 218
  desalination cell (MDCs) 159, 172, 173,
      174
  respiration 333
Micronutrients 141
Municipal 80, 185, 303, 332
  solid waste (MSW) 80
  wastewater 185, 303, 332

N

Nanotechnological methods 265
Nanotechnology 265, 275, 276, 284
  green 276
  methods 275
  processes 265
  techniques 284
Networks, wireless 278
Nutrients 119, 210
  crucial 119
  growth-limiting 210

O

Organic 143, 144, 145, 188, 190, 191, 192,
      208, 236, 238, 272, 332
  acyclic sensors 236
  pollutants 143, 144, 145, 188, 190, 191,
      192, 208, 332
  polymers 238
  vapors 272
Organic matter 4, 5, 6, 7, 17, 39, 41, 46, 47,
      48, 49, 51, 94, 187, 188, 196
  biodegradable 196
  accumulation 39, 51
  composition 5
  oxidation methods 196
Organophosphate pesticides (OPs) 235, 242,
      251, 252, 253
Osmosis 65, 138, 158, 162, 163, 167, 168,
      267, 273
  reverse 65, 158, 162, 163, 167, 267, 273
Oxidation 4, 5, 42, 83, 187, 189, 196, 265,
      268, 271, 280
  photochemical 42
  processes 268, 271
Oxygen reduction reaction (ORR) 173

P

Pest 125, 128, 312
  management 128, 312
  populations 125
Pesticides 123, 124, 252, 283
  microbial 124
  microbiological 123
  nanoencapsulation of 283
  sensing 252
Petroleum diesel 150
Photocatalysis, solar 270
Photorhabdus luminescence 127
Photosynthetic efficiency 141
Phytoremediation 78, 82, 194, 205, 206, 210,
    211, 215, 216, 217, 306, 328, 329, 330,
    331
  biochar-assisted 331
  biosurfactant-assisted 215, 216
  plant-based 210
  techniques 194, 210
Plant(s) 78, 123, 127, 194, 210, 214, 217, 219,
    271, 279, 297, 306, 314
  -based bioremediation techniques 210
  chemical 271
  dry 219
  hyperaccumulator 194, 214, 279
  -incorporated-protectants (PIPs) 123, 127
  -soil exchange process 78
  tomato 217
  transgenic 297, 306, 314
Pollutant 138, 190, 192, 193, 209, 216, 233,
    273, 275, 276, 327, 330
  agrochemicals 233
  biodegradation 209
  chemicals 327
  degradation 190, 192, 193, 273
  removal 138, 209, 216, 275, 276, 330
Pollution 57, 58, 60, 62, 78, 80, 83, 114, 116,
    183, 185, 206, 207, 211, 212, 265, 301,
    302, 324, 325
  anthropogenic 78
  health risks 211
  microplastic 83
  organic contaminant 212
  plastic 80
  problems 265
Population 171, 209, 328
  cyanobacterial 171
  degrading microbial 209

growing human 328
Productivity, influence soil 117
Products 128, 242
  biopesticide 128
  catalyzed hydrolysis 242
Properties 1, 2, 8, 11, 80, 86, 128, 129, 211,
    217, 241, 244, 245, 263, 265, 272, 273,
    274, 277, 280, 282, 331, 332
  antimicrobial 217
  catalytic augmentation 241
  electronic 274
  geochemical 277
  hydrophilic 273, 277
  magnetic 272
  physicochemical 80, 86, 263, 265
Proteins 7, 79, 85, 87, 116, 121, 127, 142,
    147, 149, 274, 283
  cryogenic 127
  glomalin-related soil 7
Pyrolysis 219, 330, 331, 332
  oil 331
  process 331
Pyrophosphatase 144

Q

QIIME software 85

R

Radionuclide contamination 193, 194
Reactions, oxidation-reduction 144
Recombinant DNA technology 327
Regulation of Biotechnology 327
Removal 5, 147, 148, 169, 170, 196, 211, 273,
    274, 277, 310, 330
  dye 147
  nutrient 148, 196
  weed 310
Resilience 12, 13, 19, 63, 94
  bolstered community 63
  climate 19
  engineering 12
  soil's 94
  thresholds 13
Respiration, cellular 271
Reverse 65, 158, 162, 163, 167, 174, 267, 269,
    273
  osmosis (RO) 65, 158, 162, 163, 167, 174,
    267, 273

photocatalysis 269

S

Sensors, electrochemical 274
Sewage 185, 187, 188, 189, 205, 217
  domestic 185, 187, 189
  municipal 188
  pollution 205
Signal transducer 240
Small-scale biochar projects 334
Smart technology 310
Social conditions 15
Socioeconomic systems 15, 308
Soil 2, 9, 10, 11, 12, 13, 14, 15, 16, 18, 37, 39,
    47, 78, 79, 81, 82, 93, 114, 297, 298,
    335, 336
  contamination 114
  degradation 2, 9, 16, 18, 93, 335, 336
  development 37, 39, 47
  diseases 78
  ecosystems 12, 13, 78, 79
  management 10, 14, 15, 18, 297, 298
  memory 11
  microarthropods 10, 11
  microbial communities 81, 82
Soil fertility 3, 50, 217, 219, 331, 333, 334,
    335
  properties 217
Soil organic 6, 14, 17, 42
  carbon (SOC) 6, 14, 17
  Matter 42
Solar radiation 35
Solid waste 80, 139, 265
Sonoran desert 34, 35, 36, 37, 38, 44, 45, 46,
    47, 48, 59, 65

T

Techniques 93, 94, 95, 101, 102, 104, 105,
    106, 107, 108, 109, 113, 121, 196, 209,
    210, 250, 281, 282, 311
  agricultural 113, 311
  bioengineering 107, 108, 109
  chromatographic 250
Thermal 163, 212
  distillation 163
  insulation 212
Toxic contaminants 207, 217
Transfer electrons 276

Transferase 144
Transport proteins 171
Treatment, aerobic 187, 188

W

Waste 189, 194, 218, 267, 271, 274, 308, 313
  industrial 267
  nitrogenous 189
  olive oil production 218
  radioactive 194
Water 16, 62, 63, 65, 188, 190, 191, 205, 207,
    208, 209
  bioremediation 190, 191, 208, 209
  conservation technologies 65
  contamination 16, 205, 207
  polluted 188
  -saving technologies 62, 63

www.ingramcontent.com/pod-product-compliance
Lightning Source LLC
Chambersburg PA
CBHW050804220326
41598CB00006B/110